History of Genetics

The MIT Press
Cambridge, Massachusetts, and
London, England

History of Genetics

From Prehistoric Times to the Rediscovery of Mendel's Laws

Translated by
T. R. W. Waters

Hans Stubbe

Originally published by VEB Gustav Fischer Verlag, Jena, German Democratic Republic, under the title *Kurze Geschichte der Genetik bis zur Wiederentdeckung der Vererbungsregeln Gregor Mendels.*

Translated from the revised second editon, 1965

This book was designed by the MIT Press Design Department.
It was set in IBM Bodoni Book
by The Job Shop
printed and bound by
Halliday Lithograph Corp.
in the United States of America.

Library of Congress Cataloging in Publication Data

Stubbe, Hans, 1902 –
History of genetics, from prehistoric times to the rediscovery of Mendel's laws.
Translation of Kurze Geschichte der Genetik bis zur Wiederentdeckung der Vererbungsregeln Gregor Mendels, rev. 2d ed., 1965.
Bibliography: p.
1. Genetics–History. I. Title.
QH431.S84413 575.1'09 73-148973

ISBN 0 262 19085 0

Translator's Preface

This edition of Professor Stubbe's book, *Kurze Geschichte der Genetik bis zur Wiederentdeckung der Vererbungsregeln Gregor Mendels*, is in large part a translation of the revised second German edition of the work originally published by VEB Gustav Fischer Verlag of Jena in 1965. However, the present edition incorporates a number of revisions and amendments together with a considerable amount of entirely new material. Moreover, an appendix containing a translation of Mendel's letter to Nägeli, written July 3, 1870, together with a short English-language supplementary reading list, has been added.

I am indebted to Professor Stubbe for many helpful suggestions and for the avoidance of a number of errors in this translation. The help of Mrs. Elisabeth Williges, who made numerous corrections and improvements in the text, is gratefully acknowledged. I am, of course, alone responsible for any errors that remain.

T. R. W. Waters
Reigate, Surrey, England
August 1971

Foreword

No branch of biology has developed in recent decades as quickly and extensively as genetics. This science seeks to investigate with ever new and improved methods the processes of heredity and development in microorganisms, plants, animals, and also man. Numerous related disciplines have helped to extend our knowledge of the nature of the material bases of hereditary factors, their reproduction and development, their variability and new formation, their transmission and the way in which they interact with internal and external environmental factors. With all its ramifications, genetics has now become a far-reaching branch of biology which, in providing a stimulus for any number of related disciplines, constantly provokes new ventures in cooperative research.

The profusion of results makes it difficult for one person to survey the entire field and to give a complete statement, based on his own experience, of these findings. It seems that the literature in this field can be dealt with only if the fundamentals of genetics, as well as its problems and results, are delineated in a series of monographs, each of which is written by a specialist who gives the reader a lively impression of the development, present state, and difficulties of each branch of the subject. Such an attempt has various advantages. The most important seems to me to be that monographs of limited size can be reprinted far more quickly than comprehensive treatises and textbooks; they can therefore do justice to the very latest developments in their subject matter.

We have decided to make such an attempt. May it be successful and show the reader, whether he is a student, teacher, scholar, or breeder, that it is worthwhile to take an active part in the science of the origin and development of living organisms.

Hans Stubbe
Gatersleben
October 1961

Preface to the First Edition

A historical discussion of an academic discipline requires an account of all its important ideas, hypotheses, and experiments including those which are later seen to be mistakes leading nowhere and which, after scrutiny by subsequent generations, are shaped anew.

It is understandable that historical preoccupations do not generally fall within the province of young scholars who, directing their attention to the essential problems of the present day, show less interest in the history of their discipline. Such an interest is usually confined to older scholars whose taste for retrospection may enable them to write with understanding on the history of their subject. In so doing they realize that the edifice that is their science was constructed during the course of human history from very different kinds of building blocks, many of which were of only temporary significance. The replacement of such blocks during the course of further scientific advances provided this edifice with ever firmer foundations.

The present discussion lays no claim to completeness in respect to the wealth of ideas that have been advanced on our subject throughout the millennia. I hope, however, that it will show all my readers that right from the early periods of human history biological phenomena excited speculation and gave rise to imaginative and mystical notions about reproduction and heredity.

Our science has developed throughout the centuries in conjunction with, and as a function of, social conditions and has gradually obtained increasingly precise results based on experiments. It is to be hoped that this study, which chiefly describes facts and barely suggests their connections with social conditions, will some day be supplemented by an account of the way in which the genetic theories and hypotheses of a given time are influenced by the social conditions of that age. The present author, however, is not competent to perform that task. The history of genetics is illustrative of the perpetual quest of human beings for the explanations of phenomena in the world of living creatures. Finally, the history of our science shows that it is only by exact experimentation that we can hope to understand the relationships between intracellular processes and the outward appearance of evolved forms.

Dr. Fritz Pohl helped me to assemble and work through the extensive literature relevant to my book; I am particularly indebted to him. VEB Gustav Fischer Publishing Company of Jena, which has published so many classical biological studies, kindly aided me in the preparation of this work, for which I am most grateful.

Hans Stubbe
Gatersleben
October 1961

Preface to the Second Edition

The considerable interest with which the historical development of our science was universally greeted meant that the first edition of this monograph was very quickly out of print. I started preparing the second edition shortly after the first appearance of my book; a number of errors were removed, and the international literature was constantly checked against important, previously overlooked, or recently published historical sources. For these reasons the text has been rearranged, amended, and amplified in several chapters; the bibliography is more extensive too. I considered it desirable that young scholars, in particular, should have a more intimate acquaintance with the great scientists of all eras. I have, therefore, added some brief biographical remarks that at least reveal the main stages of their professional careers. The additional illustrations serve the same purpose. The pictures of the Greek philosophers were retained, although there do not seem to be any authentic sculptures from that period. The picture of Epicurus has been replaced by a better one; for this I am indebted to Dr. D. Panos of Athens. I am pleased to thank the Correns family, who possess the original, for permission to publish in facsimile the complete text of a letter dated July 3, 1870, from Mendel to Nägeli. I am grateful to many friends and colleagues for their suggestions and advice given both in face-to-face discussions and in reviews of my book. Mrs. Elisabeth Williges showed great care and resourcefulness in her help both with the revision of the text and with the arduous and time-consuming study of sources. I am particularly indebted to her as I am also to my faithful secretary, Mrs. Berta Ossoba, who assisted me unflaggingly with the preparation of the second edition.

Hans Stubbe
Gatersleben
July 1964

History of Genetics

1

Prehistoric Times: The Origin of Domestic Animals and Cultivated Plants and Ancient Views on Heredity

At what point in the history of mankind did human beings first become aware of problems of heredity and of the transmission of features and characters from one generation to another? We do not know. We can assume, however, that conscious or unconscious selection of characters first occurred during the prehistoric period of transition from hunting and food gathering to tillage and animal husbandry.

Apart from chance findings in excavations of prehistoric settlements, it is chiefly discoveries in caves and then carvings and the written traditions of the oldest, slowly developing cultures which give us the first indications both of the presence of domesticated animals and cultivated plants and of the recognition of the diversity of hereditary factors which led to the formation of different breeds, races, and varieties.

It is certain that animals had been domesticated, and their products used, long before the appearance of features of domestication and the formation of those races of domesticated animals by means of which we can recognize this first diversity. Thus we are unable to specify just when the domestication of wild animals began and which is man's oldest domesticated animal. Only estimates can be made because the age of domestic animals can be studied only in those few centers of civilization where the inhabitants needed to domesticate wild animals to suit their various requirements.

We know that besides the ancestors of our present domestic animals, other types of animals were also domesticated and kept in herds to provide meat. B. Brentjes (1962) holds that herds of gazelles (*subgutterosa* and *gutterosa*) were kept in Egypt and Mesopotamia as early as the beginning of the tenth and right up to the third millennium B.C. and that they were used to supply meat prior to the domestication of the sheep and the goat. Gazelles were then supplemented, and finally replaced, by goats, which provided milk. The sable antelope (*Oryx leucoryx*), the addax, and other undefined species of antelope are known to have been tended in Mesopotamia, Syria, and Egypt during the third and second millennia B.C. It is possible that animals first began to be kept in captivity when the species of antelope which had formerly been hunted was domesticated and raised for meat supplies on the broad plains of the Near East and North Africa. It was probably because of the fact that in prehistoric times conditions for keeping animals in captivity did not favor the propagation of large herds that they were set loose again. It is known that the Nubian ibex (*Capra ibex nubiana*) was kept in Egypt during the third millennium and that the ruffed moufflon (*Ammotragus lervia*) was also kept, although briefly, during the fourth and third millennia. By the second millennium the ibex was no longer kept in captivity.

The reasons for the domestication of animals have often been discussed. Besides the need for regular and easier acquisition of meat and other products, which was perhaps especially urgent because of the diminution of the numbers of wild animals, religious considerations may have given the first impetus to the domestication of animals. Scholars who are particularly interested in the domestication of animals (Herre 1958) consider that domestication caused an increase in inherited variations and that this led to the rapid formation of different races. This idea seems to have been accepted by Charles Darwin (1868) who gives the example of the domestication of the mink (*Mustela lutreola* L.), which produced numerous breeds of different colors during the course of twenty-five generations. It seems that this matter can be accurately investigated only by means of detailed genetic analyses of wild animal populations, which have not been carried out to a sufficient extent. F. Frank and K. Zimmermann (1957) showed for the field mouse (*Microtus arvalis* [Pall.]) that genetic studies of populations of this small mammal reveal a relatively high incidence of the same mutants as occur in domestic and laboratory animals. Therefore the assumption that the mutation rate is increased by domestication can scarcely be said to be proved.

It is probable that the taming and domestication of the dog, ancestral form wolf (*Canis lupus* L.), in central and northern Europe was the first important step toward the observation of the inherited variations that caused prehistoric man to suspect—or even to know—that children resemble their parents and each other with respect to this or that characteristic.

According to Brentjes (1962), relics from the great periods of oriental culture depict several kinds of domestic dogs including the whippet, the sheepdog, the Pomeranian, and various breeds of other large dogs. In Egyptian wall paintings one occasionally sees small dogs with short bowlegs like those of a dachshund.

Endogamy and incestuous breeding among animals and human beings (sibling marriages among the Pharaohs, consanguineous marriages in ancient Israel, incestuous breeding in Greek mythology, and marriages between uncles and nieces among the Greeks) in early and prehistoric times may have helped to reveal the accumulation of identical features and characters in the offspring and may have encouraged the selection and isolation of new races, varieties, and breeds. But the observation of these facts, together with all the ideas about their causes, was permeated with superstition and mysticism which were to remain for millennia, impeding progress though giving impetus to ever new speculations about hereditary processes.

We know with certainty that domestic dogs existed in Denmark as early as the Kyökkenmödding age (c. 8000 B.C.). B. Lawrence (1967) describes the recent discovery of two skulls, one of which (found in Lemhi County, Idaho) dates back to 10,000 B.C., the other (found in southern Turkey) to 7000 B.C. From 5000 B.C. on we find domestic dogs of various races and later cattle and sheep depicted in the art of the ancient Egyptians. The domestication of the ox, ancestral form urus (*Bos primigenius* Boj.)—the primary center of domestication probably lies between Anatolia and the Nile—and the sheep, ancestral form wild sheep (*Ovis ammon* L. with various subspecies) is estimated to have taken place between 8000 and 6000 B.C. According to Reed (quoted in Nobis 1963), the sheep is the oldest domesticated animal in the Near East. Its remains were found in a cave in North Iraq, the strata of which date back to almost 9000 B.C. The domestic ox in the Swiss lake dwellings dates back to 4000 B.C. The ox species was used not only because it produced more milk as compared with the sheep and the goat; it could also be ridden and used as a beast of burden and traction. It is also known that races of dog and the goat, ancestral form bezoar goat (*Capra aegragus* Erxleben) were domesticated in the lake dwellings at about the same time.

According to Brentjes (1962), the goat was used to supply meat in Palestine, North Syria, and Iran at least after the end of the eighth millennium. It is possible that new evidence will place this date still further back so that the domestication of the goat will turn out to be as old as that of the dog. It also seems that new kinds of wild goat of the *aegragus* type were constantly captured and introduced into the already domesticated herds well into the first millennium. We have proof that goats have been kept in herds since the third millennium before our era.

From Sumerian texts and Egyptian reliefs it is known that color variations, including white, black, brown, yellow, and spotted animals, were typical characters of domestication in goats, and we also have proof that the goat supplied milk, butter, and hair from the fourth and third millennia on. Goats also seem to have been used as beasts of traction in Crete during the second millennium B.C. (Brentjes 1965). We can assume that the domestication of the goat, together with the use of wild cereals, forms an important basis for the development of farming and thus for civilization.

The sheep was first used to supply meat and milk; rams were also used as beasts of burden. A discovery in North Iraq dating back to the ninth millennium B.C. shows the sheep to be the oldest herd animal domesticated by man. This find in the elevated valleys of Kurdistan constitutes the most ancient evidence for a domesticated animal known at the present time (Brentjes 1965). The first primitive races were hairy sheep, for example, the "prickly sheep." The use of sheep for wool dates back only as far as the fourth millennium.

From about 6000 B.C. one sees the domestication of the pig—which probably took place independently in different places and at different times—whose polyphyletic descent from the European wild boar (*Sus scrofa* L.) and the Asiatic striped pig (*Sus scrofa vittatus* [Müll. Schl.]) seems definitely established. The Neapolitan pig of the ancient Romans is probably a descendant of the Asiatic striped pig, while the European wild boar led to a number of Russian, Dutch, German, and English races which, crossed with descendants of the Asiatic striped pig, gave rise to the races of pigs we have today.

The domestication of the horse, ancestral form wild horse (*Equus przewalsky* Pol.), seems to have taken place more than 6000 years ago in Elam, east of Ur of the Chaldees. At this time races with different head profiles (convex, straight, concave) and with different kinds of manes (upright, hanging, maneless) were already known (Müntzing 1958). The domestication of the horse in Europe, however, began no earlier than 3000 B.C.

According to Brentjes (1965), the oldest certain examples of domesticated horses in ancient oriental writings date back to the end of the third millennium B.C. There are authentic relics of riding horses from South Iraq from Acadian times (2400 to 2200 B.C.). Still older relics portraying horses and riders, dating back to periods at the end of the fourth millennium, were found in Susa in the West Iranian province of Chuzistan.

The ass appears as a domesticated animal in Egypt at the beginning of the third millennium B.C.; its ancestral form is probably *Equus asinus atlanticus* P. Thomas from North Africa. It appears on a seal from the Near East dated about 2500 B.C. and seems to have replaced the ox as a beast of burden in Egypt during the second millennium (Brentjes 1962).

The domestication of camels seems to have taken place between 4000 and 3000 B.C. Statuettes and bones of dromedaries have been found in Egyptian tombs of the third millennium. It was only at the beginning of the first millennium that camels became numerous in Mesopotamia and Syria. The domesticated Bactrian camel came to Mesopotamia about the middle of the second millennium, but it had certainly been domesticated centuries earlier in Central Asia. Camel bones dating back to 4000-3000 B.C. were found in North Iran and Central Asia. When the camel came to Europe is still a matter of dispute. But it is certain that the Scythian people brought the camel as far west as the Volga area during the sixth century B.C. at the very latest (Brentjes 1962).

Though the reindeer is the only cervid that was domesticated in North Europe and North Asia, other cervids are depicted on seals as early as 2500 B.C. The Hittites seem to have kept red deer. It is known that cervids were kept in Asia Minor and Mesopotamia during the third millennium. We do not have conclusive evidence for the domestication of reindeer until the last millennium B.C. Elks were also domesticated and kept for riding. Elk riders are said to have been used in the postal service during the reign of King Charles XII of Sweden (Brentjes 1962).

Among birds the pigeon, ancestral form rock pigeon (*Columba livia* Gmel.), is the oldest to be domesticated. It was kept in the Near East as early as the beginning of the fourth millennium and in Europe during the third millennium, while the hen, ancestral form Asiatic-crested hen (*Gallus gallus* L. with various subspecies), was domesticated about 3000 B.C. The domestication of the goose, ancestral form stubble-goose (*Anser anser* L.), and of the duck, ancestral form common wild duck (*Anas platyrhynchus* L.), was begun during the fourth millennium B.C. in China, though not in Europe until historic times. According to Brentjes (1962), an unsuccessful attempt to domesticate the goose was made in Egypt around 2000 B.C.

To complete our account we must also mention the insects. The domestication of bees probably took place about 4500 years ago in the Nile Valley. Apart from honey they also supplied wax for writing tablets. The silk moth (*Bombyx mori*) was domesticated in China 3500 years ago.

It is not our task to discuss the difficulties of fixing these dates. This is a matter for prehistorical studies, archaeology, ethnology, linguistics, and so on. The older the domestication process, the more unreliable are the dates. It is the same for the problem of monophyletic or polyphyletic descent.

The most exact account of the history of the origin, distribution, and domestication of a diversity of forms may be derived from a study of the rabbit (H. Nachtsheim 1929, 1949). It began to be domesticated in ancient Roman times and was first named by the historian and statesman Polybius (c. 201 to 120 B.C.). The descent of the domesticated rabbit from the wild rabbit (*Oryctolagus cuniculus* L.) is beyond all doubt. Important advances in taming and breeding rabbits were probably achieved toward the end of the classical period or during the early Middle Ages in French monasteries. The first accounts of the formation of races in the rabbit come from the sixteenth century A.D. Thus black, white, yellow, blue, and mottled rabbits are mentioned in 1548 by Agricola (the "Father of Mineralogy") in his work on subterranean animals and by the Dutchman Josef Justus Scaliger in the second half of the sixteenth century.

The first information about the existence and some of the characters of primitive cultivated plants comes from the same periods in time as for many domesticated animals. Brentjes (1964) is of the opinion that the development of cultivated cereals preceded the domestication of animals. According to Roberts (1929) and Cook (1937) the dioecious nature of the date palm was already known to the ancient Babylonians and Assyrians in 5000 B.C. The first reports of artificial pollination date from the time of King Hammurabi (2000 B.C.). Male inflorescences were commercially important at that time.

The artificial pollination of the female inflorescences of the date palm done by priests wearing bird masks (winged spirits) is portrayed on a relief that dates back to the reign of Assurnasirpal II of Assyria (883 to 859 B.C.) (Figure 1). It was known that the products of the inflorescences of the sterile (male) trees are necessary in order to obtain a rich yield of inflorescences in fertile (female) trees and that a small number of male trees suffices for the fructification of many female ones.[1] The cultivation of the date palm therefore gave the first

[1] According to W. T. Swingle (1913) date palms are today planted in a ratio of 100 female to 1 male tree.

1 Assyrian relief from the time of Assurnasirpal II (883-859 B.C.), which shows the artificial pollination of the female inflorescences of the date palm, performed by priests wearing bird masks (winged demons).

indication of the fact that plants, like animals, have two different sexes and that one could cultivate them by bringing together plants of different sexes in a special way. This is the first known case in human history of an instance of cross-pollination effected by man which gave rise to numerous varieties of an important cultivated plant and thus created possibilities for selection. H. F. Roberts (1929) reports that in four oases in the Sahara alone more than 400 varieties of date palm occur, differing from one another in the size, shape, and taste of their fruit. As no other dioecious cultivated plants were known in Mesopotamia and Egypt at that time, it was impossible to develop any other plants in the same or similar way. Similar methods of artificial pollination were perhaps used much earlier in South America for the development of primitive varieties of maize. We know nothing about this; it is, however, established that the variously shaped corncobs of primitive forms found in Bat-Cave date back to 4000 B.C.

The question of the origin of cultivated maize has been a moot point for many years. According to Mangelsdorf, MacNeish, and Galinat (1964), however, recent discoveries in the valley of Tehuacán (near Mexico City) provide a definite solution to this problem. In five caves in the valley of Tehuacán there was found a large amount of maize remains, dating back as far as the sixth millennium B.C.; these are clearly the remains of wild maize. Cultivated maize therefore developed from wild maize by essentially quantitative changes between 5000 and 3000 B.C.

According to E. Schiemann (1948) the oldest finds of emmer (*Triticum dicoccum*) and species of barley and *Vicia* in the Nile delta date back to 4000 B.C., as do the barley finds in North Syria and the discoveries of emmer and barley in North Iraq. Opinions concerning the dating of the oldest European cereal finds differ. The majority of the European finds point to c. 3000 B.C., the time at which agriculture probably first gained ground in Europe. Here too barley appears together with emmer; flax and leguminous plants also occur frequently. At the same time, moreover, bundle wheat (*Triticum vulgare antiquorum* Heer), one-grained wheat (*Triticum monococcum*), and, somewhat later, seed wheat (*Triticum aestivum*) appear; these had not been cultivated in the Orient. According to Kuckuck (1959) spelt wheat (*Triticum spelta*) was the ancestral form of many of the hexaploid cultivated wheats in the Near East (Iran, North Caucasus). It is known to have existed in Central Europe in late Neolithic times. The first finds of hard wheat (*Triticum durum*) in Egypt date back to the Greco-Roman era. Finds in Moravia dating back to the Bronze Age provide the oldest knowledge we have of rye. Oats came to Europe as a weed with emmer and barley. Millet, maize, and rice are definitely very ancient cultivated plants; sugarcane, however, is considerably more recent. The well-founded view persists that cereal grasses came to Europe from the Near East in a variety of ways.

The cultivation of many plants is reported to go back much farther, however, according to a study by H. Helbaek (1959), who reports on cereal finds in Jarmo, Kurdistan, and Iraq. It is estimated that the city of Jarmo was founded during the seventh millennium B.C., and excavations revealed grains of wild wheats of the type *Triticum dicoccoides* and *Triticum aegilopoides* as well as samples of *Triticum dicoccum.* At another location (Matarrah) in the mountains of Kurdistan samples were found which consisted exclusively of cultivated emmer; these date back to the sixth millennium B.C. The first indications of *Triticum compactum* were found in Egypt; they can be traced to the late fifth and early fourth millennia B.C. The excavations in Jarmo (7000 B.C.) reveal two-rowed barley of the type *Hordeum spontaneum*, which possessed definite features of cultivation. In Egypt and Mesopotamia two-rowed barley was replaced in 5000 B.C. by varieties containing six rows of seeds. There were lentils as early as 4000 B.C. in Egypt, and the tick bean appeared in the Mediterranean between 3000 and 2000 B.C. Flax was cultivated as early as 5000 B.C. in the mountain valleys of Kurdistan. The cultivation of the grapevine in the Near East took place around 4000 B.C. or even earlier. Olives and dates appear in Israeli and Egyptian finds from 4000 B.C.; it is established, however, that they had been cultivated before that time.

These examples of the use of cereals, leguminous plants, and other kinds of plants will suffice for the time being. It is certain that cereal grasses were cultivated far earlier than the oldest finds would lead us to believe.

We cannot here consider the details of the origin of cultivated plants, the ways in which they spread, or their primary and secondary centers of variation, and must refer the reader to the other accounts, especially those of E. Schiemann (1932, 1943, 1948).

The presence of domestic animals, cultivated plants, and above all, the appearance of new races and varieties must be considered as the first results of what were perhaps at first unconscious attempts to breed and cultivate genetically; it is certain, however, that such attempts were undertaken consciously as time went on. There is no doubt that the selection of economically valuable variants or of the products of spontaneous crossings among wild stocks (tough rachis, uniform germination, size of the grains, etc.) and their isolation and propagation, were at first the all-important considerations. In this way it was learned that desirable features as well as undesirable ones were preserved in subsequent generations; that is to say, they were inherited. It must also have been recognized in those early stages of development that this process was true of human beings as well.

In his *History of Biological Theories* E. Radl (1909) wrote:

The notion of original sin in Christian philosophy, the chastisements with which the Jewish god pursued his people into the third and fourth generation, the fate which led Aeschylus's heroes to their ruin, St. Augustine's doctrine of the predestination of man for a happy or eternally unhappy life—these are the first theories of heredity. . . . The belief in the transmigration of souls, the caste system, and hereditary privileges were other expressions of the same conviction.

It is necessary to add that hereditary caste systems are based on the idea that favorable characteristics are inherited in subsequent generations. It is for this reason that the priesthood was hereditary among the Jews of the tribe of Levi, as was the gift of prophecy in Homer's time. The practice of medicine was a hereditary privilege in the family of Asclepiades.

No doubt there are still more ancient ideas concerning the origin and inheritance of deformities or other striking features, the inheritance of acquired characteristics, the hybrid nature of fabulous creatures, and so on; these ideas, derived from mythology, religious dogmas, superstition, and philosophical speculation, were adopted during the more developed periods of prehistoric culture and continue to have influence in modern times. Thus, to give just a few examples, there is the account in Chapter 30 of Genesis of the celebrated transaction between Jacob and his father-in-law Laban. Jacob is to receive as wages all the speckled and spotted animals from Laban's flocks, and Laban is to retain the black ones. Laban removes the new spotted flock (". . . and set three days' journey betwixt himself and Jacob"). But Jacob obtained further births of spotted lambs from the black flocks by placing rods of popular, hazel, and chestnut into which he had peeled white strakes in the gutters of the watering troughs (". . . and the flocks conceived before the rods and brought forth ringstraked, speckled, and spotted").

This notion of "evil imprinting of the unborn" was very important in ancient Egypt, in India, and in classical antiquity; its effects carried over into modern times.

Besides this, however, there are religious writings which, together with epics, songs, and sayings from the oldest civilizations, contain precepts and instructions showing the recognition of the hereditary nature of sickness, health, and mental and physical features and characters. The Sutra and Veda of the ancient Hindus contained rules for a happy life. Detailed advice about how a Hindu ought to live socially and religiously has been handed down to us in the Grihya Sutra, which dates back to the last centuries preceding our era. Thus, precise instructions are given to young men on how to choose a wife. The bridegroom must be healthy in body and mind, and in the Asvalayana

Grihya Sutra we read: "The girl must not only be beautiful but must also have a good character; she must be healthy and completely free of disease." Another Sutra, the Apastamba Grihya Sutra, says: "Her health is important. She must be clear-eyed and clean-limbed. Her family history must be checked." Astangasamgraha says "that in choosing a wife a man must make sure that she has no illness which could be inherited and that her family is free of such illnesses." Those families are best, say the sutras, in which both the maternal and paternal sides have had a good reputation for ten generations because of their good character and their worthy deeds (Hiranyakesin Sutra). Rules are given for the age at which one should marry. "The man should be at least 25 and the girl at least 16 years old before they beget children, so that their progeny will be robust" (Susruta Grihya Sutra). Manu makes a point of mentioning that families in which hemorrhoids, consumption, dyspepsia, epilepsy, and leprosy occur should be avoided. In the Manu code of law it is written: "A woman always gives birth to a son who is endowed with the characteristics of his begetter [and] a man of base descent inherits the bad characteristics of his father or his mother or both; he can never escape his origin."

The Grihya Sutra prescribes various ceremonies for pregnant women in the hope that the good qualities they embody will be transferred to the expected child.

There is agreement in Hindu mythology and in ancient Indian literature that energy, strength, and mental characteristics can be inherited. An example is provided by the miraculous birth of the Pandava brothers in the story of the Mahabharata. King Pandu, who is unable to produce a son, asks his wife Kunti to appeal to the gods to fulfill their wish. She subsequently bears five sons, each of whom inherits the distinguished and illustrious characteristics of his father.

Such examples of ideas concerning the inheritance of features and characteristics in the oldest civilizations may be multiplied at will. In the Upanishads, as in many other ancient Indian works (see also J. Sen 1963, N. H. Keswani 1963, N. K. Bose 1963), evolutionary ideas are also present. Thus, in the Vishnu Purana (written at the very latest in 400 B.C.) there is a stanza that describes the order in which the highest form of life, namely man, evolved: undefined forms, sea beasts, tortoises, reptiles, birds, mammals, and finally apes. Even the number of species is given for each group. God is alleged to have come to earth ten times, first in the shapes of various animals and finally in the shape of man.

2 Reproduction and Heredity: The Hypotheses of the Greeks

Whereas the oldest traditions acquaint us with facts that refer almost exclusively to the selection of characteristics and features in the earliest domestic animals and cultivated plants, the Greeks gave particular attention to problems—in a broad sense—of evolution and heredity as they apply to man. This is particularly true of the physician Hippocrates of Cos (460-377 B.C.) (Figure 2) and of the philosopher Aristotle (384-322 B.C.), both of whom lived during the golden age of Greek culture. The speculations of the Greeks are preserved in numerous written fragments beginning with the early pre-Socratic philosophers of the sixth and fifth centuries B.C. During this period the religious myths of the Greeks were being replaced by philosophical inquiries that aimed to explain the history of the universe and the origin of human life. Although these early philosophical systems contain no concrete allusions to biological problems that we could consider evidence of a preoccupation with genetics, they do nonetheless indicate the all-embracing spirit of Greek philosophy, which strove to interpret every aspect of life and the universe. We must remember that the concept of heredity had a meaning in ancient times quite different from that which it has today. Heredity simply meant the procreation of offspring of the same kind with the same or similar features; this was not explainable in biological terms. It is for this reason that the Greeks attempted to give a philosophical account of this phenomenon. In the works of Greek poets there are frequent references to problems of heredity, and Greek statesmen introduced eugenic measures that aimed at the preservation and perpetuation of the best people, that is, those which were particularly useful to the state because of their health, intelligence, bravery, and other qualities. The concept of heredity included, in the first instance, the totality of physical and mental characters and features. It was only with Hippocrates, Aristotle, and others after them that the inheritance of specific characteristics, deformities, and illnesses was envisaged. In Greek philosophy such questions were closely bound up with imaginative ideas of reproduction and of the roles of male and female in the procreation of offspring. The development of medical—in particular anatomical—knowledge gradually clarified these ideas on the roles of male and female sexual partners in reproduction and fertilization and on the significance of sexual secretions for the creation of a new individual.

2 Hippocrates (460-377 B.C.)

2.1 The Ideas of Greek Poets, Philosophers, and Statesmen

In order to understand the hypotheses on reproduction and heredity of ancient times we must consult primarily the extant works of Greek philosophers, scientists, and physicians. Poetic and dramatic verse or the activities of statesmen are less significant, since the generalities revealed by a study of them give us hardly any information about their specific ideas on genetics. Moreover, it is not the purpose of this account of the history of genetics to discuss eugenic measures—e.g., those common in Sparta or those suggested by Plato—in any detail.

We propose nonetheless to give a brief account of how Greek poets and statesmen conceived of those questions of life, lineage, family and the preservation of the aristocracy which, in a broad sense, are directly connected with genetics (see Haedicke 1936).

A familiar feature of Homer's great epic, the *Iliad* (eighth century B.C.), is his concern with questions about the origin and lineage of the hero. The name of his family, of his forefather is of great significance, as is his divine origin; the inheritance of marks of nobility is taken for granted. The genealogical table or pedigree used by the nobleman as proof of his high birth is particularly important in the *Iliad.* The information about these different families allows us to state the kindred relationships between them; thus, a large number of genealogical trees are contained in ancient Greek poetry. In times of war the battle array was set up in terms of tribal units. A son inherited not only his father's weapons but also his strength and greatness, for it was the son who succeeded his father as the embodiment of the divine antecedents of the family. Every member of such a family could call upon the help of the god from whom he was descended and whom he asked for prosperity. For the Greeks of this era, family origin was a distinguishing mark which was conferred with the blessing of some god; it was thought of not in terms of what one got out of it but in terms of moral obligation. This concept of the nature of noble birth changed considerably in later centuries. In what is probably a later work, Homer's *Odyssey*, the aristocracy is emphasized as a social class, and wealth becomes a prerogative of noble birth. Within this social class and its noble families, physical features, strength, courage, and ability are handed down. It is just as much a matter of family duty to preserve those qualities as it is to maintain the family estate.

Hesiod (c. 700 B.C.), a poet of peasant stock, gives eugenic advice in practical terms concerning choice of a spouse, the age at which to marry, and how to determine the birth of a boy or a girl.

Of the noble poets of the sixth and fifth centuries we must mention Theognis of Megara, who enjoined the declining aristocracy to preserve the purity of their social class by keeping apart from the rising middle classes, since good could come only from good and was forever lost once mingled with the bad.

Pindar, the lyric poet from Boeotia (c. 518-442 B.C.), is the most enthusiastic exponent of the Greek notion of aristocracy. He is so convinced of the heritability of all the masculine virtues within a noble family that he does not admit to any degeneration or exception; noble glory, he claims, merely "slumbers" in undistinguished members of an aristocratic family, only to blossom forth in later generations. Innate nobility (i.e., inherited ability) is considered more important than education and instruction, for "The noble spirit of the father/ Shines forth in the nature of his son" (Pindar, 446 B.C.).

Euripides, the poet of the Greek enlightenment (480-406 B.C.), who came from the ranks of the long-established, landed upper classes, makes very frequent mention in his dramatic work of the similarities between father and son and of the son's inheritance of the father's virtues. The children of Herakles inherited their father's blazing eyes and were superior to other people because of their ancestry. Inherited ability will always be found in the genuine aristocracy. Time and time again we hear: "A noble father sires a noble son; / A base man's son is of his father's kind" (*Alcmene* fragment).

The pervading conception of aristocracy of earlier times had already changed by the fifth century B.C. In addition to the royal aristocracy of divine antecedents there developed another aristocratic class which, no longer based on birth and achievement, pretentiously sought class privileges and social advantages. This class saw the significance of ancestry not in inherited capabilities and moral obligation but exclusively in terms of its own importance and political power. It was this kind of aristocracy that Euripides lampooned in his plays.

Yet another kind of elite, frequently and sympathetically portrayed by Euripides, is that of a moral aristocracy. This can be found in a society that accords noble status only to those who have moral qualities; birth does not enter into it. In choosing a spouse, therefore, it is absurd to consider possessions and high birth rather than character. It is moral fiber and not membership in a given social classs that determines the worth of a human being; and although moral fiber may be transmitted from generation to generation in individual families, it is by no means restricted to a certain social class.

The philosopher Plato (427-347 B.C.) came from an old and distinguished aristocratic family. In some of his most famous writings (for example, the *Republic*) he proposes in no uncertain terms a number of very strict laws governing the selection of the best people in society so that his ("Ideal") state would be preserved from economic and political decay. These laws stem from the realization that the makeup of a given human being is inherited and that his father and mother both took part in transmitting the various characteristics. An important idea in Plato's theory of the state is that people are not completely equal but have different natures; it is for this reason that different people are best suited to fulfill different functions.

Drawing upon a Phoenician myth, Plato writes in the *Republic* (Vol. II, Book 3, p. 121[1]):

. . . But the god who fashioned you added gold in the composition of those of you who are to become rulers (which is why their prestige is greatest); he put silver in the auxiliaries, and iron and bronze in the farmers and other artisans. As you are all of the same stock, children will mostly resemble their parents. Occasionally, however, a silver child may be born of golden parents, or a bronze child of silver parents, and so forth.

Plato's selection procedure for the rulers and guardians[2] and their wives is based on the natural differentiation of people. This was how he hoped to create ideal people in an ideal state. But this natural inequality in men (and in what they inherit) also requires that a hierarchy of virtues be related to a hierarchy of social classes, so that each man may achieve his best according to his capabilities. He attached considerable importance to education in the development of these capabilities. Moreover, he demands not only the selection of the best but also the elimination of the bad. By "bad" Plato meant incurable disease or moral inferiority. This policy clearly presupposes the hereditability of these features. The best, however, can only pass on their good characteristics if reproduction takes place according to prescribed marriage laws. These laws caution against inebriety, licentiousness, and irresponsibility, for Plato thought that the mental, moral, and physical condition of parents during intercourse was transmitted to their offspring.

Plato's utopian concept of the state has never been realized. Nor did it stop the decline of classical Greek culture, which began with a century of domestic strife and was completed with the conquests of Philip of Macedonia (359-336 B.C.) and Alexander the Great (336-323 B.C.).

These brief examples will suffice to show the prevalence of the notion of transmission of features in the literature and political

[1] Refers to the edition of Henricus Stephanus (1578), from which Plato is usually quoted.

[2] In Plato's political philosophy, the guardian class is an elite that is drawn from the whole range of people living in the state. Members of this elite may become philosophers. Philosophers had neither family nor private property; their material needs were provided by the state. Their wives and children were held in common.

philosophy of ancient Greece. Scientific considerations, however, had little or nothing to do with the formation of this notion.

2.2 Reproduction and Heredity: Scientific Doctrines of Greek Philosophers and Physicians

It was the imposing philosophical speculations concerning the origin of the universe and of human life which led to the formulation of scientific hypotheses about reproduction and heredity among the Greeks.

The ideas of Thales of Miletus, the oldest thinker in the history of western humanity (c. 624-546 B.C. according to the ancient chronicler Apollodorus), are preserved only as part of unreliable traditions. Plato and Aristotle knew of no writings by him. He was a commercially and politically oriented philosopher who had traveled widely and was one of the Seven Sages of Greece. It was he who first brought the essentials of Egyptian geometry to Greece. He lived in the Greek commercial and cultural center of Miletus near the Lydian border. According to him, the multiplicity of phenomena is derived from a single primeval substance, water, which is capable of motion and change: He thought of the world as a small island that floats upon the endless waters of the ocean of the universe.

Anaximander of Miletus (611-546 B.C. according to Apollodorus), Thales's pupil and successor, may be said to be the founder of Greek philosophy. He sketched a theory of the creation of the universe that was based on observation and rational thought and was the first person to construct a map that showed the then known distribution of land and sea. He held that "the boundless" was the cause of all being. Primordial motion caused solids, liquids, warm and cold (that is, earth, water, fire, and air) to "separate off" from the boundless. The earth, a sphere, is suspended in the center of the boundless. Living creatures and human beings evolved from water. Anaximander claimed that human beings had first developed in the insides of fishes, from where, enveloped in a membrane, they were eventually cast ashore.

Anaximenes of Miletus (c. 585-525 B.C.) was the last of the three Milesians. Like Anaximander, he wrote a treatise entitled *On Nature.* He thought that the air in which the heavenly bodies move about the earth is the primordial element: All things are derived from air by the processes of condensation and rarefaction. Rarefied air produces fire, while condensed air produces wind, clouds, water, and earth. He knew that the moon derived light from the sun and also gave a natural explanation of the rainbow in terms of solar radiation.

The most important philosopher of the sixth century B.C., Heraclitus of Ephesus, came from a noble, and once royal, family. He lived as a hermit in the mountains and was known as "Heraclitus the Dark" because of the obscurity of his thought. In the fragments of his writings—*On Nature*—water, air, and earth are all derived from primal fire which symbolizes the principle of perpetual change. According to him everything is constantly changing—"all is flux"—and strife is the father of all things. He is extremely significant in the history of philosophy, and his doctrine of the unity of opposites makes him one of the originators of dialectics. In the tenth fragment we read: "Even Nature strives toward opposites; it is from opposites and not from unity that harmony is produced."

The Eleatics, whose school of philosophy was established in Elea (southern Italy) by Xenophanes (c. 570-480 B.C.) of Colophon (western Asia Minor), existed at the same time as the preceding philosophers. Xenophanes' thorough logical approach led him to repudiate miracles, mysticism, and the doctrine of the transmigration of souls. He cast off the popular religious notions of the Greeks, as did his pupil Parmenides (c. 510-440 B.C.), who wrote a didactic poem entitled *Nature.* Xenophanes sought to apprehend the nature of being by means of the unfettered exercise of human reason. He thought that the four elements—air, earth, fire and water—were the eternally changing primal substances of the universe. Zeno of Elea (c. 460 B.C.), a disciple of Parmenides, may also be considered an originator of dialectics, i.e., of the method of argument by antinomy for the clarification of problems in the theory of knowledge. His doctrine of the infinite divisibility of space paved the way for the infinitesimal calculus.

About the same time, however, the philosopher and physician Empedocles (492-432 B.C.), an aristocrat from Agrigentum in Sicily, was trying to clear up the quarrels between the older cosmologists in his didactic poem, also entitled *Nature.* He claimed that the four elements are all equally important primal substances. At one time they were harmoniously intermingled in the universe, which is a sphere. It was the action—in varying degrees of strength—of the forces of attraction and repulsion (Love and Strife) on the four elements that gradually produced the world, and thus animals and human beings. All life is therefore to be understood as a rearrangement of the four elements.

Because of this, Empedocles has sometimes been considered a precursor of Democritus. The fact is, however, that his scientific ideas were often pervaded by the notions of a sorcerer and mystic. Toward the end of his life he was driven into exile by groups of reactionaries. He died in the Peloponnesus.

Influenced by his teacher Zeno, Leucippus of Miletus together with his famous, versatile pupil, Democritus of Abdera (460-370 B.C.) (Figure 3), founded the atomist school, according to which the world consists of an infinite number of extremely small, qualitatively indistinguishable particles called atoms. The arrangement, position, separation, and coming together of these atoms (which differ in shape and size) are what account for the rise and disappearance of all living creatures. The materialist philosophy of Democritus is supplemented with a concept of human life in lofty moral terms. He praises cheerfulness and well-being and thinks these are best attained by the exercise of moderation. It is this quality which also leads to *ataraxia*, that is, the imperturbability resulting from self-discipline that was so important in Stoic ethics. We shall be taking a closer look at the atomists' doctrine, since it had a wide influence on a number of subsequent ideas about heredity and sex determination.

One hundred and twenty years later, Epicurus (c. 342-271 B.C.) (Figure 4) made atomism the basis of his philosophy. He came from the Attic deme of Gargettus, was a pupil of Nausiphanes, and spent his youth on the island of Samos. In 306 B.C. he set up his own school in Athens, where his teaching drew heavily on the thought of Democritus. Extensive fragments of his treatise *On Nature* were discovered in the ruins of Herculaneum. He taught that there are eternal laws which govern the course of the world. Natural philosophy was to free men from their belief in supernatural forces, from their fear of death and the gods, and from their concern about an afterlife. Reason, he claimed, led men to repose of spirit and to tranquillity—they were at one with themselves. Epicurus is the champion of Greek materialism. His philosophy inspired the Roman Titus Lucretius Carus (c. 98-55 B.C.), whose didactic poem *De rerum natura* was posthumously published by Marcus Tullius Cicero (106-43 B.C.) and Lucius Annacus Seneca (4 B.C.-A.D. 65); it was still influential during and after the Renaissance.

These brief sketches will suffice to show how pre-Socratic Greek philosophy, which embraced every aspect of Greek life, influencing poetry, music, the creative arts, mathematics, astronomy, politics, and public life, was permeated with ideas about the origin of the universe and that of life. During their long wanderings—in poverty and danger, often unaware of each other's existence, and without the help of adequate libraries or public support—the Greek philosophers gathered learned speculations of other lands. They modified them with their own ideas, and this frequently led to the promulgation of new and important doctrines. They did not teach for pecuniary gain (with the exception of the Sophists), and disciples gathered about them at every stage of their weary lives in the hope of adding to the body of knowledge on which their successors would build.

3 Democritus (460-370 B.C.)

In this period of all-embracing philosophical speculation—at the turn of the sixth to the fifth century B.C.—Greek philosophers and physicians began to ask the first concrete questions concerning the laws that govern the development of living creatures. This is extremely important for the history of our science. W. His (1870), C. Zirkle (1935, 1936, 1946, 1951), H. Balss (1936), and E. Lesky (1950) have provided excellent, detailed summaries of the doctrines of reproduction and heredity prevailing at this period. Our account is derived from

4 Epicurus (c. 342-271 B.C.)

theirs, though we single out certain points that are especially relevant to our purpose. Wherever we could, we also examined the source material in Greek or else in translation.

In the course of his dissections of living animals, the physician Alcmaeon of Crotona (sixth century B.C.) discovered the site where the nerve centers of the sensory organs enter the brain. He realized that the brain was the principal organ of sense perception. Probably the first person to inquire into the origins of human semen, he claimed that it was a part of the brain. It is, however, impossible to decide whether he came to this conclusion on experimental grounds or whether his view was merely speculative or a result of oriental influences.

Other Pythagoreans had shared this view, which was later modified by Hippo of Rhegium who postulated that semen originated in the spinal cord. This ancient account of semen was amplified into the theory of pangenesis in the writings of Hippocrates (460-377 B.C.), Anaxagoras (500-428 B.C.), and Democritus. (Democritus was here influenced by Anaxagoras, the son of a rich aristocratic family from Clazomenae in Asia Minor, whose philosophy, in turn, derives much from that of his predecessor Empedocles.) According to the theory of pangenesis, semen is formed in every part of the body and travels through the blood vessels, while testicles are merely organs through which it happens to pass.[3] We can safely say, however, that this new theory was not sharply distinguished from the old and that the influence of the older account continued strong in many places. Indeed, it was still maintained that the brain and the spinal cord were the chief places in which semen was formed, that additional semen was transferred there from the vascular system, and that the semen eventually passed by way of the testicles into the penis. Various forms of this doctrine persisted, as we shall see, until the end of the nineteenth century. The theory culminated in the "Provisional Hypothesis of Pangenesis" formulated by Charles Darwin and was finally discredited by the "Hypothesis of Determinants" proposed by A. Weismann.

In his *Timaeus*, one of the dialogues written late in his life, Plato (427-347 B.C.) suggests that semen is produced in the brain and in the spinal cord without, however, offering any detailed evidence in support of this view. This same contention was also frequently discussed in the

[3]W. His (1874) distinguished between four groups of theories of reproduction. The theory of pangenesis is characterized as an essence-type theory ("Extrakt-theorie," p. 132).

first few centuries of the Christian era. The issue was revived in the eighteenth century, as we can see in a treatise entitled *Spermatologia* (1720) by Martin Schurig, a Dresden district medical officer. In this work the opinions of contemporary English scholars, especially those of John Rogers, are quoted: "John Rogers is of the opinion that semen, as it were the very essence of the body, is chiefly supplied by the brain and carried through innumerable channels to the testicles."

Other views of Alcmaeon of Crotona about the formation and significance of semen have also been handed down. Like other thinkers and physicians of his day he believed in the existence of "female semen." He also claimed that it was "the preponderance of male or female semen which determines the sex of the child," and that equal amounts of male and female semen caused hermaphroditism. This view was also taken by Conrad von Megenberg (middle of the fourteenth century) and was held as late as the eighteenth century.

Parmenides, Empedocles, and Democritus also believed in the existence of female semen, a concept that formed part of the medical theories of ancient India. Aristotle repudiated this notion, but it was readopted by Galen and continued to be influential until modern times. K. von Baer ultimately put an end to the controversy with his discovery (1827) of the mammalian egg.

The quantitative principle, that it was the preponderance of male or female semen which determined the sex of the child, was extended (by Alcmaeon and others) to include qualitative differences in semen. Hippo of Rhegium said that "children are either male or female according to whether the semen was thick and strong or weak and watery." In Hippo's view, however, it seems that these different seminal qualities are to be found only in male semen—that is, they are not separately distributed in male and female semen—for female semen does not contribute to procreation "since it is discharged from the sexual organ."

And so according to this ancient account of sex determination, male semen alone is progenitive; its consistency (viscid or watery) is what determines the sex of the offspring. Though this account is by no means formulated in genetic terms, it seems nonetheless clear that the ideas implicit in Alcmaeon's theory of semen constitute the first attempt to view the creation of a human being as a biological process.

These ideas were modified by Empedocles, who held that the heat of the womb is decisive for sex determination: in a warm uterus semen produces males, in a cold uterus, females.

The inheritance of physical features was also interpreted in terms of Empedocles' theory of vital heat; but here the temperature of the semen and not that of the womb is the decisive factor. Censorinus has handed down the following account:

> An equal degree of heat in the semen of the parents gives rise to a boy who resembles his father; an equal degree of cold (in both) gives rise to a girl who resembles her mother. If the father's semen is warmer, and the mother's colder, then a boy is born who resembles his mother; if the mother's semen is warmer, and the father's colder, then a girl is born who resembles her father.

However, we cannot be sure that it was Empedocles who proposed this theory since it may well have been a product of later discussions on heredity.

Of all the ancient theories of generation, however, none survived longer (until modern times) than the "Right and Left Theory," according to which males come from the right side of the body, females from the left. This idea is also found in ancient Indian medical writings in which boys are associated with the right side of the uterus and girls with the left. Among Greek physicians and philosophers Parmenides and Anaxagoras were the most important proponents of this thesis. (At that time it was assumed, arguing by analogy from observations on animals, that the human uterus was divided into two parts.) The thesis was combined with the Empedoclean theory of vital heat: the right side was considered to be perfect, strong, and warm, the left to be imperfect, weak, and cold. From extant fragments, however, we learn (E. Lesky 1950) that these ideas varied considerably, were inconsistent, and in no way formed a unified system.

Anaxagoras and Parmenides probably tried to account for the formation of hermaphrodites in terms of the Right and Left Theory by assuming that normal sex formation was caused by sperm from the man's right side entering the right side of the uterus, and sperm from his left side entering the left side of the uterus. But if semen with male sexual characteristics entered the left side of the uterus, a male creature with female features was formed; similarly, if semen with female sexual characteristics entered the right side of the uterus, a female creature with male features was formed. This is to say, primary sex characteristics are determined by semen and secondary sex characteristics, by the side of the uterus.

With regard to another question, however, Anaxagoras's ideas go far beyond the Right and Left Theory of sex determination and touch on problems that were to be discussed in the Middle Ages under the guise of the theory of preformation. Anaxagoras claims that "since in all things that unite there are contained many things of all kinds, and the seeds of all things . . . men have been formed in them, and all other living creatures." He believed that a prototype of every organ and part of the future living creature was already contained within the semen. We cannot be sure, however, whether he conceived of these prototypes as ready-made, as imperceptible organs (and parts of organs), or as material elements (seeds) that only assume their final form during embryogenesis. He also made a sharp distinction between the male and female reproductive function: only males produce sperm while females have the passive role of conceiving. According to this view, moreover, female sex was determined by sperm, for males were formed from sperm from the father's right side, females from sperm from his left. Right and left side usually referred to the testicles. After puberty differences in the appearance and size of the testicles were supposed to indicate the likelihood of male or female offspring.

The Right and Left Theory was therefore an important tool for sex prognosis. It was used in a number of different ways for this purpose until modern times; superstition claimed that ligature of the left (right) testicle tended to produce boys (girls). Various "signs" during pregnancy were also used for sex prognosis both in India and in Greece. But it is not clear whether Indian physicians were influenced by the Greeks, or whether they developed the ideas underlying this practice independently.

Ancient speculations about sex determination and sex prognosis had considerable popular influence in medieval times, as is shown by the following illustration.

The earliest widely read book on natural history in German, Conrad von Megenberg's (1309-1374) *Treatise on Nature* (written about 1350), was inspired by Thomas de Cantimpré's *Liber de natura rerum* and Avicenna's *Canon medicinae.* Under the heading "On the causes of the conception of a male child" we read on page 31:

If you would know why a woman conceives a male child, and how to tell whether she will give birth to a boy, then first remark that when the sperma virile is warm and present in large amounts, such qualities prevail, and for this reason a boy is born. Moreover, if the sperm, for the most part, originates in the right testicle and attains the right side of the uterus, this constitutes a second cause, since the right side is warmer than the left, and the sperm from the right testicle more vigorous than

5 Galen (?) (A.D. 129-c. 199)

that from the left. My counsel, therefore, to a woman who will be delivered of a boy is that she forthwith lie on her right side after intercourse. Others declare that a boy is born when sperm from the right testicle enters the right side of the uterus (as I mentioned earlier), but that if sperm from the left testicle enters the right side of the uterus, then a virago is produced. However, if sperm from the right testicle enters the left side, an effeminate man results. If sperm from the left testicle enters the left side of the uterus, a girl is produced. Cold air helps to produce a boy as does a cold climate in general and also that wind, known in Latin as Aquilo, which blows out from the constellation of the Plough toward the south. For the cold drives the natural warmth of the body into the womb and increases its inner warmth. More warmth is needed to produce a boy than a girl.[4]

Aristotle (384-322 B.C.), as we shall see later in greater detail, was the first thinker to criticize these philosophical speculations on sex determination and to oppose them with the results of his own investigations. He pointed out that he had found female embryos in the right side of the uterus (and males in the left) during the course of his dissections on animals. However, he was by no means completely free of the influence of philosophical speculation.

This may also be said of Galen of Pergamum (A.D. 129-c. 199) (Figure 5), court physician to the emperors Marcus Aurelius and Severus, for despite the many advances he made in anatomy he still believed in what was essentially the Right and Left Theory of sex determination. It was this view of his which remained influential in medieval and early modern times. Galen first studied medicine in Pergamum, continued in Alexandria and other centers of medical research, and finally settled in Rome. He wrote one hundred and thirty-one treatises on medicine, of which eighty-three are preserved. Many of these writings became very popular, for they were written in a practical vein. Their pronounced religious tone commended them to Moslem scholars as well as to Christian theologians of the Middle Ages.

The atomist Leucippus and his pupil Democritus of Abdera (460-370 B.C.), and the famous physician Hippocrates (460-377 B.C.) together with members of his school propounded the doctrine that generative matter comes from every part of the body. This new doctrine gave rise to a number of quite different ways of looking at problems of heredity. When formulating his Provisional Hypothesis of Pangenesis in 1868, Charles Darwin compared it to the Hippocratic doctrine which has been known ever since as the theory of pangenesis. According to the atomist doctrine the homogeneous primal substance postulated by the Milesians was in fact decomposable into an infinite number of atoms whose shape and size, arrangement and position were just as important for the genesis of the human body as they were for anything else. Democritus thought of the procreative act as the release

[4]In popular medicine there are a large number of similar, and at times inconsistent, instructions on how to determine the sex of a future child during coitus and pregnancy. Such instructions survive to this day (see also Hovorka and Kronfeld 1908).

of a germ, preformed in all its atomic parts, from the parental body. The constellations of atoms, provided by both parents, moved at different speeds to a specific place (of great importance for the development of the germ) in the uterus. Democritus thought that the different speeds with which they moved accounted for sex formation in that the germ which arrived first determined the sex. This view was rejected by Aristotle and later by Galen, on the ground that it suggested that the hereditary factors governing sex determination operated independently of those governing somatic features. It is very difficult to say to what extent Democritus himself distinguished between these two issues since his writings and those of others who followed him are preserved only in fragmentary form.

It is, however, certain that the Democritean doctrine of pangenesis strongly influenced Hippocratic medicine. It was invoked to account for problems of heredity in all of the Hippocratic writings and gave rise to a number of new hypotheses concerning, for example, congenital defects.[5] These hypotheses, however, were purely speculative and not corroborated by facts. In contrast to Democritus's theory of the existence of atoms, their shape, and the spatial relations between them, the Hippocratic interpretation of pangenesis stressed physiologically active fluids (humors) as the bearers of hereditary traits. "Male semen is formed from all the humors in the body," we learn from one of the most important Hippocratic works *De genitura* (VII 470, 1L, from the collection in Cnidos). Elsewhere in the same work we read: "But I say that semen is drawn from all parts of the body, hard and soft, and from all the fluids in the body. There are four kinds of fluid (humors): blood, bile, water, and phlegm; all four are congenital. They are the sources of diseases" (VII 474, 5f.).

The four humors were thought to be responsible for the transmission of health and disease. "Since an organism has the same qualities as its begetters, it also has the same sorts and proportions of humors, both healthy and diseased" (*De morbis* IV, VII 542, 3f.).

Since the humors, composed of diseased and healthy elements, helped to form semen, it was thought that they could transmit those qualities from the parents to the developing child. In addition to the

[5]The Hippocratic writings of the fourth century B.C. are the oldest medical treatises of the Greeks. It seems probable that no one person wrote them but that they are a product of the whole school of medicine founded by Hippocrates. The Hippocratic school was fundamentally opposed to the medical teachings of the Pythagoreans, since it conceived of medicine not as a theoretical science but as an art or technique which drew on all kinds of new ideas.

humors, however, every particle of body tissue was used to form semen, and thus similar tissues appeared in the offspring. A predisposition to have certain kinds of tissue and organs is therefore transmitted from the parents to the offspring and fulfilled in them. It was thus not, as Anaxagoras had thought, a new, preformed creature already developed in the semen which is then discharged into the uterus, but specific material structures which were passed on and inherited.

In contrast to earlier theories, pangenesis emphasizes the functional equivalence of both sexes: they both contribute in the same way to the formation of semen. Moreover, we even read that "female sperm is present in the male in just the same way as male sperm is present in the female" (*De genitura* VII 478, 3).

Translating this statement into the language of contemporary genetic theory, we can see how closely it touches on two notions which we consider to be fundamental to sex determination: the bisexual potency of generative matter and sex-determining factors (*Realisatoren*). According to the Hippocratic doctrine, *both* sexes have a male-determining, "stronger" sperm and a female-determining, "weaker" sperm. The combined amount of both sorts of sperm from both parents is what determines the sex of the offspring. Hermaphrodites are produced by certain ratios of male- and female-determining sperm.

The inheritance of physical features is explained in the same way. In the anonymous *De Genitura* we read (VII 480, 7f.):

> . . . semen comes from every part of both the male and female body; strong semen from the strong parts, and weak semen from weak parts. It is a law of nature that it be passed on to the children in this way. If the generative matter contains more semen from a given part of the male body than from the same part of the woman the child will resemble his father more with respect to that part; and a greater amount of semen from a given part of the female body means that the child will resemble the mother more with regard to that part.
>
> But it is neither possible that the child resemble the mother in every part and the father in none, nor is the converse possible; nor is it possible for the child not to resemble one of them in some part. Rather, it is a law of nature that it resemble both in some part or other, for generative matter from both bodies helps in forming the child.
>
> The child resembles in more parts that parent who contributes a larger amount [of semen] to the resemblance, and from more parts of the body. Sometimes a daughter resembles her father more than her mother in more parts, and a boy may resemble his mother more than his father.
>
> These many facts are my proof of the doctrine I set out above: that men and women both have generative matter which determines male and female sex.

It is important to note from these pronouncements that a significant change has occurred in the Greek way of thinking about science during the period between 500 and 400 B.C. Beginning with philosophical speculation they have now come to recognize scientific laws and realized that it is semen that transmits the hereditable features of the parents to the offspring.

Whereas at first only the preponderance of either the male or the female sperm has been made responsible for the child's resembling the father, respectively the mother, the new hypothesis makes the transference of physical characteristics from both parents to the child possible and evaluates it as conforming to the laws of nature. In this it is assumed that the similarity with one of the parents increases as additional sperm is supplied by the various parts of the parents' bodies, in accordance with the tenets of the pangenesis theory. The sexual bipotency of the sperms finally supplies an explanation for the fact, not understood up to that time, that a daughter can resemble her father more than her mother.

It has frequently been pointed out that we can perceive a definite continuity in the development of such ideas in ancient thought (O. Gigon 1936, E. Lesky 1950), for old and well-established doctrines were not completely discarded but rather modified and extended. This continuity is particularly apparent in the development of problems about heredity in a single century.

Aristotle, who criticized pangenesis on a number of counts, was the first thinker to point out that the theory was unable to account for the fact that features of earlier ancestors, for example grandparents, are also handed down. He also contended that the theory could not explain similarities between ancestors and offspring in respect to voice, nails, hair, and gait, ". . . none of which help to form semen." With regard to the first issue he writes:

> Moreover [the offspring] resemble those of their ancestors who after all contributed nothing to the semen. For the resemblances recur after several generations, as was the case with a girl in Elis who married an Ethiopian: their daughter did not have the same features as the Ethiopian [what is meant here is essentially the dark skin color]; but the son of that daughter did (*De generatione animalium* I 18.722a 8f.).

It was natural for Greek philosophers, scientists, and physicians to give some thought to the inheritance of "acquired characters" in the course of their perplexing struggles to understand the nature of heredity (see also C. Zirkle 1946). The idea that characteristics acquired during life could eventually be inherited had been believed for a long time because of the influence of popular myths and legends. The best known are probably the accounts of the mythical macrocephalics or "long-heads," who bound the heads of their newly born children in order to lengthen their skulls.[6] A physician of the Hippocratic school reports on them at the close of the fifth century in his treatise *De aere, aquis et locis*:

This custom is practiced as follows: as soon as the child is born, they use their hands to shape its skull, while it is still pliable; they apply bands and other appropriate means to force the head to grow in length; it is in this way that the spherical shape of the head is distorted and its length is increased.

Thus did this practice originally determine the way in which the head developed; in the course of time, however, the head developed naturally in this way, and the practice was no longer required to make it do so. For semen is derived from every part of the body; healthy semen from healthy parts and diseased semen from diseased parts. Now, since baldheaded parents usually give birth to baldheaded children, blue-eyed parents to blue-eyed children, and squint-eyed parents to squint-eyed children, what prevents a longheaded parent from having a longheaded child?

Aristotle writes:

Children are born resembling their parents both in their whole body and in its individual parts. . . . Moreover, this resemblance is true not only of inherited but also of acquired characters. For it has happened that the children of parents who bore scars are also scarred in just the same way and in just the same place. In Chalcedon, for example, a man who had been branded on the arm had a child who showed the same branded letter, though it was not so distinctly marked and had become blurred (*De generatione animalium* I 17.721b 29f.).

As R. Martin (1928) has shown, the skull disfigurement of newly born children has a very large geographical distribution extending from South America to Indonesia. It is still practiced today by certain isolated Indian tribes in South America and is also known in Europe (southern France). E. Lesky (1950) reviews the ancient literature on this topic. He does not discuss it in terms of older Lamarckian notions but in terms of extended modifications (*Dauermodifikationen*), for the texts in question seem to suggest that artificially induced macrocephalia did not automatically occur in subsequent generations but had to be reinduced from time to time with the aid of the usual techniques.

The question of the inheritance of acquired characteristics raises the issue of looking for examples concerning the problem of heredity versus environment. All the ancient theories of heredity were based on the

[6] H. Hommel (1927) has made a study of this phenomenon in his comparison of Hippocratic and modern theories of inheritance.

fact that children resemble their parents. Exceptions were explained by reference to environmental factors of various kinds. Empedocles held that children who did not resemble their parents were born because their mother had habitually looked at statues in a loving way during her pregnancy. The belief in fright during pregnancy, so deeply entrenched in popular thought, was gradually discarded by Greek physicians, who tried to explain a lack of resemblance between parents and children in terms of internal and external influences on the embryo. Too small a uterus, pathological and traumatic deformation of the uterus, injury to the uterus caused by previous births of a large number of children, and malnutrition in the mother are cited as causes for the lack of resemblance between parents and their children. According to the physicians of the Hippocratic school, menstrual blood was used to nourish and sustain the embryo during pregnancy; any bleeding at that time was therefore thought to harm the embryo.

It is clear that the question of the relations between heredity and environment was frequently debated. Hippocratic physicians even discussed the diversity of features between various European races in these terms. (They considered the different features of European races to be more marked than those of the Asians.) They looked for the causes of these different features in environmental variations, as for example in changes in temperature and humidity. The mechanism which brought about these different features is explained as follows: Under the influence of varying climatic conditions "the developmental processes during the coagulation of the semen may change; though the people in question remain the same, their semen develops differently in summer and in winter, in dry and damp weather."

This difference in semen was thought, then, to occasion an irregular development of the embryo, which accounted for the fact that a child did not necessarily resemble its parents.

In this connection we must also note that Hippocratic physicians could distinguish between congenital disease and sickness that was contracted from environmental factors. They tried to explain constitutionally determined, i.e., congenital, disease in terms of the theory of pangenesis. In the book *On the Sacred Disease*, written about 400 B.C., epilepsy is clearly seen to be a congenital disease that occurs in those persons whose constitution is "excessively phlegmatic." The author of the work says of this disease:

> Like other diseases it arises according to hereditary factors. For if phlegmatics are born of phlegmatics, cholerics of cholerics, consump-

tives of consumptives, and melancholics of melancholics, what then prevents some of the children from contracting this disease when the parents have suffered from it? For semen comes from every part of the body, healthy semen from healthy parts, diseased semen from diseased (VI 364, 15ff.).

The study of other Hippocratic works clearly indicates that these physicians could distinguish between congenital disease—whose presence could be discovered by investigation—and other diseases caused by environmental factors. The author of the Hippocratic work *De natura pueri* mentions that there are two types of menstruation, the copious and the weak flow, both of which are constitutionally and congenitally determined. He says of them that "this circumstance, whenever it occurs, invariably was a feature of the mother's constitution, and has been handed down by her" (VII 534, 20f.).

Greek natural philosophy of the fifth century B.C. culminates in the Hippocratic doctrine of pangenesis, which was used to explain both general and specific problems of heredity. The fourth century B.C., however, sees his doctrine superseded by the work of Aristotle (384-322 B.C.), on which the theories of reproduction of the medieval scholastics were based.

Aristotle (Figure 6), a physician's son, was born at Stagirus in Thrace. He studied under Plato for a period of twenty years and later acted as tutor to the Macedonian prince who was to become Alexander the Great. He founded the so-called "peripatetic" school of philosophy, a name which arose from his custom of walking in the tree-lined avenues (*peripatoi*) in Lyceum as he taught. He was later charged with impiety and fled to Chalcis in Euboea where he died. He was the first thinker to create an all-embracing philosophical system. His analysis of rational thought includes the principle that a thought is true if, and only if, it corresponds with reality. Further, knowledge of a universal presupposes the perception of particulars.

Aristotle's hematogenic theory of semen, according to which semen is formed from the blood, owes much to the development of a scientific way of thinking which may be characterized by its teleological approach and its emphasis on conceptual analysis in dialectic terms.

Other thinkers in earlier centuries had anticipated this Aristotelian doctrine, to be sure. But it seems doubtful that Pythagoras (584-500 B.C.) ever held the view attributed to him—that semen is actually the foam formed from the richest blood. Parmenides (c. 440 B.C.), however, had already derived generative substances from the blood.

6 Aristotle (384-322 B.C.)

Although, according to pangenesis, blood was merely a means of transporting the seed which had been formed in every part of the body, Diogenes of Apollonia had described in concrete terms how blood was converted into semen. He thought that the thickest blood in the veins was absorbed by the fleshy parts of the body and that the rest was transformed into a more liquid, warm, and foamy substance, semen, as it was being carried to the sex organs.

To understand the rich contribution that Aristotle made to biology and the philosophy of nature fully, we must first review the philosophical background that gave rise to his views on the sexes, the formation and function of semen, and to his theory of development and treatment of heredity.

It is evident from his biological writings that he not only classified some five hundred and forty animal species according to their external appearance, but that, in the course of his investigations of anatomical structures (which, as he thought, were an expression of formative causes), he must have actually dissected animals of at least fifty different species. Aristotle's empirical observations and investigations open up a new era in Greek science. He discovered a number of structural interrelations in different animals and investigated the embryonic development of the chick and other living creatures. The results of these investigations enabled him to propose a theory of development.

According to him, the way an individual develops is determined by its inner nature; it is not true that its nature is determined by the way it develops. Rather, the nature of an individual determines, according to laws that are inherent in it, the developmental process in the course of which its (final) nature is realized. The dynamic principle that governs an individual's nature Aristotle calls "form" (*causa formalis*), and it is form which controls and shapes matter during the developmental process. In procreation, form is provided by the male partner and transmitted by his semen. Matter (*causa materialis*) resides in the female partner and is carried by her menstrual blood (*catamenia*). The male sex is the bearer of the principles of motion and procreation, while the female sex is the bearer of matter.

In his treatment of the formation and nature of semen Aristotle often proceeded by way of analogy with examples from the animal kingdom. He held that the testes were organs that merely regulated the frequency of coition. Semen, he thought, was formed from the blood, from that surplus (*perittoma*) of nutriment that is transformed in the body by the action of vital heat and finally concocted into semen. He held that women were unable to produce semen; in this he disagreed with Alcmaeon and other physicians of that era. In women the perittoma is constituted by the menstrual blood, which corresponds, though only in part, to the formation of semen in men. For, as the female perittoma does not have the same degree of vital heat as that of the male, it is unable to complete all three steps: nutriment, blood, semen. The Empedoclean principle of vital heat has clearly influenced Aristotle's view, for he claims that the sexes are qualitatively distinguished with respect to vital heat. The development of the embryo is initiated at the moment of conception by the male semen which endows it with form and motion. The menstrual blood is permeated by the active principles of the semen, is set in motion, and receives form without the semen actually contributing any material substance to the embryo. Aristotle tries to clarify his view of the nonmaterial contribution of semen by frequent and varied metaphorical reference to the role of artists and craftsmen. Thus he says, for example, that a sculptor who is making a statue already knows the form and shape which his finished work will have. This preconceived shape determines, as it were, the workings of his hands, which eventually create it as they fashion the material of which the statue is made. But in this process, the artist who creates the work of art does not make a material contribution. Similarly, the sole function of male semen is to convey the nonmaterial principles of motion and formative power. These change the constitution of the menstrual blood by thickening it and so preparing it as to produce an embryo of the kind determined by the semen.

According to Aristotle's doctrine, then, each and every part of a new individual is contained within the semen as nucleus and is conveyed to the embryo by the semen. For semen is formed from sanguineous nutriment which, as an organism develops, is supplied to every part of the body and indeed causes the growth of the various body parts. In this way semen, the perittoma of the blood, acquires the power to shape each and every organ. This power—and the quite specific developmental tendencies it comprises—is then conveyed to the embryo. Analogously, the menstrual blood of the woman passively contains each and every part of her body as a nucleus, which is then shaped into an organ by the action of the principle of motion of the sperm. It is then conveyed to the embryo, which is formed from the material substratum of this blood. Semen, then, brings about qualitative changes in the matter of the female organism; Aristotle compares it with the rennet that causes milk to curdle.

The heart—the formative source of the embryo, for it brings about the successive and well-ordered articulation of the remaining body parts—is the first organ to be formed in the developing embryo. In this process, however, the mother's uterine warmth (which is derived from her heart) can endow the various developing organs with specific characteristics; but it cannot determine their essential nature nor their over-all organic arrangement within the body. To achieve this, yet another nonmaterial and dynamic power is required, which Aristotle called "soul." He held that the heart was the seat of the soul, the formative power of which enabled the embryo to develop from a mere assemblage of various organs into a complete and total being.

Aristotle's views on heredity were deeply influenced by his metaphysics. His theory of heredity was quite different from those of the atomists who assumed a material basis for the transmission of physical features from one generation to another. In Aristotle's theory of heredity the menstrual blood of a woman is the material substrate for the creation of the embryo, but the decisive role falls to the nonmaterial transference of form and motion through the man's sperm. Aristotle's thoughts concerning the problem of heredity must be interpreted with this in mind. They no longer merely center on sex differences or on the extent to which children resemble their parents, as was the case with many of his predecessors. Instead, Aristotle paid particular attention to the hereditary nature of specific and generic characters and explained them in terms of the different valences inherent in the movement components conveyed by the male sperm. He attached more importance to those movements which shaped individual (rather than generic) characters, for, as he says in the fourth book of *De generatione animalium* (767b 23ff.).

> In generation both the individual and the genus are operative—the individual more so, however, than the genus, since it is the only real existence ["substance"]. For the offspring is indeed produced as of a certain kind, but also as an individual, and it is in this latter status that it has its real existence. It is therefore from the forces that the movement components are immanent in all such individuals, potentially including those of ancestors, but to a greater degree from the one who is most closely related to the individual.

This is to say that the different degrees of valence inherent in the movement components determine, in order, the characters possessed by the class, order, family, genus, species, and finally those of the individual. The formation of these different characters is successively determined during the process of embryonic development in accordance with these degrees of valence. For Aristotle, then, there is no development as we think of it in modern theory of heredity, but rather a coexistence of invariable types among which the same relations always obtain. Sex and individual characters are caused by the strongest movement components, which are the last factors to operate in the developmental process.

But how do offspring of different sex arise, and what is the cause of individual differences and variations? To begin with, the movement components, operating according to their different degrees of valence, occasion the development of the same sex and individual characters. But this pure line of inheritance is broken by the appearance of offspring that does not resemble its parents. Aristotle considered such offspring to be, in a broad sense, deformities.

For even he who does not resemble his parents is already, in a certain sense, a monstrosity; for in a way, as far as these cases are concerned, nature has simply departed from the type. Indeed, the first departure occurs when the offspring is female rather than male. This, however, is a natural necessity, for a class of animals divided into sexes must be preserved (767b 6f.).

What, then, causes such anomalies or deformities? They are due to the action of the menstrual blood on the movements conveyed by the sperm; the menstrual blood reacts in the presence of spermatic movements and may have a formative influence on them. Now this reaction can be of different potencies. If it is strong enough to prevent the movements of the sperm, which determine the male sex, from prevailing, then the embryo will be female. On the other hand, should the movements that determine the movements and the number of individual male characteristics be repressed (by the reaction of the menstrual blood), then the offspring is male but resembles the mother. Female sex in the offspring, transmission of features within the same sex, cross-sexual inheritance of individual features, and partial resemblances between parents and offspring were all accounted for in terms of the strength of the reaction of the menstrual blood to the different spermatic movements determining sex and individual characteristics.

In the case of a weaker reaction of the menstrual blood to the male movements, the valences of the latter are merely diminished and reduced by one (or more) degrees. This is how Aristotle explains the occurrence of features of male ancestors on the paternal side, for as he says in *De generatione animalium* (768a 14f.): ". . . movement components of the father are weakened [to a greater or lesser degree], then the least deviation is a change into those of the father, the next into those of the [paternal] grandfather."

And similarly on the mother's side: "In the same way, the movement of the female parent [when suppressed to a greater or lesser degree] changes into that of her mother, and if not into this (because of a greater degree of suppression), then into that of the grandmother, and so on into those of her more remote ancestors" (768a 18f.).

As E. Lesky (1950) has rightly pointed out, Aristotle thus was the first to attribute an inherent, active movement to the mother, that is, to the matter contained in her menstrual blood, for previously the principle of movement had been accorded only to the sperm. Thus the antithesis between form and matter—so important up to this point—has now become blurred.

To begin with, menstrual blood was merely the passively receiving and form-accepting principle, while semen was the actively working, form-giving principle. To explain the heredity of sex and of individual characteristics, the menstrual blood must generate a more or less powerful counteraction. The inheritance of characteristics stemming from the maternal ascendancy can only be explained, however, by an active, self-forming force inherent in menstrual blood. Aristotle confines his attempt to explain the inheritance of characteristics from paternal and maternal ancestors to cases of direct descent, for he does not offer us any explanation of how characters are inherited from the maternal ancestors of the father or from the paternal ancestors of the mother.

The essential principle of the Aristotelian theory of heredity, therefore, is that any lack of resemblance between father and offspring is caused by the suppression, in stages, of the various male movements by means of the reactive powers which reside in the substance of the menstrual blood. The perfect example of the wholesale transmission of every character of the father to the child (who would then resemble his father in every possible respect) is thus an extreme possibility. All kinds of deviation from this ideal case can occur, however, including—at the other extreme—the formation of a monstrosity, which is produced when the father's movements are suppressed to such an extent that only the poorest in quality, those which represent a sort of generalized animal, prevail. And how is it that sex and individual characteristics are determined during the operation of these processes? With twentieth century genetics research in mind, it is important to note that Aristotle already held that both internal and external environmental factors played a decisive role. Like Empedocles, he thought that vital heat had a large part in determining sex. The vital heat of the male may prevail over the female matter (which is not as pronounced), thus endowing it with movements which determine male sex; however, if it does not prevail, a female is produced.

Aristotle also adopted other views from earlier times, for he thought that viscid, strong semen would produce more males than thin, watery semen, for the former was warmer than the latter. Youths and elderly men were apt to sire females, for in the former there was not yet enough vital heat and in the latter it was declining. Similarly, a man's constitution influenced the sex of his offspring; if he had a moist and effeminate constitution he was more likely to produce females because of his lack of vital heat. These ideas were influential through the ages and were still debated in the past century. Nourishment and climatic factors were also thought to affect the constitution of the body and were therefore thought to play a part in sex determination. Boys (girls) tended to be produced if copulation took place when northerly (southerly) breezes were blowing, for the north wind invigorated and thickened the semen while the south wind enfeebled it and made it watery. The effect of nourishment on sex determination was caused chiefly by the intake of water, which is contained in all foods, and hard, cold water was thought to produce females. As Aristotle says in the fourth book of *De generatione animalium*:

For the nature of the nourishment and the bodily constitution are primarily conditioned by the tempering of the surrounding air and by the nourishment ingested, especially the water; for the body absorbs more of this than of anything else and it enters as nourishment into all foods, even the driest. Hence, hard and cold water causes in part infertility and, in part the birth of females.

Without question the procreation and heredity theories of Aristotle constitute but one element in his philosophy which brings order to his cosmic image. His achievement, in fact, was to have drawn together all the different notions of reproduction and heredity of his time and to have systematically interpreted them on teleological lines. For Aristotle, however, such teleologically interpreted developmental processes were ultimately to be explained in metaphysical terms:

Generations of people, animals, and plants will therefore exist for all time. But as they are of male and female origin, it might well be thought that for the sake of procreation each sex would contain the essential elements of both sexes. Now in accordance with its nature, the first efficient cause—embodying Idea and Form—is superior to and more divine than matter, and so it follows on teleological grounds that the higher be separated from the lower; and this is why male sex is separated from female sex in those creatures for which this is possible and to the extent to which it is possible. For the male principle of motion from which developing creatures are derived is more important and more divine; matter, however, is the substratum of the female (*De anima* II 4.415a 26ff.).

G. M. Hartmann (1959) drew attention to the fact that although Aristotle's philosophy differed sharply from the idealism of Plato, he was still greatly influenced by his teacher in his assumption of a supreme principle, the "prime mover," or divine cause of all change in true being. This dependence on a supreme principle provides the link between Aristotelianism and the scholastics of the Middle Ages.

The original, scientific, and comprehensive account of the universe propounded by Aristotle in the fourth century B.C. may be contrasted with the detailed studies in reproduction and heredity undertaken by the Alexandrian physicians who succeeded him.

Herophilus of Chalcedon (third century B.C.) is of special interest, for he investigated the problem of the origin of generative matter by dissecting the human body. (He had also discovered in this way that the brain and not the heart, as Aristotle thought, is the seat of intelligence.) He recognized the essential function of the testes and also discovered the ovaries, which he termed "female testes." We learn from Galen (*De semine* IV 596f.) that Herophilus described the ovaries as follows:

Testes which differ only very slightly from those of a man are attached to both sides of the uterus. In females, however, it is not so that both testes reside in a single scrotum, for one is attached to the right and the other to the left shoulder of the uterus, each therefore separated from the other. Now each testis is covered by a thin, sensitive membrane, is small in size and oval in shape, and has the appearance of a gland; it is enclosed within a ligamentous structure and is formed of closely knit tissue. Mares, in point of fact, have very large testes which are affixed to the uterus by several membranes together with a vein and an artery.

The testes and ovaries, however, were not thought to be the only places in which generative matter was formed, for Aristotle's spermatology influenced the teachings of his successor. Herophilus gave a good descriptive account of the testes, seminal vesicles, spermatic duct, and other organs. From Galen we learn that Herophilus considered the spermatic duct to be the principal organ of spermatogenesis, that he also mistakenly thought that it entered the bladder, and that he therefore also assumed that semen was discharged via the spermatic duct, bladder, and urethra. For Herophilus, blood still played an insignificant part in the production of semen, especially in the convoluted structures of the blood vessels in the testes which enabled it to prepare a kind of seminal liquid from which the testes produced the sperm itself.

The Stoic school, founded by Zeno (354-274 B.C.) who came from Citium, a town in Cyprus, was also concerned with questions of reproduction and heredity though he did not deal with them in a scientific way. The Stoics were chiefly interested in ethics and the good life. They sought to live in accordance with the harmonious dictates of nature and reason and investigated the nature of the human soul, a material essence that permeated the whole body and determined all of its functions. The Stoics dealt with the inheritance of spiritual features and with the essential nature of semen and its connections with the soul. The only records of this doctrine are contained in fragments and uncertain traditions. In one of these, Areus Didymus writes:

Zeno says: sperm discharged by man consists of an essence contained within the fluid; it is a part of his spirit which now resides in the mixture provided by the semen of his forbearers—a mixture drawn from each part of their souls; for this sperm contains the same formative powers as their whole beings. As soon as it reaches the womb it is absorbed by another essence—a part of the female spirit—and uniting with it, increases in seclusion. It is stimulated and set in motion by that [essence], and feeds and thrives upon the fluid.

For the Stoics, then, sperm was a special part of the essence which, possessing formative powers drawn from every part of the body, was consequently also the bearer of all those generative forces which cause the offspring to exhibit the same organs, capabilities, and spiritual propensities as its begetter. This view may contain the beginnings of the doctrine of the continuity of germ plasm. E. Lesky (1950) and others have raised the issue of whether the Stoic doctrine is merely a more advanced version of pangenesis or whether other ideas are elaborated in it. Blood was the corporeal substratum of the essence and, changing into its vaporous state, refurbished and invigorated it; it is therefore essential to consider the influence of the hematogenic theory with regard to this issue. It seems probable that the Stoic theory of semen consists of elements of both doctrines which were combined in such a way that any inconsistencies were not immediately apparent. With regard to the question of whether the sexual secretions provided by the two sexes are of equal importance, the Stoics seem to have followed Aristotle, for they thought that females were incapable of producing active sperm. Females did, however, produce a female essence that mingled with the semen but was never clearly described. On the basis of extant fragments and traditions we are therefore unable to judge whether it was Aristotle or the Hippocratics who had a greater influence on the theories of the Stoics.

Like Aristotle, the Stoics insisted on the invariability of species, for they held that sperm was only capable of producing individuals of the same species. The material generative forces that resided in the seminal pneuma were able to produce specific and generic characters and were responsible, moreover, for transmitting individual characters and features of parents and ancestors to the offspring. The generative forces of the Stoics are therefore analogous to Aristotelian efficient movements, except that the generative forces are active in both the male and the female and actually fashion the offspring. The problem of the transmission of features from earlier generations (which had caused Aristotle a number of difficulties) was dealt with more simply by the Stoics. This is (according to E. Lesky 1950, p. 170) confirmed in a quotation from Origines who says:

> Since the generative forces of both the present and previous generations reside in the father, it sometimes occurs that his own generative force predominates, in which case the offspring resembles its father; sometimes, however, it is the procreative force of the father's brother, or of his father, or of his uncle, and sometimes even of his grandfather which predominates, and the features of the offspring vary accordingly. But should all the generative forces be simultaneously excited during the convulsive movements of physical union, then it may also be the case that the one which predominates is that of the mother or that of her brother or grandfather.

It seems clear that both Aristotle and the Hippocratics proposed theories of reproduction and heredity that were more self-consistent than those of the Stoics. And so the latter cannot be said to have made any important new contributions to the solution of problems of heredity with the exception of what they had to say about the transmission of spiritual qualities. In spite of this, their philosophical system was influential for hundreds of years and, since medicine was so closely tied up with philosophy in ancient times, finally gave rise to the formation of the school of Pneumatic medicine in the first century A.D. This school was founded by Athenaeus of Attaleia, the author of at least thirty treatises (none of which have survived) dealing with every aspect of therapeutics. Adopting the form-matter antithesis, Athenaeus followed Aristotle implicitly with regard to spermatogenesis and was criticized by Galen in *De semine* for having done so. But Athenaeus could not completely disregard Herophilus's discovery that the "female testes" were able to prepare a kind of seminal liquid. However, taking his cue from Aristotle, who held that only males were able to produce semen, Athenaeus declared that the "female testes" were rudimentary organs which had no function.

It was problems concerning the origin of the hybrid offspring of two different species which interested Athenaeus most. (Alcmaeon, Empedocles, and Democritus had already commented on the sterility of certain hybrids, for example, the mule.) Aristotle had thought that animals of different species could be successfully crossbred provided that their periods of gestation were equal and that they did not differ greatly in size. Other thinkers claimed they had observed that hybrids between dog and fox and partridge and other fowl assumed, in time, the appearance of the female parent. Aristotle compared these phenomena with the changes that occur in plants when they are sown in unfamiliar ground. He writes as follows:

If animals of different kinds are crossed—which is possible only if the periods of gestation are equal and if the animals do not differ greatly in size—the first cross has a common resemblance to both parents, as the hybrid between fox and dog, partridge and domestic fowl; but as time passes, and in the following generations, the offspring assume the appearance of the female, just as foreign seeds produce plants which vary according to the land in which they are sown.

Athenaeus attributed the "dominance" of the female parent in the first hybrid generation to the nourishment that was consumed. He claimed that the hybrids he-ass × mare, she-ass × stallion, dog × vixen, and bitch × fox in all cases resembled the female parent to a greater degree than the male. He also said that the hybrids goat × ewe and she-goat × ram resembled their mother in all respects except that their coat exhibited features of their parental ancestor.

Some of Athenaeus's examples must have been cases of faulty observation since hybrids between dog and fox, goat and ewe do not occur. His claims, moreover, are inconsistent with his basic thesis (adopted from Aristotle) that the male partner alone determines the inheritance of specific characters.

Athenaeus (whose work has not survived) was succeeded by Galen, the court physician to Marcus Aurelius, who was born in Pergamum c. A.D. 129 and died in Rome in A.D. 199. The two books of Galen's treatise *De semine* form the last important work to issue from a 700-year-old philosophical debate on problems of reproduction and heredity. (See also H. Balss [1934] for Galen's theory of heredity.) Galen's anatomical studies led to modifications and improvements in current views, and these in turn entailed a revision of Aristotelian theory. Galen discovered that the oviducts inosculate with the uterus, one on each side, and not, as Herophilus thought, with the bladder. He describes their passage as follows:

Indeed the seminal vessels which spring forth from the testes clearly contain sperm, just as in the case of man; close to the testes these ducts are broad and there is a definite aperture between their walls, but further on they narrow and their apertures almost disappear; still further, as they near the head of the uterus, they broaden again and terminate in that organ, one on each side. Neither Herophilus nor Euryphon has seen that this is so; nor, indeed, has Aristotle. And I name these men not merely on account of this deficiency in their knowledge but also because they are the best authorities in the field of anatomy.

Galen's discovery led him to believe that the oviducts were the female counterpart to the *vasa deferentia*—for both sets of organs served to convey semen—and that the female semen was discharged into the uterus where it helped to form the embryo. On this point Galen is directly opposed to Aristotle and invites comparison with the Hippocratics, whose doctrine of pangenesis affirmed that semen was provided by both sexes. Using terms which date back to Alcmaeon, Galen describes the semen found in the oviducts—probably, in actual fact, the mucus secreted by those organs—as being less abundant, colder, weaker, and more watery. During the process of fertilization the main function of female semen was to nourish male sperm in the uterus. Later, the allantois was formed from the female sperm, but the chorion, amnion, vessels, nerves, sinews, cartilage, and bones were all formed from male sperm. For Galen, then, there are quantitative and qualitative differences between male and female sperm, though both are operative in fertilization. He held that the testes played a very important part in spermatogenesis; here he differs from Herophilus who saw the spermatic duct as the principal organ involved in the production of sperm. But even Galen had not completely discarded the idea that a kind of seminal liquid was produced in the blood vessels and later transformed into sperm in the testes.

Galen made a comprehensive study of the effects of castration. Herodotus (484-425 B.C.) the historian and Aristotle had both worked on this topic, and the latter had even given a detailed description of the various changes that occur in the body of a eunuch. But Galen realized that the generative matter secreted by the testes acted on every part of the body: the eunuch, in whom such action was inoperative, was neither totally masculine nor had he assumed complete femininity. Galen was the first thinker to anticipate the study of endocrinology.

With regard to sex determination, however, Galen—unlike Aristotle—had nothing to add to the ideas propounded in the Right and Left Theory and in Empedocles' doctrine of vital heat. Galen held that the degree of vital heat imparted by the pneuma to the sexually developing embryo was what determined its sex: the optimum degree of heat produced males, and lesser degrees of heat gave rise to females. His account of the inheritance of specific and individual characters was likewise unoriginal, for he was unable to formulate a new theory on the basis of his doctrine of the two sorts of sperm. Galen's eclecticism is one of the most conspicuous features of his thought, and he always tried to reconcile conflicting ideas.

For Galen, female semen played no part in the transmission of specific characters, for it was not abundant enough, nor was it vigorous in its generative power. On this point he once again invites comparison with Aristotle, though the fact that he attributes both dynamic and material principles to male semen and to the menstrual blood means that his view is an improvement on that of Aristotle. It is not the case, however, that both sexual secretions possess these principles in the same proportions. The major formative principle together with a weak material component inhere in the sperm, while the greatest material forces supplemented by a weak dynamic component reside in the menstrual blood. Galen tries to account for Athenaeus's observations on hybridization in these terms. The hybrid offspring produced by crossbreeding resemble their female parent to a greater degree because it was she in whom the material principles were most active.

Galen thought that the inheritance of specific characters depends on the material principles of the mother and that the father has the more important part to play in the transmission of individual characters. These individual differences are seen as variations or features that distinguish "one man from another and one horse from another."

But Galen's views on the inheritance of individual features are by no means clear. For although, in agreement with Aristotle, he says at one point "that any resemblance with regard to individual features arises from the principle of motion inherent in the sperm," he elsewhere emphasizes that female semen is also operative in determining individual differences. He appears to be trying to account for the fact that, with regard to their individual characters, offspring can resemble both parents despite the greater potency of male sperm.

Galen's anatomical studies in reproduction and heredity—his discovery of the actual path taken by the oviducts and his recognition of the regulative role of the gonads, especially the testes—paved the way for further advances in knowledge. But his ideas on the inheritance of specific and individual features are based partly on the Hippocratic doctrine of male and female sperm and partly on the philosophical reflections of Aristotle. Galen did not really have anything new and original to say about heredity but he utilized, in varying degrees, the ideas of his predecessors to explain the phenomena known to him.

The Greek ideas on reproduction and heredity that we have now examined in the works of their greatest philosophers and physicians turned largely on human problems; they derived far less from animal observation and rarely touched on phenomena which were noted and explained in plants.

Of the Greek philosophers and naturalists who did study problems arising from cultivation and agriculture, who examined variability and degeneration (especially as these occur in plants), Theophrastus of Eresus in Lesbos (c. 372-287 B.C.) (Figure 7) is by far the most important. He studied under Aristotle and inherited his master's library together with the leadership of the Peripatetic School, the Lyceum, and presided over its scholars from 322 to 287 B.C. He was a prolific writer, and many of his works, including *Historia plantarum* and *De causis plantarum*, have survived. We can be sure that he passed on a great deal of what Aristotle already knew about plants.

Theophrastus added to the biological research of his teacher by describing and classifying numerous species of plants. Many of the technical expressions which he introduced, for example, carpos for fruit and pericarpion for seed vessel, survive to this day in one form or another. He held that the higher plants reproduce sexually, but this fact was no longer acknowledged by the end of the classical era (see St. F. Mason 1961). His descriptions of variation in cultivated plants are of particular interest. In his *Historia plantarum* (8, 4, 1ff.) we read:

There are several varieties of both [i.e., barley and wheat], and these can be distinguished by their grain, ears, shapes, and capacities. Of barley, different varieties may have respectively two, three, four, five, and even six rows of seeds. Yet another variety is Indian barley, for it has side shoots. In some varieties the ears are large and lax, while in others they are smaller and more dense. Again, the grains themselves may be round and small or more oblong and larger; moreover, some are white and some reddish.

On inherited variations, Theophrastus writes as follows in *De causis plantarum* (lib. 2, cap 3):

Changes sometimes occur spontaneously in fruits and also, though more rarely, in whole trees. Soothsayers see omens and portents in such variations; and when they occur infrequently they are generally thought to be contrary to nature, though this is no longer the case when they are found in large numbers. Some herbaceous plants return to their wild state if they are not tended for some time. But spelt and one-grained

7 Theophrastus (c. 372-287 B.C.)

wheat when sown without their husks change into wheat after a period of three years. This recalls the variations caused by the locality in which the plants are sown; for annuals also vary in this way with time.

This passage contains the first clear reference to "transmutatio frumentorum," an issue that was hotly debated throughout the Middle Ages and even in modern times. M. Dittrich (1959) has made a detailed textual study of everything that has been said concerning this question.

In *De causis plantarum* Theophrastus makes the following remarks about the causes of variation; they apply in particular to variations which give rise to different species:

Perhaps it is different habitats that give rise to all the various different species or to some of them. For a given habitat tends to diminish differences in the crops sown in it; or, on occasion, it produces specific variations that are beneficial to the plant, as in the case of Thracian wheat. This variety of wheat sprouts late and the grain is covered with numerous glumes; both of these characteristics are caused by the winter frosts. Moreover, Thracian wheat which is sown early in other regions also sprouts late and takes a long time to mature; and wheat taken from other regions and sown in Thrace also sprouts late. In these cases, what has regularly occurred has become a part of the natural order of things.

Again, in *De causis plantarum* (4, 11, 7), he writes:

For if two or more things are able to give rise to another, then what is produced will of necessity be characterized by the distinctive features of the things that produced it; this is also the case with animals. These [variant] features which occur in the offspring are acquired from the male and from the female, from the soil and from the air, in short from whatever counts as nourishment. Features that distinguish one species from another are produced in this way; and what was once a freak of nature becomes a part of the natural order when it grows abundantly with the passing of time. All this is quite usual and occurs frequently.

Theophrastus was acquainted with several species of date palm and realized that they required the pollen from a nonfruiting palm for successful pollination. He called the flowering palms male and the fruit-bearing ones female. He says in *De causis plantarum* (lib. 2, cap. 13) that "something similar to this occurs among fish, for the male sprinkles his milt on the eggs discharged by the female." His teacher, Aristotle, had known of this phenomenon before him (*Historia animalium* VI, 13, 14).

Theophrastus distinguishes between the plants that may be raised from the fruit and those that are inferior when grown from seed, e.g., the vine, apple, fig, pomegranate, quince, and pear. Olives grown from seed are also degenerate, and the kernels of the almond give rise to a product which is inferior in taste. In chapter thirteen of *Historia plantarum* he gives a comprehensive account of differences in flowers and notes that some of them are sterile. In the cucumber and lemon the flowers that have a growth the shape of a distaff in the middle are fertile, but those that lack this growth are sterile. Jessen (1864) has pointed out that Theophrastus was the first naturalist to observe the pistil and comment on its function.

In his exhaustive study of Theophrastus's observations on degeneration in plants, W. Capelle (1949) states that the causal determinants of the cases of degeneration described by Theophrastus fall into three definite groups. The first is that of human intervention, that is, the application (or subsequent neglect) of the techniques of cultivation; the second is comprised of external influences—heat, cold, gales, continuous drought, rain out of season, and so on; the third consists of processes and changes that simply occur in the plants themselves. Such processes and changes must have seemed to Theophrastus to be "contrary to nature" since for him, and for the ancients in general, nothing occurred in nature without a cause. Such changes and variations were clearly at variance with the concept of a physis which controlled the well-ordered, purposive development of every organism.

According to Capelle, Theophrastus's scientific achievements were restricted by his unwillingness to question traditional dogmas in a serious way. For instance, without attempting to test it experimentally, he readily acquiesced in the dogma that cereals degenerate into darnel; in this, he renounced a practice on which the Hippocratics had already insisted and which they had successfully applied.

Capelle also points out that Theophrastus considered the problem of (nongenetically induced) degeneration in human beings: he asked whether the buffets of fate could transform a man's nature and whether he could be deprived of his virtue by bodily suffering.

In about 100 B.C. it was Posidonius of Apamea, second only to Aristotle in the comprehensiveness of his thought, who dealt with the wider ramifications of the problem of degeneration long after Theophrastus. Posidonius's detailed investigations into degeneration and its causes did not merely consist of examples taken from the plant and animal kingdoms, but they also included phenomena drawn from his study of whole races and tribes.

3

Reproduction and Heredity: The Hypotheses and Practical Knowledge of the Romans

The Romans added little that was new and original to man's understanding of the theoretical foundations of reproduction and heredity. They adopted the fund of ideas developed by the Greeks, in particular those of Hippocratic medicine and Aristotelian philosophy. For example, the ideas contained in *De rerum natura*, the didactic poem written by the famous Roman philosopher-poet Lucretius (Titus Lucretius Carus c. 98-55 B.C.), are ultimately derived from the Greeks. Based on the world view of Lucretius's master Epicurus (who was deeply influenced by the atomism of Democritus), the poem reflected the moral philosophy of Epicurus, who taught that men should limit their needs and desires and seek to acquire tranquillity and repose of spirit; men should delight in philosophical knowledge and in the sovereign power of the mind.

Lucretius wanted to free his fellowmen from the fear of death and the fear of the gods. Following Epicurus, he proclaimed a materialistic concept of the universe; his poem is considered the most consummate exposition of Epicureanism preserved from antiquity. The most vigorous and consistent materialism of ancient times had been expressed in the atomic theory of Democritus. It is this hardy materialism, adopted by Epicurus, that was set forth in Lucretius's poem.

De rerum natura, which leans on the ideas implicit in the entire Greek materialistic tradition, is the most important philosophical work of the ancient Romans. Of Lucretius's successors, none was equal to him. In teaching that matter was indestructible and in perpetual motion, he provided the foundations for one of the most important laws of modern physical science. The contrary movements of the atoms, their separation and union, are what cause the origin and decay of material bodies.

The poem is a mixture of truth, erroneous ideas, and naïve conceptions. Lucretius takes account of Aristotle's observations on the inheritance of features from earlier ancestors by writing:

> It often happens that children resemble their grandparents or have features similar to those of their great-grandparents. This occurs because the parents' bodies often retain large quantities of semen [atoms] in many different configurations which are derived from ancestral stock and handed down from one generation to another.

From this semen Venus brings forth a varied assortment of characteristics and reproduces ancestral traits of expression, voice, or hair; these features—as well as our faces, bodies, and limbs—are also determined by the specific semen of our kin (IV, 1218-1226).

What Lucretius had to say about development and selection is equally important, as seen from the following passages:

Nothing remains forever what it was; everything is in a state of flux. Nature transmutes everything and forces all things to assume new forms. When one thing, consumed with age, declines and decays, another emerges from ignominy and waxes strong. Age produces changes in the nature of the universe in this way. The earth also passes through successive phases and, at one stage, can no longer bear what it could, while at another it can bear what before it could not (V, 830-836).

In those days many species must have died out completely for they were unable to reproduce their kind. Each different kind of creature that now enjoys the breath of life has been protected and preserved from the beginning of its days by its prowess, cunning, or speed. Many of them survive under human protection because their usefulness has made us want to care for them (V, 856-862).

Superstition and mysticism were rampant in the age of Roman supremacy. Pliny the Elder (Gaius Plinius Secundus, born A.D. 23 and died during the eruption of Mount Vesuvius in A.D. 79) reports at great length on these aspects of the civilization of ancient Rome. A Roman commander and superintendent, he was also the author of the *Naturalis historia* in thirty-seven books. In it he comments on the origin of the universe, the geography of the earth, the state of medical knowledge, and other biological and agricultural questions of his age. In the seventh book we read:

It often happens that healthy parents have deformed children, and deformed parents healthy children or children with the same deformity, as the case may be. Certain kinds of marks, moles, and even scars also reappear in the offspring. A birthmark on the arm can be observed in Dacians of the fourth generation.

The Romans were aware of the hereditary nature of individual characters and deformities. Pliny writes that the Lepidus family gave birth to three children with a membrane over the eyes. Extravagant ideas abounded to account for the fact that children are like their parents. For example, the mental images of parents during coition, thoughts suddenly passing through their minds, abrupt changes in their moods, and the diversity of their intellectual powers were all thought to occasion differences and resemblances between themselves and their children. Misshapen births were attributed to a number of different causes. Pregnant women who ate salty foods would give birth to children without nails. Failure to hold the breath made the delivery more difficult, and yawning during childbirth supposedly produced an infant who was stillborn. Sneezing after copulation was thought to cause abortion. The smell of lamps being extinguished could also occasion the birth of a malformed child.

The Romans had some very curious ideas on the origin of new animal species and on hybridization.[1] Three of the greatest Roman men of letters, Varro, Virgil, and Pliny, all give accounts of the origin of bees from the carcasses of oxen.

Marcus Terentius Varro (116-27 B.C.), one of the most learned authors of ancient Rome, wrote a treatise on agriculture entitled *Rerum rusticarum libri tres* in which he says: "As for bees, some are descended from bees themselves, and some from oxen, for they can be born from the rotting carcasses of those animals. This is why Archelaus, in an epigram, calls them 'winged offspring of the putrescent ox.' He also says, '. . . wasps are born of horses, and bees of calves. . . .' "

It is Virgil (Publius Vergilius Maro, 70-19 B.C., born in Andes near Mantua and buried in Naples), the author of the *Georgics,* a poem on agriculture, who has given us the most detailed account of the reputed origin of bees from cattle. Pliny refers to this account in his *Naturalis historia* (11, 20, 23) and says:

When entirely lost, bees can be restored by smearing the carcass of an ox, but recently dead, with mud or, according to Virgil, the carcasses of young bullocks. Wasps and hornets are born from horses' bodies, and beetles from those of asses in the same way; nature can change some things into others. But all these creatures are seen to pair, and nonetheless their offspring possess almost the same nature as that of bees.

For the Romans, then, one species could give rise to another by a process of metamorphosis. There was, in fact, a rich heritage of ideas, often reminiscent of fable and legend, concerning hybridization and exceptional births. Pliny gives some examples in book seven of *Naturalis historia*: Alcippe gave birth to an elephant; a maidservant was delivered of a snake during the Marsian war; the Emperor Claudius had written that a hippocentaur had been born in Thessaly. Aristotle, and later Pliny, relate that various wild beasts meet together at the waterholes in Africa where they copulate, one kind with another, and produce new species of animals. The writings of the Greeks and

[1] Well-founded facts jostle with fanciful ideas in the numerous quotations from classical authors cited in J. D. Hofacker (1828). This study is a versatile account—with particular reference to horse breeding—of the characters inherited in animals and in man.

Romans contain numerous references to sexual relations between men and beasts. The Minotaur resulted from the union of the wife of Minos with a bull. Pindar (518-442 B.C.), Herodotus (484-425 B.C.), and Strabo (63 B.C.-A.D. 19) all write about sexual relations between women and goats in an Egyptian city where the god Pan was worshipped. It was emphatically forbidden by the Mosaic Law for men to copulate with animals, but nevertheless this practice was associated with religious sects in Egypt. According to Mode (1960), ritual copulation seems to have been practiced a great deal in ancient India. C. Zirkle (1935) has made a thorough study of early concepts of hybridization.[2]

Though the Romans cannot be said to have contributed much to the theoretical foundations of reproduction and heredity, they did develop the study of applied genetics to a quite remarkable degree. Many of their ideas on selection, breeding, fruit grafting, and so on, are still valid today. But these valuable ideas appear side by side with a host of inaccurate observations together with fanciful accounts of events which could never have occurred in the way they are described. Knowledge of these times is made available to us in the works of the great Roman authors, but we can be sure that many of the facts they described were already known to the Greeks and were merely adopted by the Romans. They kept pedigrees of their domestic animals, especially their horses and doves. The quality of animals selected for breeding purposes was all-important; age, season, pedigree, and hereditary capacities were also carefully considered.

Both Columella and Pliny wrote at great length on vegetative propagation and on the methods and results of grafting. Columella, who came from the Spanish town of Cádiz, became a landed proprietor in Italy after finishing his military career. About A.D. 60 he wrote the twelve books of *De re rustica* which contain a comprehensive account of agriculture, the neglected state of which Columella deplored. He was

[2]It must not be forgotten that notions involving curious hybrids were still being worked out in the late Middle Ages and were often entertained in modern times. For example, the hybrid produced by crossing a horse (or an ass) with a cow—known as a jumar—first appears in the middle of the sixteenth century. C. Zirkle (1941) has made a detailed textual study of the material pertaining to this creature. A perusal of the literature reveals that the jumar was thought to exist for about 250 years—until the beginning of the nineteenth century. The term "jumar" was first mentioned in Conrad Gessner's *Historia animalium.* Scaliger, John Locke, Ferchault de Réaumur, Buffon, Bonnet, Voltaire, and von Haller were among the many who wrote about this hybrid until the impossibility of the cross was eventually proved.

familiar with the techniques of slit grafting, bark grafting, and budding which even today are practiced in much the same was as he described them. It was widely held that grafting produced new species. Pliny writes (*Naturalis historia* 15, 11, 10): "Grafting ordinary quinces on to the strutheia (sparrow apple) has produced a new species, the Mulvian quince; this is the only species of quince which is also eaten raw."

Pliny gives several examples of this kind and makes frequent mention of the fact that new characteristics appear in the shoots that have been grafted. He also remarks on differences in the taste of the fruit: "The tree which is most receptive to every kind of graft is generally thought to be the plane, followed by the hard oak; but both of these spoil the flavors of the fruit."

The Romans were familiar with a very large number of different varieties of plants. In *De re rustica*, Columella refers to several varieties of wheat, though some of these may actually have belonged to a different species, as for example winter wheat (*Triticum monococcum*). Four varieties of spelt were known; they were distinguished by their color, quality, and weight (*De re rustica* 2, 6).

In the fifteenth book of *Naturalis historia*, Pliny names and describes the characteristics of fifteen different kinds of olive, four kinds of stone pine, four quinces, seven peaches, twelve plums, thirty apples, forty-one pears, twenty-nine figs, eighteen chestnuts, and eleven nuts. The whole of the fourteenth book is devoted to the vine, in particular to the fifty varieties from which noble wines can be produced; in all, ninety-one species (in fact, subspecies or varieties) are distinguished. Columella says (*De re rustica*, 3, 2, 29), that the varieties of the vine are as numerous as the grains of sand in the Sahara. He knew that many varieties lose their distinctive character when transplanted to other habitats.

With regard to domestic animals, Pliny names different breeds of cattle, but none of horses and dogs. Sheep are classified according to the region from which they come and the quality of their wool. A hundred years before the time of Pliny, Varro had already distinguished hunting dogs from watchdogs. Columella, a contemporary of Pliny, refers to black watchdogs and white shepherd dogs, and describes their qualities. Many different breeds of chickens and pigeons were known, and some of them had a considerable market value. Columella and Pliny refer to bantams and other breeds whose cocks are particularly suitable for fighting.

Pliny also describes bees (*Naturalis historia* 11, 18, 19):

But there are also forest bees, which have a bristly appearance and are much more irascible, though they are more industrious as well. Domesticated bees are of two kinds: the best are short, speckled, and round in shape, while the inferior ones are long and look like wasps; the worst ones of this kind are hairy. In Pontus there are white bees which make honey twice a month. In the neighborhood of the river Thermodon two more kinds are found: one that builds nests in trees and another that builds underground. With their triple-tiered structure of combs they are very productive.

These examples of the Romans' knowledge of different species, varieties, and breeds could be multiplied at will. They undertook procedures of cultivation and selection which show that they were acquainted with the basic principles of heredity and their applications to plant and animal breeding.

Varro, Virgil, and Columella all gave instructions for the care and handling of seed. In *Rerum rusticarum* Varro says: "The finest ears must be threshed individually in order to obtain the best seed for sowing."

Columella writes in *De re rustica*:

The next instruction I shall give is that when the crop has been harvested and is on the threshing floor, you should at once provide for the sowing that is to follow. For, as Celsus has already observed, when the harvest is of an average quality you should gather up all the best ears, and so provide for the future. But if the harvest is exceptional, then the threshed-out grains should be shaken about in a container, and those which settle at the bottom should be kept for seed. It is clear that strong seed produces seed that has strength, and weak seed produces seed that is weak.

Virgil gives a number of important instructions pertaining to seed and says, "If the best grains are not handpicked every year the crop can deteriorate, even though tended with great care, for I have observed this to be so."

In *Naturalis historia* (18, 24, 54) Pliny writes:

The best seed is that of last year; two-year old seed is poor, three-year-old is even worse, and seed from before that time is barren. There is a general rule which applies to every kind of seed: The first grains to settle at the bottom of the threshing-floor should be kept for sowing, for they make the best seed since they are the heaviest; there is no better way of selecting them. The ears on which the grains are separated by gaps must be discarded. The best grains are reddish in color, and they keep this color when crushed by the teeth; those which are white inside are inferior.

The Romans were most diligent and attentive in their work and observations, as the following passage from Columella's *De arboribus* shows (quoted from H. O. Lentz 1859):

In order to procure shoots from good vines, those vines which have borne large, sound, ripe fruits are marked with a mixture of ochre and vinegar at the time the grapes are gathered. This is not washed away by rain. The procedure is continued for three or more years, as long as the vine continues to bear sound fruit, and provides evidence that the stock is excellent and that the quality and quantity of the grapes is not due to some favorable vintage.

The Romans were careful observers of the processes of plant cultivation and animal breeding and modified their breeding practices in the light of what they saw. Concerning sheep rearing Pliny says (*Naturalis historia* 8, 47, 72):

In this breed, particular attention is given to the mouth of the rams, for the wool of their progeny has the same color as the veins under the tongue of the parent ram; if these are of several colors, the lamb will be piebald too. Changes in the water they drink can also cause character differences. There are two main breeds of sheep: the jacketed sheep and the common sheep. The former have softer wool, and the latter are more fastidious in their pasture; jacketed sheep sometimes feed on brambles.

This passage also appears in Varro's writings (Zirkle 1951). Columella tells of another case (*De re rustica* 7, 2): "Rams of a marvelous color were once brought to the city of Gades where my uncle bought some of them. He tamed them and mated them with ordinary sheep. They produced lambs that retained the color of their sires, and this was also the case in subsequent generations. . . ."

Horses and asses of both sexes were frequently crossed, and the resulting hybrids were described by the Greeks (Homer, Hesiod, Plato) as well as by the Romans (*Naturalis historia* 8, 44, 69). The cross between a she-ass and a stallion was known as a hinny (*hinnulus*), and the hybrid that issued from the union of a he-ass with a mare was called a mule (*mulus*); both were known to be sterile. (According to Zirkle [1951], this is documented in the writings of Empedocles, Democritus, and Aristotle.) But Herodotus and Varro report that fertile hybrids were said to have occurred occasionally as well. Varro also claims that the Romans used to capture wild asses and mate them with domesticated varieties to improve the breed. Pliny tells of crosses between wild and domesticated pigs. There is no end to such examples which show how carefully the Romans observed the workings of nature. Many of their studies were widely circulated in manuscript form until the fifteenth and sixteenth centuries. During the Middle Ages they had a wide influence on the development of the applied natural sciences.

4 The Middle Ages

4.1 The Legacy of Ancient Natural History and Its Influence during the First Centuries of the Christian Era

There were but few contributions to the study of reproduction and heredity during the early centuries of the Christian era. The supremacy of the Church, the political unrest, and the ravages of war which accompanied it not only seriously impeded the development of the scientific studies which had been undertaken in the schools and academies of the Greeks but also allowed the cultural legacy from classical times to be forgotten during a period of a thousand years. By a circuitous path, however, these ideas were reintroduced into the West.

With the exception of Clement of Alexandria (c. A.D. 200) and St. Augustine (A.D. 345-430), the philosophers, naturalists, physicians, and Church Fathers of the first three centuries of the Christian era make no mention of the teachings of Hippocrates and Aristotle. St. Augustine's reflections on reproduction and development are scattered throughout his many writings (A. Mitterer 1956). In the fourth century Vindician—who came from Africa as did St. Augustine—wrote the *Gynaecia.* This study contains an account of the gynecological, physiological, embryological, and anatomical doctrines of ancient times. It also refers to the Right and Left Theory as it pertains to reproduction, sex determination, and sex prognosis. The theories of the Greek philosophers were also preserved in some of the great encyclopedias of later centuries, as for example in the *Etymologiarum sive origines libri XX* of St. Isidore of Seville (570-636). In this work Isidore compiled a lexicon of classical learning and of the teachings of the Church Fathers. Several hundred years were to elapse before the ecclesiastical centers—in particular the monasteries—took up the important task of preserving and interpreting Greek philosophy. As late as 1209 the Synod of Paris proscribed the scientific works of Aristotle; they were not allowed to circulate until Pope Gregory IX issued a special dispensation in 1231.

According to J. Stur (1931), Michael Psellus, a product of the civilization of Byzantium (b. 1018 in Constantinople, d. 1079), ranks with those great thinkers who recorded the theoretical ideas and practical knowledge of their age in speeches and letters, and in countless works on natural science, mathematics, astronomy, jurisprudence, archaeology, grammar, and history. In his miscellany, *De omnifaria doctrina*, he makes several references to problems in the study of reproduction. For example, he considers the issue of fixing the point in time at which conception occurs; he talks of resemblances (and the lack of them) between parents and their children; he discusses the possibility of giving birth to boys or girls as required; he comments on the reasons for sterility and on the birth of twins and triplets, and asks how the embryo is nourished and whether it is, in fact, a living creature. But he added very little that was new and original, and the influence of the Greeks is evident throughout his work.

The following passage, quoted by E. Lesky (1950, p. 173), shows how closely Psellus follows the Greeks, especially the Stoics:

> Semen is secreted by men as well as by women. Moral qualities, character traits, and bodily features are all determined by the procreative forces that inhere in both kinds of semen, in that of the woman and in that of the man. If both sorts of semen are present in the same proportions, then the child resembles its father and its mother to an equal degree. But if the male sperm preponderates over that of the female, then the children bear a greater resemblance to their father; and if the female sperm is present in greater quantities than that of the male, then the children recall the features of their mother.

We are greatly indebted to the Arabs for the preservation of numerous works of Greek philosophy and science. From the remains left after the downfall of Greek civilization in Alexandria and Asia Minor they rescued much that would otherwise have been lost to posterity and translated it from the Greek original or from Syriac into their own tongue. This material then proceeded by way of Mesopotamia and Egypt to Spain where it was yet again translated into medieval Latin.

The Arabs were content merely to preserve and pass on the philosophy and learning of the Greeks; they did little to further the study of the biological sciences. On the other hand, they were excellent horse breeders and were no doubt aware of the importance of selecting parent animals; according to ancient legend, the processes of artificial insemination were familiar to them.

The Arabs wrote commentaries on the writings of the Greeks; for example, Avicenna (Ibn Sina of Bokhara, A.D. 980-1037) (Figure 8) commented on the complete works of Aristotle. In his *Canon medicinae*, the first book of which contains a Galenical account of human physiology and anatomy, he expounds the classical Right and Left Theory of reproduction. At the end of the fifteenth century this book remained the most comprehensive treatment of the subject.

8 Avicenna (980-1037)

When the Arabs retreated from northern Spain, their universities and libraries fell into the hands of the Spaniards, and the works of the Greek philosophers, especially those of Aristotle, were translated into

Latin. From that time ancient philosophy and science played a major role in the definition of theology and Christian philosophy in the monasteries. By the thirteenth century, the height of the scholastic age, this influence had resulted in a schism between natural science and philosophy on one hand, and theology on the other, and thus determined the character of subsequent European thought.

Until the end of the Middle Ages, scholars were content to interpret the philosophy of Aristotle in an endeavor to reconcile it with the revelations of the Holy Scriptures and the views of the Church Fathers. But it was precisely the influence of Aristotle that encouraged the empirical outlook that began to replace faith and speculative thought; people felt it to be their duty to make observations and conduct experiments; these are the only techniques that make it possible to determine the facts and discover the truth. We must not think that superstition and mysticism had now been forgotten. Unsubstantiated ideas, like that of the inheritance of acquired characters, are often found in the writings of the Middle Ages and still occur in those of modern times (C. Zirkle 1946).

The knowledge of that age was set forth in voluminous encyclopedias compiled by the friars of the great monastic orders: the Benedictines, the Franciscans, and the Dominicans. Between 1225 and 1240, for example, Bartholomaeus Anglicus, an English Franciscan, wrote *On the Properties of Things*, a widely read work which was also translated into a number of European languages. It contains a number of scientific facts along with the creatures of legend and fable. The encyclopedic work entitled *Speculum majus* (*The Great Mirror*), written about 1250 by the French Dominican friar Vincent of Beauvais, gives a more comprehensive account of the knowledge of his day. It falls into four parts, one of which called *Speculum naturale* (*The Mirror of Nature*), contains descriptions of trees and plants, of birds, fish, and sea monsters, of draft and brood animals, of wild beasts, and finally of man. Most important of all, however, was the encyclopedic treatise *De natura rerum*, compiled by the Dominican friar Thomas of Cantimpré (c. 1210-1276). Writing of man at the outset of his work, Thomas describes animals, plants, minerals, and metals together with the astronomy and meteorology of his age.

The great encyclopedias of the thirteenth century also discuss Greek ideas concerning the origin of semen; they contain theories that are largely variations of the ancient doctrine of pangenesis of Anaxagoras,

Democritus, and Hippocrates. A detailed account of these views may be found in C. Zirkle (1946).

4.2 The First Naturalists of the Early Middle Ages

These compilers of scientific knowledge were succeeded by men who, distinguished by their superior talents for observation, by their freedom from prejudice, and by their steadfast attempts to perceive new relationships, assumed the role of the true naturalist and increased the body of biological knowledge. But though these pioneers of exact biological research may well have made original contributions to botany and zoology, they had little to add to the stock of ideas that had been handed down to Western culture by Hippocrates and Aristotle as far as studies in heredity and reproduction are concerned. Yet their refusal to accept Church dogma, superstition, and mysticism paved the way for a new era in science. Albertus Magnus, one of these pioneers, is remembered as one of the greatest naturalists of medieval times. Albertus (Count Albert von Bollstädt) was born in 1193 (or 1207) at Lauingen on the Danube and died in Cologne in 1280 (Figure 9). He studied medicine and philosophy at Padua, entered the Dominican order, and studied theology at Cologne. After teaching at various Dominican monasteries (Hildesheim, Freiburg, Strasbourg, and Regensburg where, twenty-five years later, he was to return as prince-bishop), he was sent to the University of Paris and appointed Regent Master of Theology. He was later recalled to Cologne to take charge of the *studium generale* and was entrusted with numerous important missions of the Church to different European countries.

Albertus was an extremely versatile thinker. He wrote numerous theological and philosophical studies in addition to a lengthy series of commentaries on Aristotle. He was the first great Christian Aristotelian of early medieval times. It was he who made the philosophy of Aristotle as well as the works of other classical naturalists and physicians available to Western scholasticism. His method was to observe and describe the great variety of individuals who make up the organic world without, however, propounding any new scientific theories at the same time. In the field of botany, he wrote *De vegetabilibus libri VII*, a commentary on an ancient text dating back to Nicholas of Damascus (b. 64 B.C.) and containing the views of Aristotle and Theophrastus. Albertus takes issue with some of these classical ideas, and his study is made all the more valuable by the addition of his

ALBERTVS MAGNVS EPISCOP' RATISPONE.
Magnus es, at major fieri dum ALBERTE, recusas
Dispercam, si quid majus hic orbis habet.

9 Albertus Magnus (1193-1280)

own observations. Problems of heredity and of variability in the offspring are considered in the fifth book of the commentary where he writes of the union of one species of plant with another and of the transformation of one kind of plant into another. The issue of whether rye can change into wheat is taken up again, for Albertus held that rye could turn into wheat two or three years after it had first been sown. Similarly, wheat could degenerate into spelt or rye (a belief adopted from Theophrastus and still held today!). Albertus thought that such cases of metamorphosis were due to the fact that plants receive their nourishment from the soil. Material changes in the composition of the soil—and therefore in the nourishment—could easily occur, and this made it possible for one plant to change into another. He also thought that the aspen and the birch could spring from the stumps of oaks and beeches and that fungi and grasses might grow from putrefying matter. Other changes could be observed in cases of grafting: e.g., when grafted onto the willow tree, the plum produced fruit that had no pits; vines grafted onto the cherry, apple, or pear tree matured at the same time as the fruits of the parent tree. The transformation of a wild plant into a cultivated variety was accounted for in terms of superior environmental conditions, e.g., nutrition, cultivation of the soil, conditions of the habitat, sowing season, and so on. For Albertus, problems concerning variations that were inherited during the process of cultivation did not exist, though he did note that deformities were sometimes inherited. He also held that the inheritance of acquired characteristics was possible both in animals and in man. Albertus wrote at length on the origin and nature of semen. In *De animalibus libri XXVI* he advances the view that semen is formed in the whole of the body and accounts for sex determination and resemblances between parents and children on the lines of the theory of pangenesis. Broadly speaking, he follows Hippocrates, even though he thought—along with Aristotle—that women did not produce any sperm. In the tenth book of his treatise, Albertus also talks in terms of the Right and Left Theory. He says, for example, that a pregnant woman who feels the weight of her child bearing on the right side of her womb will probably be delivered of a boy, for males are usually formed on the right.

Albertus's famous pupil, the Dominican friar Thomas Aquinas (1225-1274) (Figure 10) was born in Aquino just outside Naples and died in the monastery of Fossanuova near Rome.[1] Like his teacher, Thomas tried to reconcile the doctrines of the Christian Church with the philosophy of Aristotle. A. Mitterer (1947) has given a detailed account of his views on human reproduction and development, on the formation and special properties of male and female semen, and on how these two kinds of semen are secreted and combine. In one of his most important treatises, *Summa theologica*, St. Thomas considers the question of whether semen is, in fact, produced from surplus food. (A residue of nutriment, we remember, was conveyed to the blood and, in

[1]In this study we confine ourselves to a brief note on St. Thomas's spermatology and do not consider the development and importance of his philosophical system as a whole.

10 St. Thomas Aquinas (1225-1274)

men, produced semen, in women, menstrual blood, or female semen.) St. Thomas decides that semen is formed from the substance of the begetter, for the son is like his father in respect of that which he receives from him. But if it were true that semen is produced from surplus food, he could never receive anything from his grandfather nor from more remote ancestors, for the same food had never existed in any of them. St. Thomas assumes a virtue or power, emanating from the soul, to be operative in the semen. It is this virtue that is passed on from the ancestors through the father to the son.

Like Aristotle, St. Thomas held that a child who did not resemble his parents was a monstrosity. In a similar vein, the birth of females was attributed to the inhibiting influence of the terrestrial factors of generation. "The birth of a girl must be at odds with the purposes of the terrestrial factors of generation and does not occur contrary to the designs of the divine power (God, celestial spirits, heavenly bodies), for these are also causes of this monstrous circumstance" (E. Lesky 1950). Female sex was considered to be a deficient formation which occurred only when the normal formation, that of a male, was unsuccessful. St. Thomas therefore agrees with Aristotle, who thought of the male as a perfect creature who, as the bearer of the principles of form and motion, was responsible for the reappearance of individual, sex, and species characters in his progeny of the same sex as himself.

Roger Bacon (1214-1294), the English philosopher and naturalist, was also an advocate of the Hippocratic theory of pangenesis. Although befriended for a while by Pope Clement IV, Bacon was an adversary of clericalism and suffered a great deal because of his scientific ideas. He made a sharp distinction between theology and secular studies and opposed the claim that the study of the natural world was subordinate to theology. Experience, experiment, and mathematics were the three mainstays of knowledge. With regard to pangenesis, he joins with Albertus and St. Thomas in criticizing the Aristotelian doctrine that semen derives from the blood secreted from surplus food. Bacon seems to think that the nutritive substances are themselves transformed, for he says:

And when enough of it is produced, according to Aristotle and Avicenna in *De animalibus*, a third faculty, which may be called the reproductive faculty, begins to operate. Now the surplus of nutriment is absorbed by this third faculty which conducts it to the place of reproduction where it is separated out or transmuted, so that it can give rise to an individual that is similar to itself (from *Communium naturalium* [1268] "*de generatione hominis,*" see Zirkle 1946).

Albertus Magnus, St. Thomas, and Roger Bacon are only the first in a series of important medieval naturalists who, deeply influenced by the writings of the ancients in their treatment of reproduction and heredity, must be considered advocates of the Hippocratic doctrine of pangenesis. A study of the subsequent literature reveals the quite remarkable fact that this doctrine survived, in many different guises, for some 2300 years. The fundamentals of the doctrine were adopted by almost all the celebrated biologists and doctors of the Middle Ages and were the basis for the formulation of a number of far-ranging hypotheses as late as the nineteenth century. The intellectual hegemony of the Greek philosophers endured for centuries. With regard to the history of thought, it is interesting to note that it was only with the dawn of modern times that the experimental results of inductive research began to lead to the solutions of biological problems that had occupied mankind for nearly 2500 years.

There are very few extant fourteenth-century records that contain references to pangenesis. However, the debate about the origin of semen was resumed, particularly in medical treatises, in the fifteenth and sixteenth centuries. Still captive to the dominating influence of Hippocrates, Aristotle, and Galen, semen was yet again thought to derive from surplus food, or alternatively, from the active participation of different bodily parts. It was chiefly the brain that was held to function in this way: semen was formed in the brain or else taken from the blood and deposited there and was then conveyed through the spinal cord to the testes. But the heart, lungs, kidneys, and even the eyes were also thought by some physicians to be organs where semen was formed.

Even in the notebooks of that all-embracing, revolutionary spirit Leonardo da Vinci (1452-1519), Aristotle's doctrine is mentioned and apparently accepted, for there are no critical remarks of any kind. Unlike many of his contemporaries, Leonardo understood that the black people of Africa are not whites who had been burned black by the sun, and that both parents participate in the transmission of features, for he says:

> The blacks of Ethiopia are not the products of the sun; for when black gets black with child in Scythia, the offspring is black; but if a black gets a white woman with child, the offspring is grey in color. Now this proves that, with regard to the embryo, the seed [*semenza*] of the mother has the same vigor as that of the father (*Quaderni* III 8v).

The great physician, alchemist, and philosopher of the sixteenth century, Paracelsus (Bombastus von Hohenheim) was born on November 10, 1493, in Einsiedeln, Switzerland, and died on September 24, 1541, in Salzburg. His attempts to unify the study of medicine, theology, and philosophy resulted in another quite distinct version of pangenesis. In his *Opera omnia* II we read:

> Sperm, or the seminal fluid of a man that is visible to the eye, does not constitute the true semen, for semen is, in fact, a semimaterial principle which is contained within the sperm, or the *aura seminalis* conveyed by the sperm. The physical substance called sperm is a secretion from physical organs, but the *aura seminalis* is produced by, and emanates from, the *liquor vitae*. It springs from the *liquor vitae* in the same way as fire originates from wood, for though wood contains no fire, fire and heat can be generated from it. This emanation or separation occurs as a result of a kind of decomposition and inner heat which can be produced in a sexually mature man when he is thinking of

a woman, or by his proximity to her, or by contact with her, in just the same way as a piece of wood can start to burn when it is exposed to the concentrated rays of the sun. Every organ of the body, together with all their energies and capacities, contribute equally to the formation of semen; the *liquor vitae* is the extract from all of them. In the fetus, the organs and capacities are formed from the *aura seminalis*, the purest and most perfect part of the *liquor vitae*. They are contained within the seed of the seminal fluid, which is necessary for the formation of the new human organism. Semen is, as it were, the extract of the human body which contains all the organs of that body in an ideal form (see Zirkle 1946).

Paracelsus, one of the greatest pioneers of modern medicine despite the fact that his medical and scientific ideas were larded with pantheism and mysticism, argued furiously with many of the medical authorities of his time, with the notable exception of Hippocrates. He spent many years as an itinerant surgeon who wandered throughout Europe with his pupils or with soldiers, and thus we are indebted to him for a description of syphilis and the plague. He introduced a number of chemical medicines and new drugs (e.g., opium) and lectured on medicine in the German vernacular so as to inform the common people. Denouncing feudalism and excessive taxation, Paracelsus was condemned as a rebel and a heretic, but his plans for establishing a new social order make him one of the first German social reformers.

In *De morbis mulierum curandis* (1542), Nicholas de la Roche expounds the theory of pangenesis as it pertains to the female body:

For the *vasa spermatica* draw out the material substance of female semen from every part of the female body and convey it to the testes (ovaries?). When they have received it and transformed it (concocted it by the action of vital heat), they transfer it to the *parastatae*. These vessels, surrounding the uterus on both sides, grow larger as they leave the testicles and finally terminate in the uterus (from chap. I, "*Testes foeminarum*").

In *De morbis muliebribus* (1597), Martin Akakia suggests that misshapen fetuses could be due to incompatibilities among the seeds taken from each part of the male and female body; for if they were unable to fuse properly so as to mature, malformations or double formations could result.

Anaxagoras and the atomists had thought of the procreative act as the release of a germ, preformed in all its atomic parts, from the parental bodies. Moreover, a given configuration of atoms taken from a certain part of the parental bodies constituted the same bodily part in the offspring. In *Commentarii et animadversiones . . . de causis plantarum Theophrasti* (1566), Julius Caesar Scaliger opposed this view and pointed out that deformed men, even though they lacked certain organs and other parts of the body, had produced sound children.

By the end of the sixteenth century we see the first detailed description of an unforeseen hereditary variation in one of the higher plants. It was provided by Sprenger, a Heidelberg apothecary, in 1590. In his garden laboratory, Sprenger discovered a new form of greater celandine, *Chelidonium majus*, which possessed deeply incised, laciniate leaves. He named this new form *Chelidonia major foliis et floribus incisis*. Gaspard Bauhin (1596) gave a detailed description of it, and Clusius (1601) named it *Chelidonium majus laciniato flore* in *Rariarum plantarum historia.* In *Historia plantarum universalis* (1651), Jean Bauhin named the new form *Chelidonium folio laciniato,* while Chabraeus characterized it as *Chelidonium majus folio laciniato ac florum itidem foliis incisis* in *Stirpium Sciagraphia* (1666). Guy de la Brosse described this new form in *Description du Jardin Royal des plantes médicinales* (1636). In *Schola botanica sive Catalogus plantarum quas ab aliquot annis in Horto regio parisiensi studiosis indigitavit* (1689), J. P. Tournefort listed three species of *Chelidonium*: *Chelidonium majus vulgare*, *Chelidonium majus foliis quernis*, and *Chelidonium majus foliis et flore minutissime laciniatis.* Morison (1680), Linnaeus (1753), Miller (1731, 1760, and 1768), and many others described or briefly mentioned this new form, which had attracted wide attention. The earliest statements made about it referred to Sprenger's stock in Heidelberg, from where the form spread over most of the botanical gardens in Europe. In the first edition of his *Gardener's Dictionary* (1731), however, Miller reported that the new form had been discovered growing in a wild state in Wimbledon, Surrey. This statement is doubtful and moreover was subsequently in part withdrawn by Miller himself.

From the beginning, the new form was constant. Every botanist who examined it found that it displayed the same characteristic features. Miller observed it, annually raising it from seed, for a period of more than forty years and discovered no variation whatsoever. G. Bauhin (1620) was the first to report on the variability of the new type. A different form, in which the petals and foliage were even more deeply incised, was discovered in the cultures of the Jardin Royal in Paris. The aforementioned description, *minutissime laciniatis*, given by Tournefort (1689), also refers to this new variety. Morison (1669) names it *Chelidonium majus foliis tenuissime dissectis.* He held the new form to be a degenerate variety of the old and writes (1680):

A plant from the seed of a Chelidonium with oaklike leaves is believed to be degenerate, because a seed collected from this Chelidonium with very finely dissected leaves, and sown by me, produced an oakleaved Chelidonium with laciniated flowers, and reverted to its original form and shape in all its parts—which shows that it is only a sport of nature; for to constitute a distinct species, it should maintain itself always in the same condition.

The new form was generally considered to be variable (Miller 1760). E. Roze (1895) gave an account of *Chelidonium majus laciniatum*, in which he maintained that this form was, in fact, a new species which, under the influence of external conditions, could give rise to the varieties *fumariaefolium* and *crenatum*. Finally, we must mention that the laciniate variations of *Chelidonium majus* L. are discussed in a study by F. Widder (1953).

4.3 The Discoveries of the Seventeenth Century

In the seventeenth century, natural science took a big step forward thanks to the work of the English anatomist William Harvey (1578-1657) (Figure 11). As Professor of Anatomy and Surgery at the Royal College of Surgeons in London he enlarged our knowledge of physiology by his discovery of the dynamics of the circulation of the blood (1628). With regard to embryology, we are indebted to Harvey for his correct explanation of the origin of the chick from the yolk of the egg. He also knew that fertilization by the male was a necessary feature of the procreative processes in higher animals. From his observations on the chicken egg he boldly concluded that mammals also come from eggs, even though he had never seen a mammalian egg. However, Harvey did not yet conceive of an egg as a differentiated structure of a specific type. For him the concept of an egg characterized an undifferentiated mass—a mass which nonetheless represented a definite stage in the development of animals, even the higher species. For Harvey, then, the notion of an egg included not only the eggs of birds but also the larvae and pupae of insects and finally the supposed *primordium* from which the embryo developed in man. Of insect larvae, for example, Harvey writes: "Such, too, are the seeds of insects, called worms by Aristotle, which, though imperfectly made at first, seek food for themselves and are nourished by it; in this way they grow from a larva into a pupa—from an imperfect into a perfect egg or seed" (*Exercitationes de generatione animalium* II, 4).

In the chapter entitled "An Egg Is the Common Origin of All Animals," Harvey writes:

11 William Harvey (1578-1657)

[Animals] which bring forth in fact [*actu animal pariunt*] are called viviparous, while those that bring forth potentially are oviparous. For we, like Fabricus, hold that every *primordium* that lives potentially ought to be called an egg; moreover, we make no distinction between what Aristotle called a worm [i.e., a larva] and an egg, both because there is no difference to the eye, and because the identity is in conformity with reason. . . .

Harvey uses the term "contagious" to describe the action of the semen on the egg. The inheritance of features from the paternal side was therefore intelligible to him. He also draws the following parallel: a woman's brain is capable of understanding the world—that is, of producing ideas in the likeness of objects; similarly, her uterus has the power to form its eggs—the "ideas" of the uterus—in the likeness of the man who fertilizes her. King Charles I of England had placed at Harvey's disposal hinds and does from the Royal Deer Park so that he could study the nature of fertilization. At intervals, Harvey dissected the uterus and ovaries of these animals after they had been mated, in order to answer the old problem about the origin of semen and the role it played in fertilization, but he found nothing to support the views that had prevailed since antiquity.

Harvey presented two possibilities for the development of the egg after it had been fertilized by semen: either the complete material was already present and merely needed to be shaped, as a statue would be fashioned from stone, or the material had to be assembled and was differentiated as it was produced. The former theory of development is known as metamorphosis, the latter is called epigenesis, and according to Harvey, both kinds of development occur in animals. Some spring from a ready-made mass; their different organs are then formed by the metamorphosis of this mass. In others, however, the organs are assembled one after another and grow and assume a definite shape at the same time.

As always, whenever new research methods lead to a deepening of knowledge, its advocates stand between two worlds. In many respects they are still tied to traditional concepts, but in other respects they enter virgin territory.

The development of the microscope gave a great impetus to scientific progress in the second half of the seventeenth century in Italy and in England as well after the founding of the Royal Society. Robert Hooke (1635-1703) contributed a great deal to progress in microscopy as did the Dutchman A. van Leeuwenhoek (Figure 12), the first microscopist to examine specimens by means of transmitted light. Van Leeuwenhoek was born October 24, 1632, in Delft and died there August 27, 1723. He was a merchant for whom biology and optics were merely an avocation, but from the lenses which he ground he produced some 400 microscopes with magnifying powers of up to 250-fold. Van Leeuwenhoek and his contemporaries, the Italian Malpighi (1628-1694) (see Figure 13), Professor of Medicine at the universities of Bologna, Pisa, and Messina, and the Englishman Nehemiah Grew (1641-1711), all made important contributions to the studies of plant anatomy and physiology. In 1674 van Leeuwenhoek described the red blood

12 Anton van Leeuwenhoek (1632-1723)

corpuscles of mammals, amphibians, and fish. Three years later, at just about the same time as van Ham and Nicholas Hartsoeker (1656-1725), he discovered the human spermatozoon and later, spermatozoa in rabbits and in dogs. The first information about this discovery was contained in a letter written in November 1677 to the president of the Royal Society, Viscount Brouncker. A few years earlier, Rainier de Graaf (1641-1673) had found vesicular lumps in the ovaries and thought he had discovered the mammalian egg. Others proved him

13 Marcello Malpighi (1628-1694)

wrong, but the structures he discovered are known as the Graafian follicles to this day. Van Leeuwenhoek, van Ham, and Hartsoeker all assumed that a spermatozoon was a preformed organism—infinitely small, though complete in all its parts—that was nourished in the womb and eventually developed into an embryo.

For a long time, however, the function of these newly discovered spermatozoa was unclear, and what was in fact the vital element in the seminal fluid remained a controversial issue. Did the fluid contain hordes of parasitic "worms," or were these worms themselves the fertilizing agents? Pierre Dionis discussed this question in his *Disserta-*

tion sur la génération de l'homme (Paris, 1698). He rejected the theory that the egg is fertilized by a single sperm with the comment that it would be absurd to produce millions of sperms when only one was necessary. As late as 1868 even Darwin thought that the seminal fluid in its entirety was the fertilizing agent. He also held that the extent to which the fetus resembled its father depended on the amount of semen contributed by him.[2]

With regard to the origin of semen, however, Dionis says that it is composed of a number of similar elements that are probably separated and filtered out from the blood by the testicles. The blood then flows back to the heart from where it is redistributed and is replenished with new sperm from every part of the body. Dionis was also an advocate of pangenesis.

During the seventeenth century relatively few observations were made of the inheritance of features in plants, animals, and man.

From the collected letters of van Leeuwenhoek, M. J. Sirks (1959) has published the following remarks, made in 1683, on dominance in rabbits:

Many people in our country keep rabbits, some for pleasure, others for profit. Usually these rabbits are very large, white in color, and have long ears, considered to be a special mark of beauty. In order to cause these white rabbits to have grey offspring, so they can sell these in the spring for wild ones, they mate a grey buck with white does. This grey buck is caught, when still young, in the dunes where all rabbits are grey. It is mated not only with white, but also with piebald, blue, and black does, and all the young which issue from such unions take their father's grey color. Indeed, it has never been seen that any such young rabbit had a single white hair, or any hair other than grey. Moreover, they will never grow to the size of the mother, nor have such long ears as she; also, they will never be as tame as the mother but will always retain something of their wild nature.

E. O. von Lippmann (1911) quotes the following passage from *Physica subterranea*, written by J. J. Becher in 1669:

Similar results may be observed in other species of animals, especially in pigeons. For example, it usually happens that the young of a black cock and a white hen are either completely black or completely white. It is only later, when these blacks are mated with blacks, and the whites with whites, that black and white speckled birds appear. In arboriculture [*in arborum insitione*] nature achieves similar results, for if a tree bearing red fruit is crossed with a tree that bears white fruit, it is only after the second cross that mottled fruit appears. Examples of this kind can also be seen in human beings. If, for instance, an elderly,

[2]It was only in 1841 that A. Kölliker proved that spermatozoa are not parasites but true reproductive agents.

dark-complexioned Spaniard or Portuguese marries a woman with a fair complexion, then—should the mother's nature predominate—her children will also be fair and have their mother's nature. But if these children then marry persons who are similarly constituted, then they will have grandchildren whose complexion and temperament recalls that of their grandfather. The selfsame qualities have also occasionally appeared in the third generation, as shown by van Helmont in his *Alphabet of Nature.*

5 Early Modern Times: The Eighteenth Century

5.1 The Theory of Preformation

Looking at the egg yolk through his microscope, Malpighi (Figure 13, p. 74)—in contrast to Harvey—observed what were already well-defined structures. He held that these first stages of embryonic development constituted the actual structure of the egg, like the germ of an unripened seed which at that time was thought to be the equivalent of an egg. Malpighi was convinced that the new organism was already present and preformed in the egg and gradually became visible as the preformed parts grew in size. We have already encountered Harvey's notion of epigenesis and Malpighi's idea of preformation in our study of the ancients. Empedocles and Democritus were partisans of the theory of preformation, while the dynamic ideas of Aristotle represented a variant of epigenesis. Malpighi's doctrine of preformation was adopted by other famous biologists of the seventeenth and eighteenth centuries, including Spallanzani (1727-1799),[1] von Haller (1708-1777), and Bonnet (1720-1793). According to Malpighi, the preformed egg was the starting point for all further reflection on reproduction and heredity. His theory contrasts sharply with the preformationism of van Leeuwenhoek, Hartsoeker, and Boerhave (1668-1738) who all were of the opinion that the future adult was enshrined within the preformed sperm. Perhaps the most important result of the battle that raged between ovists and spermists for more than one hundred years was the realization of the important fact that the female's contribution to her progeny could be traced back to material structures and was not based on some undefined role of her menstrual blood. For the time being, the question of the origin of semen from the body remained unchallenged by the controversy, and preformationism did nothing to dislodge the theory of pangenesis.

[1]It is worth mentioning that Lazzaro Spallanzani, a professor in Pavia, also conducted a number of experiments (1780-1785) which were the first attempts at artificial insemination. This work was clearly at odds with the moral views of the Church. He found that frogs' eggs immersed in fresh water produce tadpoles only if the seminal fluid of the male is added. He also tested the viability of seminal fluid and was familiar with the effects of chilling on the life-span of semen. Moreover, Spallanzani was the first to conduct experiments with different quantities of the seminal fluid and in different dilutions, though he failed to take into account the presence of individual sperms in the seminal fluid. Finally, he successfully inseminated a bitch by artifical means. Sixty-two days after the injection, she gave birth to a litter of three healthy pups. He also recognized the future importance of these experiments "for breeding different kinds of mules (hybrids) and for making further progress in research on the reproductive processes" (cf. R. Götze 1949).

Gottfried Wilhelm Leibniz (1646-1716) was also an adherent of preformationism. In his *Monadology* he writes:

Philosophers have been greatly perplexed about the origin of forms, entelechies, or souls. Today, however, when exact researches on plants, insects, and animals have revealed the fact that the organic bodies of nature are never produced from a chaos or from putrefaction, but always from seeds wherein there was doubtless some preformation, we conclude not only that the organic body was already existent before conception, but also that there was a soul in this body; that, in short, the animal itself was present, and that by means of conception it was merely prepared for a great transformation, so as to become an animal of another kind. . . . .

One of the last advocates of eighteenth-century preformationism was the Swiss Albrecht von Haller (Figure 14). He was born near Bern on October 16, 1708, and studied medicine in Tübingen for two years. In April 1725 he traveled to Holland to pursue his medical studies at the University of Leiden where he received his doctorate in medicine in 1727. After further studies in London and Paris, he returned to Switzerland in 1728 and lectured on anatomy as an assistant at Basel University. Following a tour of the Alps with Johannes Gessner in 1729, he wrote his famous poem *Die Alpen,* which describes the life of the mountain peasantry in Switzerland. In 1736, after six years of medical practice in Bern, he was appointed Professor of Anatomy, Surgery, and Botany at the University of Göttingen where he remained until 1753. Here he set up a botanical garden, founded the "Sozietät der Wissenschaften," and established an institute for obstetrics, the first maternity clinic attached to a university in Germany. For his great achievements in science he was given a hereditary title in 1749. After his return to Switzerland, von Haller was appointed manager of the salt works at Roche (Waadt) and held this post from 1758 to 1764. During these years he was also actively working on forestry and agriculture. His work *A Catalogue of the Wild Trees and Shrubs Which Grow in Helvetia* (Bern, 1763-1772), together with numerous other treatises on the improvement of agricultural output like *A Treatise on Cattle Pest* (Bern, 1772), are the accomplishments of a true pioneer. He died December 12, 1777, in Bern.

Von Haller was the founder of experimental physiology. He set forth his views on preformation in his *Grundriss der Physiologie für Vorlesungen* (Lecture Outline for Physiology), in which he discusses the views of both spermists and ovists. He writes, for example (p. 652):

Some have attributed everything to the father, especially now that—with the help of magnifying glasses—the celebrated worms that

14 Albrecht von Haller (1708-1777)

are found in male semen are well known; and it was indeed justified to observe that their shape and figure conforms to that of the incipient embryo.

Or again (p. 653):

. . . others, no less famous, whose work deserves to be believed by those who follow them, have taught that the future individual is present in the maternal ovary; the semen from the male gives it life and movement, and also transforms it in different ways; but it is already present in all its parts.

Von Haller's own view is best summed up in a sentence from his most important work, *Elementa physiologiae corporis humani*: "There is no process of development; no one part of the animal body is made before any of the others, and all are produced simultaneously."

Von Haller also assumed that semen was produced from particles of food in every part of the body. With regard to the regeneration of organs, however, he developed a special theory of pangenesis. He did not attribute to any given organ merely the formation of the seeds that would develop into the organ alone, but he held that the head could produce seeds for the tail (and conversely) and that in the event of the absence of a given organ, regeneration could take place from the surplus of seeds provided by some other organ. He also maintained that semen was absorbed by the blood and flowed throughout the body and that it was responsible for all the changes that could be observed in a pubescent youth (smell, hair, voice, and so on).

Von Haller's great contemporary, the biologist Georges Louis Leclerc de Buffon (1707-1788) (Figure 15), adopted a critical stance toward preformationism but was a confirmed advocate of pangenesis. His treatise *Histoire naturelle* (1749) contains his ideas on both these topics. He claimed that human growth resulted from the absorption out of the atmosphere of specific living particles for each organ and that the body contained no superfluous particles until growth had been completed. At this point, when the body had less need for new living particles, those that were supernumerary in any part of the body were conveyed to reservoirs, that is, the testicles and the seminal vessels. These are replenished with surplus food (see Aristotle) at puberty, as is shown by the phenomena that accompany this stage of development. According to de Buffon, such living particles are taken from every part of the body and stored in the testicles and seminal vessels of a man and in the ovaries of a woman. The seminal fluid is therefore a kind of extract of every bodily part. If the seminal fluids of both sexes are mixed, the same or similar particles unite because of their affinity to form a small, structured body which then is to grow in the mother's womb. If the particles provided by the man are more numerous than those from the woman, the offspring is male; the converse is also true (see Democritus of Abdera). During von Haller's and de Buffon's lifetime similar ideas were expressed by two other remarkable thinkers of the eighteenth century, René-Antoine Ferchault de Réaumur (1683-1757) and Pierre-Louis Moreau de Maupertuis (1698-1759) (Figure 16).

15 Georges Louis Leclerc de Buffon (1707-1788)

Réaumur is well known to us as a great physicist and important zoologist. He was the author of an excellent work on the life and development of insects and was the first to discover that the society of bees is composed of the queen, drones, and workers. In *Art de faire éclore et d'élever en toute saison des oiseaux domestiques de toutes espèces* (1749) he pointed to the problems involved in the idea that the

16 Pierre-Louis Moreau de Maupertuis (1698-1759)

different particles, taken from every part of the body, unite in the way necessary to form new organs. What power is it that is able to overcome the chaos, to arrange the parts that belong together, to form the organs, to join them one to the other, and finally how is the seed formed, which is so small that one cannot get a really good view of it even with the best microscopes? Réaumur's comments are probably best seen as a critique of the philosophy of Descartes (1596-1650) and his mechanistic ideas on the development of living creatures, which Réaumur rejected.

Maupertuis, however, is better known for his quarrel with Voltaire than for his scientific work. This philosopher, mathematician, and geometer had far-reaching interests in biology, particularly in genetics. G. Hervé (1912) and B. Glass (1947) have written about his importance as a geneticist.

Pierre-Louis Moreau de Maupertuis was born September 28, 1698, in Saint-Malo. A sensitive child, he was brought up by the Abbé Coquaud and then studied for two years at the Collège de Marche. In 1718 he entered the army which he left with the rank of captain of dragoons. After a brief trip to London in 1728 to study under Newton, he journeyed to Basel, where he studied mathematics, made friends with Bernoulli, and attended his lectures on integral calculus. By this time Maupertuis had already become interested in biological problems, and in 1736 he led an expedition to Lapland to measure the meridian. His observations there confirmed the theory that the earth was flattened at the poles. In 1740 he accepted an invitation from King Frederick the Great to go to Berlin, but at Voltaire's suggestion the invitation was withdrawn. In 1746 Frederick appointed Maupertuis president of the *Académie Royale des Sciences et Belles-Lettres* (now known as the German Academy of Sciences) in Berlin, a post which he retained until 1756. He died July 27, 1759, in Bernoulli's home in Basel.

Maupertuis attacked preformationism in chapters twelve and thirteen of his treatise *Vénus physique.* His most powerful argument was to ask how first-generation hybrids which display a blend of the qualities of their parents (white × black or horse × ass) are possible if the offspring is preformed in either the spermatozoon or in the egg. According to the spermists or animalculists the offspring would have to resemble its father alone, while according to the ovists or ovulists it would have to resemble its mother; in point of fact, however, it had the defects, habits, propensities, and mental qualities of both. In the fourteenth of his *Lettres sur divers sujets* (1753) Maupertuis writes:

> But the system of eggs and that of the animalcules are equally discredited by *Vénus* as well as by the work of Buffon: the supposed observations of those who saw eggs in the tubes, entirely formed fetuses inside the eggs, and fetuses in the seminal fluid of the male; all these are mythical and do not deserve our attention.

He gives a detailed account of his views on reproduction and development in *Système de la nature.* In *Vénus physique* he had already written that the seminal fluid contained elements that were capable of forming the heart, head, entrails, arms, and legs. These elements (*élémens*) having a great affinity with the corresponding elements in the seminal fluid of the other sex, joined to form the embryo (cf. Anaxagoras).[2]

In *Vénus physique* he writes (pp. 89-90):

One should not believe that the two seminal fluids contain only the number of particles required to form one fetus, or the number of fetuses that the female is to carry: each of the sexes undoubtedly provides many more than are needed. But once the two particles that are to touch are united, a third, which could have effected the same union, is unable to link up and is not used. It is thus, by these repeated operations, that the child is formed from particles supplied by its father and its mother, and often bears visible marks of being part of both.

According to Maupertuis, these *élémens* are the smallest units of matter from which the different organs of the embryo can be formed. Above all, they also transmit characteristics from the parents. If there are either not enough or too many, a monstrosity results. In *Vénus physique*, Maupertuis writes further (p. 90):

If each particle is joined to those which should be beside it, and only to those, a perfect child will be born. If some particles are too far away, or do not have a suitable form, or are too weak to be capable of joining with those to which they should be united, a monster is born because of these deficiencies. But if it should happen that superfluous particles should find a place and are joined to those particles whose union is already sufficient, the excess of these particles also produces a monster.

In the second part of *Vénus physique*, Maupertuis comes to the following conclusions:

1. The seminal fluid of a given animal species contains an immense number of *élémens* that, because of their resemblance, are particularly suited to produce animals of the same species.
2. The *élémens* likely to produce character traits similar to an individual (parent) are those that are generally present in greater numbers and which have a greater affinity.
3. Each part of an animal is probably formed from *élémens* provided by the part in question.

But how does it happen that many creatures resemble their ancestors rather than their parents? Maupertuis explained this well-known phenomenon by saying that since there is an excess of *élémens* in the seminal fluids, different *élémens* derived from earlier generations will preponderate in a later generation and thus make the children resemble distant ancestors rather than their parents.

[2]W. His (1874) called Maupertuis's reproductive theory of *élémens* the "theory of formative forces" (p.137).

The composition of elementary particles accounts not only for physical features but also for mental and spiritual characters. Such qualities are perfected by education when, in the course of several generations, the children are trained in the occupation of their father.

It is clear the Maupertuis anticipated some of the basic principles elucidated by Darwin. He writes, for example ("*Essai de cosmologie*," in *Oeuvres*, Vol. I, pp. 11-12):

> Chance, one might say, had produced an infinite multitude of individuals; a small proportion of these was so constituted that the animals' organs could satisfy their needs; in an infinitely greater number of them there was no adaptation, nor order; these last have all perished. Animals without a mouth could not live; others which lacked reproductive organs were unable to reproduce their kind. The only ones that survived are those in which there were both order and adaptation; and these species which we see today are only the smallest portion of what a blind destiny had produced.

Maupertuis also conducted experiments. He crossed different breeds of various animals and observed the characteristics that appeared in the hybrid offspring. The most comprehensive account of his detailed investigations into a case of human polydactylism may be found in his *Lettres* (p. 307). He traced the genealogy of the Ruhe family in Berlin and studied cases of the appearance of extra toes in chickens and in dogs.

In his *Lettre sur le progrès des sciences*, Maupertuis, encouraged by Réaumur, prescribes that such costly experiments be subsidized by the state and systematically conducted: "To make natural history into a true science we should apply ourselves to research that would reveal to us not the individual aspect of one animal or another but the general process whereby nature reproduces and conserves" (p. 418).

There is no doubt that Maupertuis believed in pangenesis. In *Vénus physique* (pp. 120-121), he writes: "As for the substance from which in the seminal fluid of an animal the particles are formed that resemble that animal, it would be a bold, yet not implausible, conjecture to hold that each particle supplies its own germs."

Concerning the further development and mode of operation of the *élémens*, however, he has little to say. We can therefore assume that the elements that are destined to form the heart, the head, or some other organ are not real organs in themselves, which merely need to grow, but must be undefined, information-bearing patterns of growth.

Preformationism, however, which believed that the completely developed new individual was already contained either within the male sperm or in the female egg, was compelled by logic to assume that the germs for subsequent generations were already contained within any given germ. Charles Bonnet (1720-1793), a Huguenot from Geneva who discovered parthenogenesis in plant lice, held an extreme form of this view. According to his theory of "emboîtement," or encapsulation, the germs from which the second generation was formed already contained the germs for the third generation, and so on—that is, within the first member of the species were contained all the members of that species, preformed, each generation within the previous one, until the end of the world. L. Spallanzani also believed in this theory of encapsulation. Such extravagant reflections, however, finally made way for an epigenetically oriented theory of development.

5.2 Epigenesis

It was the work of the embryologist Caspar Friedrich Wolff (1733-1794) that encouraged the idea of epigenesis. The son of a master tailor, he was born in Berlin, where he studied at the Faculty of Medicine and Surgery and was trained by the famous anatomist, J. F. Meckel, Sr. At Halle University, which he attended in 1755, he made a number of original observations on the development of animal and vegetable organisms, and in 1759 presented his dissertation, entitled *Theoria generationis.* A greatly expanded German version criticizing Bonnet's work appeared in 1764 under the title of *Theorie von der Generation.* According to Wolff the nourishment and growth of plants depended on an essential force, or *vis essentialis,* which had the power of assembling new organs (in no way previously existent) from the blobs or bubbles of a clear, vitreous, homogeneous substance. In the case of the embryonic chick, in which the various vessels and organs were assembled from undifferentiated ampullae, he thought that development was also initiated by the influence of the *vis essentialis.* These ampullae are cells from which all organisms develop. Wolff also realized that the sap of plants deposits substances that produce turgid cell structures, whereas the corresponding structures in animals are softer and more pliable. Because of these observations, he may be regarded as one of the originators of the cell theory—eighty years before Schwann. He had a rather odd view of fertilization, however, for he thought of pollen and male semen as a kind of "purified food" necessary to stimulate the development and growth of the germ (see also A. E. Gaisinovich 1956-1957).

Three basic principles of cell theory must be attributed to Wolff:

1. Cellular structure is an essential mark of all living creatures.

2. Plant and animal cells are similar with regard to their essential parts.
3. The remaining morphological formations (vessels, fibers, and so on) are modifications of cells (from B. E. Raikhov 1947, chap. II).

Wolff proved that the components of the developing organism are not preformed but arise epigenetically during the course of development. Although he had undertaken a critical analysis of the theories of Bonnet and von Haller, in his *Theorie von der Generation*, von Haller remained unconvinced for he was afraid that any further acceptance of the idea of epigenesis would undermine faith in the Church.

In 1761, two years before the end of the Seven Years War, Wolff was summoned to the field hospital at Breslau. He also gave lectures on anatomy and first met Cothenius, who was the senior medical officer there. In 1763 he left Breslau for Berlin, where he received permission from Cothenius to give public lectures. Though he taught with great success for a period of three years, he incurred the displeasure of the academicians, who prevented him from taking up an official academic career. Discouraged by intrigues as well as by his vain attempts to obtain a permanent appointment in Berlin, Wolff accepted an invitation from Empress Catherine the Great to the Academy in St. Petersburg in 1766. Universally admired, he was able to continue his work for twenty-seven years until his death. According to Gaisinovich (1961), Wolff had been recommended for this post by Leonard Euler, an advocate of epigenesis who had read Wolff's dissertation, knew him personally, and correctly assessed his abilities as a scientist.

Wolff and his wife lived a secluded life in St. Petersburg. He was head of the Department of Anatomy of the Kunstkammer and later became director of the Botanical Garden of the Academy. He held the position of prosector and studied the monstrosities of the anatomical collection which had been started in 1718 by an edict of Peter the Great. In Wolff's lifetime, the St. Petersburg collection contained forty-two monstrosities, while the one in Moscow had sixty. He interpreted these teratological phenomena as evidence against preformationism. Kölreuter, who was the naturalist most likely to have understood Wolff, had retired from the Academy a few years before the latter's arrival in St. Petersburg.

Wolff's study, *De formatione intestinorum* . . . , appeared as early as 1768. It contained a report of his investigations into the formation of the intestinal canal in the developing chick, emphasizing once again the untenable character of the theory of preformation. Wolff raised epigenesis to the status of a general law for all living things.

In his opinion, it was not merely the stomach and intestines that arose from laminated, undifferentiated structures, as described in his study; he also thought that all the other organs were similarly developed from "germ layers," an observation further pursued by Christian Pander (1794-1865) in 1817 and conclusively confirmed by K. E. von Baer.

Karl Ernst von Baer (Figure 17) was born February 17, 1792, at Piep, the family estate in Estonia. He studied medicine at Dorpat, and

17 Karl Ernst von Baer (1792-1876)

in 1814, after he had graduated, he went to Vienna. A year later he traveled to Würzburg where he studied anatomy and embryology under Döllinger, and he also took part in the investigation of Döllinger and Pander into the developing chick. In 1817 von Baer was offered the post of prosector at the Institute for Anatomical Studies of the University of Königsberg and was appointed Associate Professor in 1819. The following year he set up a zoological museum and was granted a full professorship in zoology in 1822. During the winter of 1826-1827 he was the first to discover the mammalian egg contained within the Graafian follicle of the ovary of a bitch. In 1829 he went to St. Petersburg for a brief stay. In 1834, however, he moved there for good, having been elected to the Academy. His most important work, *Über die Entwicklungsgeschichte der Tiere*, contained an account of numerous pieces of original research and gave a great impetus to the development of embryology. In Russia he held a number of important posts. From 1850 on he devoted himself to anthropological, prehistorical, geographical, and ethnographical problems. In 1867 he returned to Dorpat, where he died November 16, 1876.

For a long time, Wolff's treatise on the alimentary canal was ignored. It was rediscovered in 1812 by J. F. Meckel (1781-1833), an anatomist from Halle, who translated it into German. This provided Wolff with the recognition he deserved as a pioneer of modern embryology. In connection with his work *The Metamorphosis of Plants*, Goethe called him an "excellent precursor" for his investigations. Albert Kölliker (1817-1905) wrote in his work on embryology: "Wolff was the first to study an organ from its first beginnings to its completion; what is even more important, he showed that the development of such a complex mechanism as the intestine originated from simple, laminate, undifferentiated structures."

Our knowledge of the history of genetics is greatly increased by the contribution of B. E. Raikhov (1947, 1952), the Soviet historian of science. His first report after examining the posthumous manuscripts of C. F. Wolff presents a new aspect of Wolff's work, which enhances his importance as a figure in the history of biology. From Raikhov's studies it is clear that Wolff did a great deal of work on the problem of the transformation of species. Raikhov himself calls him an "important precursor of Lamarck and Darwin." Stability and variability are the two great biological problems that preoccupied Wolff. He distinguished between two kinds of variability: One was brought about by the influence of light, heat, air, humidity, and nourishment. Such instances of variability caused by environmental influences he called *variations.* He knew that plants that were taken from St. Petersburg to Siberia gradually changed their appearance, and he was also aware that the variant Siberian variety assumed its original form when returned to St. Petersburg. He was therefore familiar with the concept of modification and realized that such modifications were not inherited: "Variations are not hereditary; the outward form and its internal structure would be constant if they did not change because of modifications of nonessential and nonhereditary qualities."

In addition to "variations," however, Wolff was also familiar with other changes that were continued hereditarily in subsequent generations and thereby influenced the formation of species. He observed deformities, such as cases of hexadactylism in man that reappear to a certain extent in the descendants and persevere in conserving their forms and structures.

The religious aspect of the question whether it was possible for animals to exist in any form other than the normal form presented a problem. Wolff answered that it seemed not impossible for external factors to penetrate structures, nor was it impossible for these structures to manifest different forms. The abruptness with which such new structures could appear, however, was puzzling, for stable organisms could change and form the beginnings of new and equally stable ones. It followed that a given structure was far from being a primordial product of nature which had received its existence directly from the hand of God. Wolff acknowledged the existence of stable species and genera, though he pointed out that sometimes new ones could be suddenly produced which did transmit characters from generation to generation. In other words, he realized that the principle of invariability was coexistent with the contrary principle which was responsible for the origin of new species and genera.

For Wolff, the constancy of species and genera was derived from the specificity of a structured substance that reproduced itself and was therefore hereditary. But only those external factors which intruded upon the qualified substance and modified it could cause hereditary changes.

Despite the errors he made from the standpoint of present-day knowledge, Wolff deserves a distinguished place in the history of eighteenth-century biology for having fought against the theory of preformation and for having anticipated the ideas of Lamarck[3] and Darwin. The embryologists who followed him relied on his methodology and work. He anticipated the cell theory and the theory of development.[4]

The philosopher Johann Gottfried Herder (1744-1803), a contemporary of Wolff, also commented on genetic problems in his *Outlines of a Philosophy of the History of Man* (1785-1792). He made a clear distinction between changes that occur in man due to environmental influences and those resulting from internal causes. He regards the genetic force as the mother of all formations and also examines the effects of climate and the occurrence of artificially induced deformities in man:

For centuries, different nations have molded their heads, pierced their noses, bound their feet, or extended their ears. Nature remained true to herself; though forced for a while to take a course against her will and to send fluids to the distorted parts, she proceeded on her own way as soon as she was able and produced her own more perfect image. But if the deformity was genetic, and effected in the natural way, the case was quite different, for it was hereditary, even in particular parts. Let it not be said that art or the sun has flattened the Negro's nose. The formation of this part is connected with the structure of the whole skull, the chin, the neck, and the spine; the proliferous spinal marrow is, as it were, the trunk of the tree on which the thorax and all the limbs are formed. Comparative anatomy gives adequate proof that this characteristic has affected the whole figure and that none of these solid parts could be changed without an alteration of the whole. Thus the Negro form is transmitted in hereditary succession, and it may be changed only by genetic means. Transport the Negro to Europe and he will remain as he was; but let him marry a white woman, and a single generation will effect a change which the fair-complexioned climate could not have produced in centuries. So it is with the shapes and forms of all nations: regions alter them very slowly; but after a few generations of intermarriage with foreigners, every Mongol, Chinese, or American feature vanishes. . . . Lastly, the most exquisite means employed by nature to unite variety and stability of form in her genera

[3]We will not discuss here problems in the theory of evolution as they were developed in *Philosophie zoologique* (1809), the most important work of Wolff's great contemporary, the botanist, zoologist, and philosopher Jean-Baptiste Lamarck (1744-1829).

[4]Raikhov and Gaisinovich (1961) disagree widely in their interpretations of Wolff's contribution to cell theory and the theory of evolution. Gaisinovich is essentially of the opinion that Wolff did not have the clear ideas about cells and tissues, their function and relationships, that Raikhov supposes. Again, according to Gaisinovich, Wolff had no inkling of the genetic principle, the basis of cell theory, and had only incomplete and disjointed views on genetic phenomena, inherited and noninherited variations.

is the creation and union of two sexes. With what wonderful delicacy and spirit do the features of the two parents unite in the countenances and forms of their children—as if their souls had been transfused into them in different proportions and the multifarious natural powers of organization had been divided between them! That diseases and features, indeed that propensities and dispositions are hereditary is known to all the world; in some wonderful way even the forms of long-departed ancestors frequently return in the course of generations. Equally undeniable, though by no means easy to explain, is the influence of the bodily and mental affections of the mother on the unborn child; many lamentable examples of the effects of this influence have been borne till death.

5.3 The Great Botanical Discoveries of the Latter Part of the Seventeenth and of the Eighteenth Centuries

In the seventeenth century the emphasis in discussion had been placed upon the clarification of the origin of human and animal semen and on the role of the sexes in reproduction, fertilization, and heredity. While the controversy between preformation and epigenesis continued to flare up without being definitely solved, the problem about sexuality in plants came very close to a solution in 1692, thanks to the work of a London doctor, Nehemiah Grew (1641-1711).

In 1694 Rudolf Jakob Camerarius (1665-1721) published a short work entitled *De sexu plantarum epistola.* Beginning with ancient times, it summarized all the known facts relevant to this problem and included a large number of his own observations. Camerarius (Figure 18) was born in Tübingen, where he studied at the university. He became Associate Professor of Medicine and director of the Botanical Garden in 1688. Seven years later, after the death of his father, he was appointed full Professor. He was the author of numerous works on medicine and botany, the most important of which is his *Letter on the Sex of Plants,* which describes the structure of flowers, the male and female organs of plants, hermaphroditism, monoecism, dioecism, and doubling, and gave particular attention to the part played by pollen in the formation of the seed.[5] He writes (p. 25):

In the vegetable kingdom . . . seminal reproduction, that gift of perfect Nature for the preservation of the species, will not take place unless the plant itself has previously been prepared by the anthers of

[5] About fifty years later, J. G. Gleditsch (1749), the director of the Botanical Garden in Berlin, gave a detailed account of his pollination experiments on palms (*Chamaerops humilis*). Anticipating the work of Kölreuter, this study was an important contribution to the recognition of sexuality in plants.

18 Rudolf Jakob Camerarius (1665-1721)

the flower. It therefore seems reasonable to confer a more noble name upon these anthers and to say that they function as male sexual organs; their capsules are the containers in which the seed itself, that most subtle element of plants, pollen dust, is secreted and accumulated, and are the vessels from where it is later released; for when it has been duly strained and refined, it reaches the tip of the plant where it has its greatest effect and can be propagated. Now, while the anthers of the plant produce male seed, the stigma, or style, corresponds to the female reproductive organs, where the young germ is conceived and nurtured with maternal care. . . . But these parts which together serve the purposes of reproduction are short-lived; for each year Nature is compelled to produce new organs of generation for the new germs. As Malpighi remarks, a given uterus does not continue to function: for in the year in which it appears, each branch of a plant possesses its own generative organs which remain fertile only for a short time; the rest of its life that branch is sterile, and we may therefore ask with Theophrastus (*Hist. plant. lib.* 4, *cap.* 14) whether it should be regarded as the same plant or a different plant.

In his *Letter*, Camerarius draws numerous parallels between the sexuality of animals and that of plants. In both, male and female sex products must meet for the embryo to be produced. But the function of sperm and pollen was still unclear:

To solve this difficult question it would be most desirable to hear from those whose optical instruments have given them the eyes of a lynx what the pollen grains from the anthers contain, how far they penetrate into the female structures, whether they remain intact until they reach the place where the seed is received, and what comes out of them when they burst. We could hardly expect to find a germ in globular pollen itself or in the seminal vesicles of the plant, i.e., in the fertilizing agent or in the as yet unfertilized eggs. With regard to this question, some say that the fruit is produced by the male and is merely cared for and nourished by the female, as it might be by the soil; but others maintain that the generative substance of the male is not introduced into the uterus, as the seed of the plant into the earth, but that the male seminal fluid must be compared to the anthers of the plant, and the organ of conception of the animal to the seed of the plant (p. 30).

The part played by the wind in pollination was known to Pliny (*lib.* 16, *cap.* 25); it was demonstrated in a precise way by Camerarius, who also points out that many trees and shrubs can produce seed without having been fertilized, though such seeds were not viable; that is, that fruit could develop without pollination but these are then seedless. Many of his experiments were inconclusive or raised new doubts. It was Camerarius who realized the existence of sexuality in plants and its function. He found that pollen was transmitted to the stigma in self-fertilizing plants and that wind pollination caused cross-fertilization. We are indebted to him for all of these discoveries, though his study met with little public approval at the time. The English botanist Philip Miller (1721) was the first to describe the part played by insects in pollination, which was later clearly recognized by J. G. Kölreuter. It was the botanist C. K. Sprengel, the rector of Spandau (b. September 22, 1750, in Brandenburg; d. April 7, 1816, in Berlin) who gave a thorough account of this process in his famous book *Das entdeckte Geheimnis der Natur im Bau und in der Befruchtung der Blumen.* Sprengel's contemporaries ignored this important discovery, however, and it was left to Darwin to acknowledge it in 1859.

The seriousness with which Camerarius tackled his work is shown by a quotation from the physicist Robert Boyle (1627-1691) which appears at the end of his *Letter*: "Experiments designed to prove theoretical or practical hypotheses must be repeated with the greatest of care, for an isolated experiment does not prove an hypothesis" (p. 50).

In the eighteenth century, an age of great progress in the natural sciences, the increasing number of new species and varieties made systems of classification necessary and again raised the question of the fixity or variability of species. At the same time, the discovery of sex and heredity in plants gave an ever-growing impetus to experimental investigations into the possibilities of artificial crossbreeding. We have seen that already in antiquity some early conceptions of the significance of the sexes and the role and development of sex products arose and that in myths and histories there exist monstrous combinations of human and bestial forms, that new species sprang suddenly out of decay and putrefied matter, but also spontaneously out of weeds and cultivated plants. The appearance of new varieties was observed but no satisfactory account for their existence could be provided, particularly one that did not cast serious doubts on the story of creation.

The importance of explaining sexuality in plants and the attendant possibility of producing plant hybrids is also made evident by the fact that during the eighteenth and nineteenth centuries some of the great European scientific academies arranged open competitions designed to solve both problems.

This discovery of the sexual nature of plants culminated in the work of the great Swedish doctor and botanist Carl von Linné (Figure 19), usually known in English by the Latinized version of his name, Carolus Linnaeus. The son of a Lutheran pastor, Linnaeus was born May 23, 1707, in Råshult in southern Sweden and died January 10, 1778, in Uppsala. Having studied medicine at the universities of Lund and Uppsala, he undertook an expedition to Lapland in 1732, which supplied him with the material for his *Flora Lapponica* (1737). He combined his gift for careful observation with lofty artistic inspiration and an exuberant talent for description. While still a young man influenced by Camerarius (*De sexu plantarum epistola*) and Vaillant (*De sexu plantarum*), he drew up the beginnings of a coherent system of nature in which he expounded his ideas on the sexuality of plants. These ideas formed the basis for his artificial system of classification, in which species were subsumed under genera, and genera under classes. For Linnaeus the different species had been created as ideas of God.

In 1735, when he was twenty-eight years old, Linnaeus went to

19 Carl von Linné (1707-1778)

Holland, where he published the first edition of his *Systema naturae*. The description and classification of plants and animals which it contained covered only ten folio pages, whereas the last edition from his own hand (1766-1768) comprised 2300 pages! The Linnaean system of classification embodied a binomial nomenclature of the kind first introduced by Bauhin. For Linnaeus nomenclature was required to clarify the process of classification and was therefore as important as classification itself. He had already laid down the basic principles of the binomial system in his early work *Critica botanica*, and this system had been thoroughly established in his most important treatise on botany, *Species plantarum* (1753). He discarded the names then in use, and invented new ones, immortalizing famous botanists with some of his new generic names: for example, Rudbeckia, Pisonia, Commelina, Dorstenia, Hernandia, and so on. His *Fundamenta botanica* (1736), *Genera plantarum* (1737), *Hortus Cliffortianus* (1738), and *Classes plantarum* (1738) also appeared in Holland.

Linnaeus returned to Sweden in 1738 and practiced medicine for a time. In 1741 he was appointed Professor of Medicine at Uppsala, and a year later he became Professor of Botany at the same university.

The task of characterizing the influence of the Linnaean system of classification on his contemporaries and of showing in detail what made him the founder of taxonomy belongs in the realm of a history of biology. In such a work, his name and the era named for him will always be sure of special appreciation. In view of the growing knowledge about the multiplicity of the organic world, how did his views on the fixity of species come to change during the course of his life? Inspired by religion in his early study *Fundamenta botanica*, Linnaeus had written the famous statement that was later made a precept in *Philosophia botanica* in 1751: "Species are as numerous as there were created different forms in the beginning."

Linnaeus first began to doubt the truth of this proposition in *Critica botanica* during his stay in Holland, where he had had the opportunity of observing monstrosities and artificially induced variations in Clifford's garden at Hartecamp. But for Linnaeus such works of nature in a sportive mood enjoyed only a transitory existence. Nor did he consider the host of different varieties that had been observed to be significant; they had been produced by chance, were conditioned by climate, by temperature, and by the soil, and reverted to their original form when restored to their former environment. In 1742, however, a *Linaria* with actinomorphic flowers was found. Linnaeus described it in his *Peloria* (1744):

> Nothing can be more wonderful than what has happened to our plant: the deformed offspring of a plant that used to produce flowers of an irregular form have now reverted to a regular form. This is not merely a variation with regard to the maternal genus, but an aberration in terms of the whole class; it provides an example unequaled in the whole of botany, which may now no longer be thought of in terms of the differences between flowers. What has happened is indeed no less wonderful than had a cow given birth to a calf with the head of a wolf.

This peloria was fertile and produced similar offspring, and Linnaeus was obliged to draw the following conclusion:

That it can happen that new species of plants can be produced; that genera that have different sexual organs can be derived from the same source and have the same nature; indeed, that there are different sexual organs in individuals of a given genus—these discoveries would seem to subvert the fundamental principles of fertilization in plants—the principles, that is to say, of the whole of botanical science; the natural classes of plants would appear to be brought to confusion.

In 1759 the Imperial Academy of Sciences in St. Petersburg announced an open competition with the following words:

"Sexum plantarum argumentis et experimentis novis, practer adhuc jam cognita, vel corroborare, vel impugnare, praemissa expositione historica et physica omnium plantae partium, quae aliquid ad foecundationem et perfectionem seminis et fructus conferre creduntur." (To establish or discredit the sexuality of plants by means of new arguments or experiments apart from those already known, with an introductory historical and physical exposition of all the parts of a plant that are believed to contribute something to the fertilization and development of a seed and of a fruit.)

Linnaeus's earlier qualms about the fixity of species are taken up again in his entry for this competition, *Disquisitio de sexu plantarum*, for which he was awarded the prize on July 6, 1760. In that treatise, however, he no longer doubted that new species could be produced by crossbreeding, for he had succeeded in obtaining, by artificial means, a hybrid between two species of *Tragopogon* or goat's-beard (*T. pratensis* x *T. sporrifolius*). He therefore concluded that the different species of a given genus had originally been members of a single species and had subsequently come into being by means of crossbreeding. In a letter to Bäck written in 1764 he wrote: "We may assume that God made one thing before making two, two things before making four; that he first of all created simplicia, and then composita; first a single species from a genus, and then mixed the different genera so that new species would develop. . . ." (quoted from Hagberg 1946, p. 229).

The ancient principle of the invariability of species was thus abandoned. Linnaeus himself expunged from the last edition of *Systema naturae* the sentence that there are no new species.

From that time onward, all biologists were preoccupied with the problem of the constancy of species. In a lecture given in 1749 Johann Georg Gmelin, Professor of Chemistry and Botany at the University of Tübingen, raised the question whether it was possible for additional species, other than those created by God, to come into being. He considered Linnaeus's peloria to be an example of a new species, provided it was fertile and retained its characteristics, and concluded by saying that years of experimental research would be required to furnish a definitive answer to this question.

Beset with doubt and misgivings throughout his life, Linnaeus had painstakingly won his way to the conclusion that species had not, in fact, been created and fixed for all time. But it would be quite wrong to think that his work anticipated evolutionary theory as we understand it today. In point of fact, he rejected ideas of this sort in his *Discourse on the Increase of the Habitable Parts of the Earth* (1743).

Still, biologists were no longer fettered by the dogma of the invariability of species. Experiment and analysis enabled them to prepare for the triumphs of the nineteenth century.

Massive progress was made toward this breakthrough thanks to the work of Josef Gottlieb Kölreuter (1733-1806) (Figure 20). We have already mentioned that he studied sex in plants and discovered the significance of insects to pollination in our treatment of the work of Camerarius and Sprengel. The son of an apothecary, he was born April 27, 1733, in Sulz on the Neckar. After having studied at Tübingen and Strasbourg, he went to St. Petersburg in 1756 where, at the Imperial Academy of Sciences, he spent three years classifying the specimens of the ichthyological collection. In 1759, however, he undertook his first studies of hybridization in plants. Two years later he returned to Sulz and then moved to Calw in Württemberg. In 1763 he was appointed director of the Ducal Gardens in Karlsruhe and Professor of Natural History, but six years later he relinquished this post, irritated by the garden staff who interfered with his experiments and ruined them. Under conditions in which it was difficult to obtain scientific accuracy he continued his experiments in his own small garden until 1790. He died November 11, 1806, in Karlsruhe.

Hybrids formed from different species of plants had been known for a long time, but they had been found by chance, having been reported by gardeners who had no particular interest in science. We shall mention only a few of Kölreuter's many predecessors, the most important of whom were the botanists and gardeners of England at the beginning of the eighteenth century. C. Zirkle (1932, 1935) has given a thorough survey of experiments in hybridization up to the time of Kölreuter.

Cotton Mather (1663-1728) reports on his observations in a letter written September 24, 1716, to James Petiver. These observations were subsequently included in Mather's book, *Religio Philosophica*; or, *The*

20 Josef Gottlieb Kölreuter (1733-1806)

Christian Philosopher (1721). He describes spontaneous crosses in maize, crosses between different varieties of "Indian corn" having red and blue grains and the usual variety, which had yellow grains. The extent of this spontaneous cross could be seen from the diffusion of red
and blue grains and the usual variety, which had yellow grains. The

observations on maize, and J. Logan (1736) demonstrated by controlled experimentation that pollen is essential for the formation of seed in maize (cf. Zirkle 1932).

Mather also reported on spontaneous crosses between *Cucurbita pepo* var. *ovifera* and *C. pepo* var. *condensa* or *C. pepo* var. *maxima*. The flowers of *C. pepo* var. *ovifera* when pollinated with the pollen of *C. pepo* var. *condensa* produced bitter fruits.

Thomas Fairchild (1667-1729) produced the first artificial hybrid from *Dianthus caryophyllus* and *D. barbatus*, which was named for the first time in a gardening journal in 1717. This sterile hybrid was reproduced by vegetative propagation for more than a hundred years. Fairchild, a gifted market gardener with far-ranging interests, was known for having introduced many foreign plants into England. Later there was some doubt about whether the hybrid had occurred spontaneously or whether it had been obtained by artificial means. Zirkle (1935) has made a detailed study of this question. By the end of the eighteenth century progressive cultivators in England had begun to carry out systematic experiments in hybridization on cultivated plants in order to obtain increased productivity.

Using information provided by the Liège botanist Morren (1858), E. Lehmann (1916) wrote of a little-known predecessor of Kölreuter, the Parisian postal clerk N. Guyot, who was one of the first who "deliberately produced hybrids on the basis of knowledge about sex in plants." He had a clear understanding of the transmission of pollen and its consequences. He cultivated variously colored plants of the same species (*ranunculus* and *auricula*) close to one another and obtained offspring in which the colors were mixed.

Kölreuter, however, who was skeptical about all these statements about successful crosses, was the first to undertake systematic hybridization experiments in plants. Linnaeus's *Tragopogon* hybrid also seemed doubtful to him. To start with, he succeeded in obtaining a cross from *Nicotiana rustica* × *N. paniculata* and described it in 1761. He realized that it was intermediate between the parent species and that it was sterile in both sexes. From this successful experiment he drew the following conclusion:

With great pleasure I saw that they [the hybrids] maintained precisely the mean between both the natural genera, not only merely in the extension of their branches and in the position and color of their flowers in general; they also displayed an almost geometrical proportion—compared with their natural parents—in all the parts of the flower, with the exception of the anthers alone. These facts completely justify the ancient Aristotelian doctrine of reproduction by means of two sorts of seed and also refute the doctrine of animalcules together with that which assumes that embryos or germs are originally present in the ovaries of animals and plants and merely need to be stimulated by male seed.

The description of the characters of this hybrid revealed that its nature was determined, in equal proportions, by both parents. At the beginning of his first *Vorläufige Nachricht von einigen das Geschlecht der Pflanzen betreffenden Versuchen und Beobachtungen*, written in 1763, Kölreuter makes this quite general inference:

Two different, homogeneous liquid substances are necessary for the procreation of every natural plant. These two substances, which are ordained by the Creator of all things to unite with one another, are male and female seed. It is easy to understand that the virtue inherent in the one differs from the virtue that resides in the other, for the two substances are different in kind and have dissimilar natures. The combination and union of these substances, which conform to certain proportions and occur in a most intimate and orderly way, produces a third substance which is intermediate between the first two. The virtue it possesses, formed from those two simple virtues, is therefore of a composite and intermediate nature, and it may be compared with the intermediate type of salt produced from an acid mixed with alkali (p. 42).

With regard to the sterile hybrid which results from *Nicotiana rustica* × *N. paniculata*, however, Kölreuter concludes:

A hybrid that is perfect, yet quite sterile in both sexes, is produced in just the same way as any other natural plant. It displays the same competence in its form and development. Following its growth from an embryo into a plant whose flowers are almost completely formed, the keenest eye will discover no lesser perfection than in a natural plant. Yet the hybrid is imperfect, and I can perhaps say without exaggeration that its deficiency is the most important of all imperfections, for it is sterile; this fact would never occur to the greatest philosopher who saw such a plant for the first time (p. 43).

For Kölreuter, however, it was the conversion of the hybrid into the paternal species from which it was derived that constituted the all-important advance in the transformation of species by experimental means; for example, when the hybrid derived from *Nicotiana rustica* × *N. paniculata* was transformed into *N. paniculata* by means of repeated back-crossing with the paternal species. In this connection Kölreuter says (1764): "In the eyes of those able to discern the essential value of things, I believe that my discovery has accomplished, if not more then at least as much as if I had transmuted lead into gold or gold into lead" (p. 87).

Even though he was successful in obtaining hybrids from experiments with several other genera and species (*Dianthus*, *Matthiola*, *Hyoscyamus*, *Verbascum*, *Hibiscus*, *Datura*, *Cucurbita*, *Aquilegia*, *Cheiranthus*, and so on), Kölreuter's hybrids and his "transformation of species" by back-crossing met with little approval from his contemporaries. F. J. Schelver (1812) and A. Henschel (1820) doubted the authenticity of Kölreuter's experiments. They were opposed to the idea of sexuality in the vegetable kingdom and sought to explain hybridization in a different way. But Sageret (1826), a great authority in his field, said of Kölreuter: "Having reproduced several of his experiments, I became more and more convinced of his accuracy and of his veracity; I therefore believe that he deserves our confidence."

It can be demonstrated in the history of any science that few important discoveries were appreciated in their full significance by contemporaries. In the history of genetics the rejection of the works of Camerarius, Kölreuter, and Sprengel (see for instance F. J. Schelver's *Kritik der Lehre von den Geschlechtern der Pflanze*, 1812, 1814), who created the basis for the development of plant genetics, represents one of the great blunders of an epoch in which the dogma of nonvariability of the species had not yet been overcome.

In the eighteenth century, the *Linaria* with peloric flowers which had caused Linnaeus to doubt the doctrine of the fixity of species was not the only case of a variation known to have suddenly occurred in a higher plant. In 1715 Jean Marchant (1719) observed a laciniate form of *Mercurialis annua* which he called *Mercurialis foliis capillaceis.* One year later four plants displaying identical variations appeared in the same part of the garden together with two other plants that differed slightly from the laciniate form. In 1761 M. Duchesne discovered a variant in some seedlings of *Fragaria vesca* at Versailles; instead of the usual ternate leaves, it had simple, oval verging on cordate, leaves with strongly serrated margins. The flowers were normal, the sepals slightly enlarged and also toothed, and the fruits contained small seeds. Of eighty descendants of this variant, all but four were constant. For purposes of cultivation, however, the new form was unserviceable, for it was too delicate and very susceptible to frost. In his *Histoire naturelle des fraisiers*, Duchesne (1766) describes a number of other changes that occurred in the characters of *Fragaria silvestris* and *F. hortensis.*

The sudden appearance of these new forms always led to lively discussions about their classification. When one or several of the characters of a new form varied hereditarily and showed no tendency to revert to the ancestral form, as was frequently the case, it was usually characterized as a new species. But the difficulties inherent in the concept of a species began to be apparent: the same form was described as a new species, as a variety, or a race. However, there is no exact definition of the concept of a species. Having described a number of variants in his *Histoire naturelle des fraisiers*, Duchesne concludes:

How must it be categorized, I then asked myself? Is it a species? . . . New ones develop from it: is it merely a variety? . . . How many varieties are there in other genera which are considered to be species? For a long time I wavered between these two alternatives. To resolve my difficulty, I examined what botanists said about the characters that distinguish species from varieties and compared their ideas with what I had observed in nature concerning strawberries, in particular with that strange phenomenon of which I was a witness, and of which I previously spoke. It seemed to me that something was wrong with the ideas I had conceived, but the fact that different authors used the same words to refer to completely different concepts was the main cause of the confusion.

H. Hoffmann (1881) writes in the introduction to his *Rückblick auf meine Variations-Versuche von 1855-1880*: "As for the concept of species, I must confess that I have lost it during the course of these investigations; there are no sharp boundaries for this concept, and there is no one feature which it must possess."

Roze (1895) and Marchant (1719) decided to consider as species certain new forms of *Mercurialis* and *Chelidonium* they described. Anticipating subsequent discussion, we shall now refer to the particularly striking example of the Lombardy poplar, *Populus pyramidalis*, which many botanists considered a perfectly genuine new species. S. Korschinsky (1901) has given a detailed account of its origin; he maintained that *P. pyramidalis* is not to be regarded as an authentic species but rather as a variety of the black poplar, *P. nigra*. This view is still held today. *P. pyramidalis* differs from *P. nigra* not merely in its growth form, but also in the shape of its leaves. The variation suddenly appeared in a single male plant of *P. nigra*. Its place of origin is not known with any certainty, but it is now generally thought to be in Asia Minor. From there it probably made its way to Italy and was then propagated throughout Europe by Napoleon. Of the few female specimens, six have been described. They are probably best regarded as hybrids derived from other species, for they all differ with regard to characteristic features from the male plants of *P. pyramidalis*.

In animals, too, deviations from the normal form were increasingly observed. L. C. Dunn and W. Landauer (1934, p. 240) quote a passage from Nathaniel Highmore's *History of Generation* (London, 1651) where we read (p. 31): "Myself also have seen a kinde of poultry without rumps, which breeding with their own kinde, still brought forth chickens wanting that part. If with others sometime had rumps, sometime but part of a rump."

Felix d'Azara (1801) reports that a male animal that was found to be hornless was born into a South American herd of horned cattle in 1770. This hornlessness was hereditary; a "mocho" (that is, hornless) breed was produced and widely propagated.

In 1791, a ewe of S. Wight's flock gave birth to a male lamb that became the paternal ancestor of the Otter sheep, also known as the Ancon sheep (ancon = elbow). Colonel D. Humphreys (1813) reports on this to the Honourable Sir J. Banks: "When both parents are of the Otter or Ancon breed, the descendants inherit their peculiar appearance and proportions of form."

This new variety was distinguished by its long body and very short legs. The animals developed more slowly when compared to those of the original form, for they could not run and jump as well as their normal siblings, but they were greatly prized for their inability to clear high fences. Later, however, drawbacks became apparent: it was difficult to drive them to market, and they did not fatten up as quickly as their normal siblings. Little by little the breed was discontinued.

In 1919 the Ancon breed appeared for a second time on a farm in Gudbrandsdalen in eastern Norway (Wriedt 1925). Crossbreeding revealed that this characteristic was simply recessive. Landauer and Chang (1949) crossed Ancons with normal sheep and obtained animals (seventeen) of the usual kind with no Ancons in F_1. Ancon × Ancon always produced Ancons (fifty-three animals). Back-crosses between heterozygous ♀ × Ancon ♂ produced thirty-six Ancons and forty-three normal sheep. Both authors were able to undertake skeletal mensuration on these animals of different extractions. On the basis of the differences they found, it seems probable that these variations were phenotypically similar and genetically distinct.

Many other nineteenth-century examples may be added to these early discoveries of the sudden appearance of new forms in animals and in plants. Most of these, however, pertain to hereditary variations in domesticated animals and in cultivated plants; reports of the appearance of new varieties of wild animals and plants become less frequent. Such reports were always chance discoveries, whereas a systematic quest for hereditary variations in cultivated plants was shortly to be undertaken.

6 The Nineteenth Century

6.1 The Great Plant Breeders of the Nineteenth Century

Kölreuter's investigations had again raised problems concerning the nature of those delicate processes that characterize fertilization. These had not yet been explained, for the techniques of microscopy had to be considerably refined before they could reveal pollen germination, the growth of the pollen tube, and the further development of the pollen grain. However, progress in scientific studies of hybridization was preceded at the end of the eighteenth and beginning of the nineteenth centuries by the work of plant breeders, especially those in England, who conducted experiments in plant hybridization in order to achieve increased productivity. The work of these planters includes a number of important discoveries in genetics together with a wealth of significant critical observations.

Thomas Andrew Knight (1759-1838), for example, was convinced that crossbreeding could bring about vast increases in the yield from animals and plants. A man who combined scientific intuition with considerable practical experience, he was president of the Horticultural Society of London from 1811 until his death in 1838. He was particularly interested in the vine and in apples, pears, and plums and tried to cultivate frost-resistant and productive strains. But it is his experiments in crossing the pea plant (1799-1823) that are of special interest to geneticists. Knight (like Mendel after him) was aware of the advantages of the pea as an experimental object: it has numerous varieties with distinct features, is rigorously self-pollinated, is easy to cultivate, and its offspring can be isolated without difficulty. An important part of Knight's experiments was his careful castration of the flowers used for crossing and his control observations on the flowers that had not been pollinated after castration. Exuberant growth and vitality in the hybrids was observed time and again. He was the first to discover that the gray color of the seeds was dominant to the white. Backcrossing the hybrid with its parental form (which had white seeds), he obtained both gray and white seeds in the subsequent generation, though he did not determine the numerical ratios in which they appeared.

The famous Italian naturalist and pomologist Count Giorgio Gallesio (Figure 21) obtained similar results from his work on hybridization in carnations. He was born May 25, 1772, at Finale in Liguria and died November 29, 1839, in Florence. He studied jurisprudence at the

21 Count Giorgio Gallesio (1772-1839)

University of Pavia, from which he graduated in 1793. He held a number of high offices of state and was secretary of the delegation from Genoa at the Congress of Vienna. In 1811 Gallesio published his *Traité du citrus*; in 1816 a further book appeared, entitled *Teoria della riproduzione vegetale* and containing his latest observations on plant cultivation. There we read: "I fecundated some carnations having white flowers with the pollen taken from carnations bearing red flowers, and conversely; the seeds I gathered from them gave me carnations with flowers of mixed colors." He continues: "It follows that the crossing of these, not being natural, produces varying effects which bear the imprint of their different sources in the proportion to which a given one is dominant" (pp. 76 and 79).

Martini (1961) pointed out that this is the first use of the term "dominant." Roberts (1929), in his *Plant Hybridization before Mendel*, claims that Sageret (1826) was the first to make use of the concept of "dominance." It is now clear, however, that Gallesio had used it ten years earlier. We do not know whether Mendel was familiar with Gallesio's work; Darwin, however, had read both of his studies. He frequently and enthusiastically quotes from them in *The Variation of Animals and Plants under Domestication.*

William Herbert (b. 1778), a contemporary of Knight, made a detailed analysis of the concept of species by examining whether the fertility of hybrids could be used as a criterion for membership in the same (or different) species. He decided that the fact that hybrid offspring, whether fertile or sterile, are produced establishes that both individuals used in the cross have a common origin in the same genus and that there was no really sharp dividing line between varieties and species. Moreover, since climatic and edaphic influences affected fertility and sterility, there was indeed no strictly formal way of classifying such hybrid offspring; the decisive factor in a successful cross was that environmental conditions happened to suit the constitution of the partners in question.

A high-ranking dignitary of the Church of England (he was dean of Manchester Cathedral), Herbert was obliged to come to terms with a rebuke to which he had been subjected for holding this progressive opinion. He writes (1847, p. 6):

> The only thing certain is that we are ignorant of the origin of races; that God has revealed nothing to us on the subject; and that we may amuse ourselves with speculating thereon; but we cannot obtain negative proof, that is, proof that two creatures or vegetables of the same family did not descend from one source. But we can prove the affirmative, and that is the use of hybridizing experiments, which I have invariably suggested; for if I can produce a fertile offspring between two plants that botanists have reckoned fundamentally distinct, I consider that I have shown them to be one kind, and indeed I am inclined to think that if a well-formed and healthy offspring proceeds at all from their union, it would be rash to hold them of distinct origin.

Herbert was convinced that crossbreeding could produce new and valuable forms. While Knight was primarily interested in crossing different kinds of fruit, Herbert obtained extensive practical knowledge about horticultural plants and some experience of agricultural varieties. Ignorant of the literature on hybridization, Herbert worked alone for many years before he came across Kölreuter's work. He comments: ". . . and as I have found nothing in his reports to the best of my recollection opposed to my own general observations, it is unnecessary to state more concerning his mules than the fact that he was the father of such experiments."

John Goss (1820) chose varieties of peas as the subjects for his experiments in crossbreeding as Knight had done. Goss informed the secretary of the Horticultural Society of London of his results in a letter written in 1822.

He crossed the "Blue Prussian" with the "Dwarf Spanish" variety and found that the seeds of the first filial generation all displayed the white color of their male parent. In the second generation he found "that these white seeds had produced some pods with all blue, some with all white, and many with both blue and white peas in the same pod."

When these blue seeds were planted separately from the white ones, it was seen that the blue seeds produced plants that contained only blue peas, while the white seeds gave forth plants that contained only white peas as well as such that had blue and white peas in the same pod. After publication, the secretary of the Society supplemented Goss's letter with a remark drawing attention to the work of A. Seton who had also reported on crosses in peas at a meeting of the Society held the very same year. The latter (1824) had crossed the green "Dwarf Imperial" with a white variety and obtained a pod containing four peas that were all green in color; but the plants produced from these seeds were intermediate with regard to their size and the time it took them to mature. Moreover, they had pods with green or white peas. As Seton wrote, "they were all completely either of one colour or of the other, none of them having an intermediate tint" (Vol. I, p. 237).

Like Knight, Goss and Seton had observed dominance and segregation; but they did not examine later generations, nor did they study the numerical ratios in which the characters were transmitted

A few decades later, Thomas Laxton (1866, 1872) carried out further experiments in crossing peas. They confirmed the facts that had been discovered by Knight (1823) and Goss (1822): dominance of a parental character, for example, color and shape of the peas; and segregation of characters in the second filial generation. Laxton's experiments were not carried out on a scale large enough to admit of numerical analysis, but the different forms were harvested separately and replanted. Laxton then obtained a result that showed—probably like none other in botany—the grouped segregation of two *pairs of factors* in the Mendelian sense. He says (1872, p. 13):

I have noticed that a cross between a round white and a blue wrinkled pea, will in the third and fourth generations (second and third years' produce) at times bring forth blue round, blue wrinkled, white round, and white wrinkled peas in the same pod, that the white round seeds when again sown, will produce only white round seeds, that the white wrinkled seeds will, up to the fourth or fifth generation, produce both blue and white wrinkled and round peas, that the blue round peas will produce blue wrinkled and round peas, but that the blue wrinkled peas will bear only blue wrinkled seeds.

An experienced plant breeder, Laxton knew that maximum variability appeared in the third and fourth generations after the cross and that it was not possible to obtain fixed new varieties any earlier, though a reversion to parental characters could occur in earlier generations.

The work undertaken from 1856 to 1860 by the great French plant breeder Louis de Vilmorin (1816-1860) showed an awareness of segregation, though not experimentally proved, in the flower color of *Lupinus hirsutus.*[1] His son Henry de Vilmorin (1843-1899) subsequently (1879) reported on this work. In selecting varieties of *L. hirsutus* that would remain constant with regard to the color of their flowers (blue or pink), Vilmorin observed segregation in the offspring of parent plants whose flowers were blue or pink; this suggested spontaneous crossing between flowers of each color. Crosses between blues and pinks repeatedly segregated in the ratio of three pink to one blue (if the hybrid had been pink); others segregated in the ratio of three blue to one pink (when the hybrid had been blue). This observation needed to be further tested since the experimental material had been of limited proportions.

In his later experiments in crossing peas, H. de Vilmorin (1890), like Goss (1822), found segregation of characters, different combinations of characters, and differently colored peas in the same pod.

The famous breeders of the nineteenth century paid a great deal of attention to spontaneous variations when selecting new varieties for cultivation. K. Rümker (1889) gives some excellent examples in his *Anleitung zur Getreidezüchtung*. Le Couteur seems to have been the

[1]Louis de Vilmorin deserves credit for introducing into cultivating practices the idea of examining the offspring separately. Generally speaking, this was seen to be absolutely necessary for clarifying problems about the inheritance of individual characters and features.

first to select unfamiliar varieties from hybrids in a systematic way. In 1841 R. Hope (according to P. Shirreff, 1873) described a Fenton wheat that had developed from an isolated plant. The well-known breeder Patrick Shirreff, who came from Haddington in Scotland and lived until the late 1870s, cultivated a number of new varieties by selecting spontaneous variations in oats and wheat (1873). His experiments involved:

1. Mungoswells wheat: a single specimen was discovered in 1819; it was constant after the fourth generation.
2. Hopetoun oats: a single specimen was discovered in 1824.
3. Hopetoun wheat: a single specimen was discovered in 1832.
4. Shirreff oats, origin unknown.

5. Shirreff's bearded red wheat	
6. Shirreff's bearded white wheat	put on the market about 1860
7. Pringles wheat	
8. Early Fellow oats	
9. Fine Fellow oats	put on the market about 1865
10. Long Fellow oats	
11. Early Angus oats	

There was no lack of experiments designed to cause spontaneous variations to breed true in rye. Martini (1871) attempted to fix the ternate character of three-flowered spicules. Blomeyer tried to achieve the same thing with "Leipzig rye." Neither attempt was completely successful, even though the character of three-flowered spicules was increased to 75 percent in Leipzig rye. Echinate rye and lax-eared rye, both cultivated by Wollny (1885), were obtained from spontaneous variations in rye.

In 1883 Rimpau discovered a ramose ear in a two-rowed variety of barley which gradually became constant after repeated selection for this trait. But this new variety was not commercially produced, for it had flat and partly misshapen grains. He also described ramose ears in rye (1899), but these were not extensively cultivated either. Selecting for a red-glumed, nonaristate landwheat, Rimpau (1877) also discovered three new forms of ear: (1) aristate; (2) white-glumed; (3) two plants with extremely dense ears on a shorter and more stocky stem. All three forms remained constant when cultivated, though the first two had no market value.

Beseler's Anderbeck oats is a well-known example of a spontaneous variation that was valuable for purposes of cultivation. It was grown as a muticate variety by selection from the aristate "Probstei oats." He also found spontaneous variations in wheat that remained constant when cultivated. From a spontaneous variation Heine cultivated Emma spring wheat in Emersleben; this turned out to be a valuable variety. And Drechsler found an aristate variety of squarehead wheat in the experimental field in Göttingen.

According to Darwin, Metzger (1841) paid a great deal of attention to the striking variations in maize. He described twelve subspecies with numerous subvarieties whose differences in height ranged from 16 to 18 inches to 15 to 18 feet. He stressed the variability of the cob both in the shape and in the form and color of the grains.

While all these nineteenth-century breeders more or less adopted a practical outlook in their work, some of the academies were anxious to explain the phenomenon of hybridization.

6.2 The Open Competitions of the Academies in the Nineteenth Century

The inductive approach to botany that had been initiated by Camerarius, Linnaeus, and Kölreuter inspired the Physics Section of the Prussian Academy of Sciences in Berlin to arrange another open competition sixty years after that of the Academy in St. Petersburg. This was suggested in 1819 by the botanist Link. At first, however, no response was forthcoming. In 1822 a subject for open competition was offered again: "Is Hybrid Fertilization Possible in the Vegetable Kingdom?"

Only one contribution was entered in 1828, by the apothecary and botanist A. F. Wiegmann (1771-1853) from Braunschweig. He had been a member of the Imperial Academy of Naturalists (Leopoldina) since 1821. The paper he presented was entitled *Über die Bastarderzeugung im Pflanzenreiche* (1828). Though it contained a number of methodological errors—absence of castration and isolation, together with other inadequate safeguards against cross-pollination—it nonetheless made some important advances and was awarded half of the prize by the Academy. Wiegmann studied the genera *Verbascum*, *Dianthus*, *Nicotiana*, *Brassica* (cross-pollination by means of insects), *Allium*, *Phaseolus*, *Pisum*, *Vicia*, and *Avena*. From his results he drew the following important conclusions:

1. Hybridization takes place within the vegetable kingdom; this provides a valid proof of the sexual nature of plants. The influence of the male parent which transmits pollen to the stigma is apparent in the characteristics of the first filial generation.

2. The pollen that has reached the stigma is absorbed by the secretions of the latter into the cellular tissues of the vessels of the pistil and thereby conveyed to the conceptacle. The fertilizing moisture is carried off and assimilated, and is seen to be formative and activating in the albumen, the product of the female parent.

3. Some hybrids revert to the features of their male parent to a greater degree, while others maintain an intermediate condition between both parents. More frequently, however, hybrids resemble their female parent, while also having some of their characters in common with their male parent and others with their female parent, though they cannot resemble both parents with respect to a given characteristic. Again, other features (for example, the shape of the sepals in *Nicotiana*) assume an intermediate condition. Some hybrids occasionally show an intermediate condition with regard to the color of their flowers, though in many cases the color of the hybrids is the same as that of the male or female parent. In Wiegmann's own words (p. 22):

> I own that it is a mystery why they sometimes resemble their male and sometimes their female parent to a greater degree, and why in some parts they bear the stamp of their father, in others that of their mother; but can we explain the same phenomenon in the animal kingdom where sexual function is more potent and more significant? Should we refuse to call the hybrid products of the vegetable kingdom hybrids because we are unable to understand those intimate processes whereby they are formed? Kölreuter, who assumed the existence of male and female seminal fluid in plants as well, declares that the frequent resemblance of a hybrid to one or the other of its parents to an extreme degree was due to the preponderance of the nature of one of these seminal liquids over the nature of the other with which it had been artificially mixed. (Kölreuter's *1. Fortsetzung der Nachrichten über das Geschlecht der Pflanzen*, 1763, p. 16.)
>
> Moreover, such resemblances seem to proceed from the quantity of pollen that is assimilated by the stigma, for the different stages of development of the seeds are derived solely and exclusively from the weaker or stronger agency of the foreign pollinating substance and are proportionate to the quantity of foreign pollen which is absorbed.

4. Wiegmann rejects Linnaeus's view that hybrids resemble their mother with regard to their organs of fertilization and their father with regard to their foliage and general appearance.

5. When hybrids are formed by crossing different varieties and species, their seeds are fertile; whether the same is true for hybrids formed by crossing plants from different genera remains doubtful.

6. Individuals produced from a seed and a capsule taken from the hybrid plants often differ considerably from one another with regard to fertility and the shape of various parts; sometimes they resemble their

mother and sometimes their father to a greater extent. Mixed pollinations composed of equal proportions of pollen from the plant in question and that taken from a foreign plant do not produce hybrids. 7. The closer the parents are related, the easier it is to produce a hybrid. Therefore it is easiest of all with different subspecies and varieties, next, with many species of the same genus, and more difficult with plants of different genera. As Wiegmann says: "The most important point is that the different plants do not differ greatly from one another with regard to their natural constitutions and that their secretions are not too heterogeneous; for when this is the case, the fertilizing substance, pollen, is not absorbed by the stigma" (VI, 1, p. 27).

It is of great historical interest to note that despite Wiegmann's lack of experimental investigation, his work produced a wealth of new observations which would today be designated as the dominant, recessive, and intermediate inheritance of individual characters. He also observed the fact of segregation, though the polygenic differences in the species he used in his study did not admit the formulation of any laws.

In the same year that Wiegmann presented his essay to the Prussian Academy, Sageret (1826) published a brief note in France which, oddly enough, was little quoted although it contained results and methodological particulars that could have been useful to subsequent researchers.

M. Sageret (1763-1851), a naturalist and a practical farmer, was a member of the Société Royale et Centrale d'Agriculture de Paris. Brogniart named the genus *Sageretia* after him.

His experiments in crossing members of the Cucurbitaceae family (1826) are of great interest. For the first time in the history of plant hybridization, he arranged the characteristics of the parents in sets of opposed pairs. In this way, he distinguished the following characteristics when crossing two different varieties of melon (*Cucumis melo L.*):

Melon cantaloup brodé (♀)	**Melon chaté (♂)**
1. Yellow pulp	White pulp
2. Yellow seeds	White seeds
3. Reticulated skin	Glabrous skin
4. Costae very prominent	Lightly traced costae
5. Sweet taste	Bittersweet taste

From this cross Sageret obtained two fruits that showed him that the characters were not blended or intermediate, but that they were quite clearly identical with those of one or the other of the parents—that each of the characteristics was dominant to another in the hybrid. He introduced the concept of "dominance" in a clear way (p. 302):

Thus, to conclude, it seemed to me that, in general, the resemblance of the hybrid to its two parents consisted not in the intimate fusion of the various characters belonging to each of them, but rather in the even or uneven distribution of those same characters; I say "even" or "uneven" because it is far from being identical in all of the individual hybrids having the same origin, for there is great diversity among them. These facts are confirmed by a huge number of my own experiments.

He continues (p. 307): "the bitter flavor of the chaté variety was found together with the characteristics of the common melon and the serpent melon; in others, the character of the common melon was dominant. . . ."

The independent distribution of characteristics and the dominance of one over another are the two important contributions made by Sageret. The union and distribution of characteristics which can be combined in different ways form the source of the countless different varieties observed in nature—particularly in the melons studied by Sageret (p. 304):

One cannot admire too much the simplicity of the means by which nature gave itself the power to vary its products endlessly and to avoid monotony. Two of these methods, the union and distribution of characteristics combined in different ways, can bring these varieties to an infinite number.

The transmission of individual characteristics and their combination led him to the realization that in the new forms produced by hybridization characteristics were found that had long since disappeared, characteristics that had been transmitted from ancestors "whose germs still existed, though their development had never been favored."

The important new developments contained in this work—the juxtaposition of pairs of characteristics and their possibilities of combination—were apparently not further pursued by Sageret in the subsequent generations of his experimental subjects since there are no further statements about the distribution of characteristics in the progeny.

We do not know what prevented Sageret from publishing the results of his later investigations. When he was fifty-six years old he undertook the administration of an estate, and it seems probable that his practical agricultural duties made it impossible for him to continue with his scientific work.

Such work, however, was now being undertaken in a number of different countries. In his *Geschichte der Botanik vom 16. Jahrhundert bis 1860*, Julius Sachs (1875) described the situation of that time as follows (p. 461):

Those who took a serious view of this matter in the [eighteen hundred] twenties and thirties would have welcomed further comprehensive debates on the problem of sexuality in plants. As early as 1819, following Link's suggestion, the Berlin Academy of Sciences had attempted to stimulate new investigations into the core of the sexual question by announcing the open competition: Is Hybrid Fertilization Possible in the Vegetable Kingdom? The only entry—the one presented by Wiegmann as late as 1828—was thought to deserve only half of the prize.

Later, in 1830, the Dutch Academy in Haarlem was more successful. At Reinwardt's instigation it changed the question, relating it to practical considerations of plant culture. Gärtner (1772-1850) presented his entry in 1837, having been delayed by external circumstances, and was awarded the prize together with an honorable mention.

Carl Friedrich Gärtner (Figure 22) was born May 1, 1772, in Calw, Württemberg. The son of Joseph Gärtner, a well-known professor of botany and a member of the Imperial Academy of Sciences in St. Petersburg, he completed his father's three-volume treatise, *De fructibus et seminibus plantarum* (1788-1807). After attending the grammar school in Calw and the Friars' School in Bebenhauşen he became apprenticed at the Royal Dispensary in Stuttgart. After two years of studying pharmacology, he began courses in medicine at the Karls-Akademie in Stuttgart. In 1794 he went to Jena, one year later to Göttingen, and finally to Tübingen where he revised and completed his dissertation, which dealt with chemical physiology. In 1796 he returned to Calw where he practiced medicine for six years. He then set out on a journey to France, England, and Holland to broaden his medical and scientific knowledge. In France he met Cuvier, Jussieu, de Candolle and other celebrated scholars; in England he saw Dryander, Lambert, and Banks. One year later he returned to his native town. He declined an offer of the curatorship of the Botantical Gardens at Tübingen. In 1805 he completed the third volume of the treatise on carpology which included investigations made by his father and many of his own observations. In the end he devoted himself exclusively to botany—especially to the study of sexuality and hybrid fertilization in plants—and in 1826, Gärtner published the first treatise, *Vorläufige Nachrichten über die Befruchtung der Gewächse*. In 1837 he entered the competition arranged by the Dutch Academy of Sciences. In 1846 on

22 Carl Friedrich Gärtner (1772-1850)

the occasion of the fiftieth anniversary of Gärtner's doctorate, the Margrave of Württemberg appointed him a Knight of the Order of the Crown, and his native town of Calw accorded him the freedom of the city. Gärtner died on September 1, 1850, in Calw.

The subject of the open competition of the Dutch Academy of Sciences in Haarlem was: "What does experience teach about the production of new species and varieties by the artificial fertilization of flowers of the one with the pollen of the other, and what economic and ornamental plants can be produced and propagated in this way?"

Gärtner entered a brief note for this competition relating how he had conducted his extensive experiments over many years. To the Academy, however, it was by no means clear how he had arrived at his results. Moreover, he had not appended any specimens, nor had he given the sources from which he had derived his results. He was requested to remedy these deficiencies and submitted an amended manuscript together with 150 pressed specimens of hybrid plants. It was for this that he was honored with the prize plus an honorable mention in 1837, though his revised study, entitled *Versuche und Beobachtungen über die Bastarderzeugung im Pflanzenreich* was not printed in German until 1849.

In contrast to the earlier investigations of other writers, this study shows great progress in experimental methodology and in comparative descriptions of living hybrids and their parent species. Above all, it recognizes well-defined regularities in crossbreeding. Gärtner carefully evaluated more than 9000 experiments and drew the following conclusions of general importance:

1. Experiments in hybridization require that the species in question be well defined and carefully raised. All fruit that is gathered must be stored separately, all seeds must be resown, and all seedlings must be raised.

2. If a flower is fertilized with a mixture of pollen (its own and foreign pollen), no blending of characteristics appears in the offspring, as is the case, for example, when trivalent salts are produced from many chemical mixtures. Uniform fertilization by one of the kinds of pollen always occurs, namely, by the one which has the strongest sexual affinity with the ovaries. Successive mixed pollinations of the halves of the stigma of *Nicotiana rustica* with pollen taken from *N. paniculata* and *N. rustica* produced either pure specimens of *N. rustica* or genuine *rustica-paniculata* hybrids. Each grain of pollen therefore operates separately and independently of every other, so that no modification of the one by the other takes place in the offspring. Two or more paternal types never blend with the maternal type, nor are two different types of embryos formed from the same ovule. Imperfect hybrids, or semihybrids, are not the result of a mixture of their own pollen with foreign pollen (as Kölreuter had assumed), but are produced in the second generation.

3. Supporting Kölreuter's work, Gärtner observed that the same hybrid forms are always produced by crossing pure species. Hybrid types produced by crossing the same two species are therefore not

indeterminate and variable, but constant and producible according to specific laws of formation. As Gärtner himself says (p. 235):

We have therefore found it to be an invariable law of hybridization that the seed taken from an original cross between two pure species produces nothing but seedlings of the same form, and that however often such a cross with the selfsame species is repeated, the resulting hybrids always have the same form.

4. In simple hybridization, the maternal and paternal factors of two different species of plants are both active; both sets of factors possess their own nature and formative power which determine a specific development and formation of the characters in the offspring.

5. In hybridization, the individual characters are modified, blended, and crossed, and in part cancel each other out. It follows that a general law of hybridization, in plants as in animals, is that parent characters will not be delivered pure and unaltered to the given partner.

6. We do not know the criteria according to which the characters of the parents are blended in hybrids, nor do we know why it is the whole general appearance that is altered in a given hybrid, while in others it is only certain parts, such as leaves, flowers, fruits, or seeds. Hybridization is therefore not a chemical process, as Kölreuter assumed, but a process analogous to animal reproduction; it is capable of producing variants or varieties in the subsequent progeny of both animals and plants. In Gärtner's own words (p. 428):

The fertilization of fertile hybrids with their own pollen provides a source of numerous varieties as is evident from Kölreuter's experiments with *Mirabilis*, and as the flowers of carnations and auriculae, the seeds of *Zea mays* show; the knowledge of this fact has long since been exploited by flower lovers. The fertilization of mixed hybrids or of varieties obtained in this way does not produce uniform results: rather, both of the fertilizing processes appear to fluctuate irregularly when the embryos are formed; this produces embryos with different developmental forms in a single fertilization and in a single ovary. Kölreuter comments: when such hybrids are self-fertilized, the results will be irregular and diverse.

7. The production of new forms from the elements and characters of the parents by hybridization is equally important for plant physiology and taxonomy. For taxonomy, experiments in hybridization raise the question whether there are stable species or whether they are subject to further development and change with the passing of time.

From his experiments, Gärtner concluded that the appearance of new varieties after crossing showed that plant species were fixed within limits beyond which they could not vary.

Despite the merits of his work, he received adverse criticism from the man who finally, seventeen years later, succeeded in explaining the fate of hybrids in subsequent generations. With Gärtner's work specifically in mind, Gregor Mendel wrote in the introductory remarks to his *Versuche über Pflanzenhybriden* (Experiments in Plant Hybridization, 1865):

The fact that, so far, no generally applicable law governing the formation and development of hybrids has been successfully formulated will not surprise those who are cognizant of the magnitude of this task and can appreciate the difficulties with which experiments of this kind have to contend. A final decision can be made only at such time when we possess the results of detailed experiments on plants belonging to the most diverse families. Those who survey the work done in this field will conclude that of all the numerous experiments made, not one has been undertaken to such an extent and in such a way for it to be possible to determine the number of different forms in which the offspring of hybrids appear, or to classify these forms with certainty according to their separate generations, or to ascertain their statistical relations.

In a letter to Carl von Nägeli, published together with others by Carl Correns (1905), Mendel also wrote that he had been unable to get a clear picture of Gärtner's experiments because detailed descriptions of the individual experiments were lacking, adequate diagnoses of the different hybrid forms had not been made, and none of the statements about the characteristics of the hybrids was sufficiently precise. Correns agreed with this view. But Olby (1966) has referred to some correspondence by Gärtner which appeared in *Flora* in 1827. There Gärtner describes the offspring of a cross between a dwarf variety of maize with yellow seeds and a tall variety of maize whose seeds had red stripes. In F_1 the seeds were all yellow like those of the maternal plant; in F_2, however, four plants each yielded a cob, but only two had exclusively yellow seeds, though they were somewhat larger than those of the maternal plant. One of the other two remaining cobs had 64 more or less red and gray seeds out of a total of 288; the other cob had 39 similarly colored seeds out of a total of 143. This gives a segregation ratio of 3.18 yellow to 1 colored. Since yellow endosperm is dominant to red, Gärtner had observed a clear case of segregation in the Mendelian sense, though he was unable to interpret it.

Preceding the publication of the results of Mendel's work, the Paris Academy of Sciences announced an open competition in 1861 on the subject of "the study of plant hybrids from the point of view of their fertility and the permanence or impermanence of their characters."

In 1863 two essays were entered for this competition: that of D. A. Godron (University of Nancy) was entitled "The Study of Plant

23 D. A. Godron (1807-1880)

Hybrids Considered from the Standpoint of Their Fertility and the Permanence or Impermanence of Their Characters" and Charles Naudin's entry entitled "New Investigations on Hybridization in Plants." Naudin was awarded the prize.

The study by Godron (1807-1880) (Figure 23) was primarily concerned with answering one of the secondary questions of the competition—whether the characters of hybrids propagated by self-fertilization for several generations remained constant or whether such hybrids reverted back to the original forms. He investigated species hybrids from the genera *Verbascum*, *Primula*, *Nicotiana*, *Digitalis*, *Antirrhinum*, *Linaria*, and *Aegilops* and discovered that species hybrids are sterile and that their sterility is mainly due to a lack of pollen, or rather to its impotency. In his experiments he repeatedly observed exuberant growth in the hybrids of the first filial generation (for example, in *Verbascum*). The hybrids only became fertile by means of a second fertilization with the pollen of one of their parents or of some other appropriate species. By this process, which can be successively reiterated, the hybrids gradually reverted back to one of the parent forms. Godron drew these conclusions from his work:

1. Absolute fertility from the first filial generation onward characterizes hybrids produced from two stocks or varieties of the same species.
2. The sterility of simple hybrids which are isolated from their parents proves that they are genuine species hybrids.
3. Hybridization is not possible between species belonging to different genera.

The possibility of attaining hybrids in conjunction with the degree of their fertility provided Godron with a criterion for recognizing stocks and varieties of the same type and for defining species and natural genera.

In his experimental work, he does not give any numerical data on the frequency with which the various characters appear in subsequent generations. This deficiency is also apparent in Naudin's work; in many ways, however, the latter's study is more advanced and comes closer to achieving its objectives than that of his compatriot.

Unhappily—from the standpoint of present-day knowledge—the work of Charles Naudin (1815-1899) (Figure 24)[2] was also mainly concerned with species hybrids. He studied the following genera: *Papaver*, *Mirabilis*, *Primula*, *Datura*, *Nicotiana*, *Petunia*, *Digitalis*, *Linaria*, *Ribes*, *Luffa*, *Coccinia*, and *Cucumis*. His investigations also suffered from a lack of experimental material; various accidents, frosts, droughts, and pests prevented important plants from maturing or destroyed their fruit. Most important of all, he did not have the space to conduct large-scale experiments. For all that, he obtained a number of important results and discovered segregation of characters in the

[2]For a while, Naudin was an assistant under Decaisne at the Paris Museum. At the age of 62 he was appointed director of the "Jardin d'Acclimatation" in Cap d'Antibes, an estate that had formerly belonged to the botanist Thuret. Naudin's personal life was full of misfortune and anxiety: his children died and his eyesight failed him. He died alone at the age of 84, completely blind.

24 Charles Naudin (1815-1899)

hybrids, though he failed to classify the hybrid types numerically. Full of promise, Naudin stood on the very brink of the discovery made by Mendel only a few years later. The most important results of Naudin's investigations in 1863 are these:

1. Uniformity in the first filial generation and the identity of reciprocal crosses. Naudin himself writes: "In the hybrids whose origin was well known to me, I have always found a great uniformity in the appearance of the individuals of the first generation which were derived from the same cross, no matter how numerous they were." In the same report he added: "but what I can assert is that all the reciprocal hybrids I obtained, both from closely related species and from unrelated species, have resembled one another as if they had come from the same cross."

2. First-generation hybrids usually assume an intermediate condition with reference to their two parent forms. However, it is often seen that some of the characters of the hybrids resemble the male or female parent to a greater degree. "All hybridologists agree that hybrids—and we are still speaking of first-generation hybrids—are mixed forms, intermediate between those of their two parent species. . . . It has also been observed that hybrids sometimes have a greater resemblance to one of the two species in some aspects, and to the other in different aspects; this is also true, and we have seen an example of it. . . ."
3. In contrast to first-generation hybrids, those of the second generation manifest an exceptional and disorderly variation. As Naudin says: "Beginning with the second generation, the general appearance of hybrids changes in a most remarkable way. Very often the perfect uniformity of the first generation is followed by an extreme diversity of forms, some coming close to the specific type of the father, others to that of the mother, while yet others suddenly and completely revert back to the one or the other. At other times the steps leading back to the parental types are taken slowly and by degrees; sometimes one sees the whole collection of hybrids tending toward the same side."
4. Beginning with the second filial generation, all hybrids show a tendency to revert back to their original forms. This reversion is caused by the separation of the two "specific essences" contained within the pollen and in the ovules of the hybrids. In the hybrid two different essences come into contact, each of them determining its own kind of growth and principle of final form. Each tries to dissociate itself from the other, but various quantities of both essences are intermingled in the organs of the plant. Naudin pictures the hybrid as a living mosaic whose discordant elements are optically indistinguishable when they are in this intermingled state.
5. The separation of the elements takes place in the pollen and in the ovules of the plant: "The facts warrant the conclusion that pollen and the ovules—especially the pollen that is situated at the tip of the male flower—are precisely the parts of the plant where the separation of the specific essences is made most forcefully; this hypothesis becomes even more probable from the fact that these organs are at the same time highly differentiated and very small, a double reason for making the localization of the two essences more perfect."

Both the pollen cells and the ovules contain in part purely paternal, and in part purely maternal, essences. If a pollen grain which has completely reverted back to the paternal species meets an ovule that also possesses the characters of the male parent, fertilization gives rise to a plant that has also reverted back completely to the paternal species. The union between a pollen grain and an ovule which have segregated toward the maternal side produces a plant that resembles the original maternal species. But when there is a combination of a pollen grain and an ovule that have differentiated in contrary directions, a hybrid results. These considerations account for the multiformity of hybrids, some of which resemble the female parent, others the male, while yet others assume an intermediate character. The number of these different forms of second-generation hybrids is decided purely by chance. With regard to the "purity of the gametes" and to the origin of individual hybrid forms in accordance with the laws of probability, Naudin's work comes close to Mendel's discoveries; but Naudin was working with what were considered to be good species and was therefore unable to determine the numerical proportions of the different forms.

In 1864 he presented the Academy with a second paper which confirmed his earlier results: i.e., the uniformity of first-generation hybrids, the identity of reciprocal crosses, and the exceptional and disorderly variation that appears in the second hybrid generation. This kind of variability, however, does not introduce anything new, for it ranges within boundaries which it cannot cross; rather, it is a blending of forms already present in the parent types.

Naudin also examined the question whether new species could be produced from hybrids. He rejects the question with a note on the problems associated with the concept of species together with the remark that the natural hybrids of *Salix* and *Rubus* repeatedly segregate, so that the essential characteristic of a new species, constancy, was lacking in them.

M. Wichura (1865) revealed his practical knowledge of hybrid fertilization in species hybrids of the willow tree when he published, one year later, his detailed study on natural and artificial hybrids in the genus *Salix*. He produced simple and complex hybrids from numerous species and found, like Kölreuter and Gärtner, that the hybrids often showed exuberant growth: greater height, broader ramification, longer life-span, resistance to cold, and more luxuriant and earlier flowering. More frequently, however, Wichura found that the hybrids underwent a deficient and weak development and had a low degree of fertility or were completely sterile. He felt that the intermediate condition apparent in some hybrids was most important, for he says:

The secret of reproduction depends on the fact that it is possible for two different cells to fuse together and become a unified whole. Once this process is accepted as a fact, we must recognize it as being natural and necessary for the union of the two cells, when they belong to differently constituted individuals, to produce a more or less intermediate form whose character does not change, no matter whether it was the ovule or the pollen cell that was taken from species a or b. Each of these two cells, the germinal vesicle or the pollen tube, bears within itself the type of the individual from which it was taken, and each of the two species supplies a numerically identical part, namely, a single cell, to the new individual. Even in reciprocal crosses; the union of both must give rise to one and the same intermediate formation which contains the same proportions of both species. . . . Since the sexual union of quite differently constituted individuals—i.e., hybridization—always yields a reproductive product which observes the mean between the types contained within the egg and pollen cell, we may regard this as a law that is valid for all cases of reproduction and therefore also applies to the production of varieties (pp. 87-88).

Wichura observed the appearance of "varieties" in the progeny of these intermediate hybrids and explains the fact of the new combination of characters as follows (p. 88):

The appearance of a variety therefore proves that the egg or the pollen cell (or both) from whose union it was produced must have belonged to a type that varied from that of the original parent plant.

The sex cells (of the hybrids in question), which is the term I propose to use to designate the germinal vesicle and the pollen tube, do not therefore merely have the function of propagating the individual but also have the faculty of producing new variant forms.

We have observed in hybrids that the force that forms varieties resides principally in the pollen and to a lesser degree in the ovules; this is probably also the case in genuine species.

Wichura's work does not contain anything that is fundamentally new and original; his choice of experimental material was unlikely to produce such results. The discoveries he made in annuals were not automatically applicable to perennial plants.

6.3 Gregor Mendel and the Inheritance of Individual Characters

As has been mentioned on p. 120, those who survey the nineteenth-century studies on the production of hybrids

will conclude that of all the numerous experiments made, not one has been undertaken to such an extent and in such a way for it to be possible to determine the number of different forms in which the offspring of hybrids appear, or to classify these forms with certainty according to their separate generations, or to ascertain their statistical relations. Courage is indeed required to undertake such extensive labors. But they would seem to constitute the only right way of ultimately achieving the solution to a question which is of inestimable

25 Gregor Mendel (1822-1884)

importance in connection with the evolutionary history of organic forms.

Gregor Mendel (Figure 25) wrote these words in the introduction to his celebrated essay (1865) *Experiments in Plant Hybridization.* He first presented this study at the monthly meetings of the Brünn Society for the Study of Natural Science held on February 8 and March 8, 1865. Artificial fertilization undertaken in ornamental plants in order to produce new color varieties had prompted him to embark upon this work. The striking regularity with which he had repeatedly observed the appearance of the same hybrid forms—provided the fertilization had been made across the same species—had led him to undertake further experiments designed to solve the problem of the transmission of parental characters to the offspring.

After the rediscovery of Mendel's work, his life, his importance, and the fortunes of his studies were frequently described. These matters form an important topic in a number of historical publications (Correns 1922, Iltis 1924, Richter 1924-1941, Krumbiegel 1957, von Tschermak-Seysenegg 1956, and others). J Křiženecký (1965) has published a number of texts and sources on Mendel's life and influence; he was interested in all branches of agriculture, was a fruit grower and a beekeeper, and compiled meteorological data for a period of many years. N. Vavra (1965) has undertaken a geneticopomological analysis of Mendel's system for cultivating apples and pears. Weiling (1966) has also written about Mendel as a fruit grower.

The son of a farmer, Johann Mendel—he assumed the "religious" name Gregor upon taking holy orders in a monastery—was born at Heinzendorf near Odrau (the present-day Czechoslovak village of Hynčice) in what was then a German-speaking part of Austrian Silesia. He attended grammar school in Troppau and Olmütz to study for the priesthood. In 1843 Mendel became a novice at the Augustinian monastery of St. Thomas in Brünn where, in 1847, he was ordained a priest and undertook pastoral duties for a time. After studying to be a schoolmaster as well as teaching grammar school in Znaim, Mendel was not able to pass the University of Vienna's teaching-certificate examination. He then turned to the study of science; from 1851 to 1853 he took courses at the University of Vienna, attending Franz Unger's lectures on plant physiology, those of Andreas von Ettinghausen on mathematical physics, and a course given by Christian Doppler on experimental physics, all of whom deeply influenced him. Returning to Brünn, he obtained a teaching post in physics and natural science at the German Technical Institute in 1854. He was able to discharge the duties of this post for fourteen years while still remaining at the monastery. In 1855, taking the examination for a teaching certificate the second time, he failed again. Recent discoveries made by J. Křiženecký (1963) in the Mendel Archives of the Moravian Museum in Brünn reveal, however, that Mendel did not, in fact, fail his second examination but was obliged to withdraw his application during the course of the examination on grounds of illness. It was during this time that he conducted the experiments that led to the discovery of the laws named after him. In 1868 he was elected abbot of his monastery and entrusted with the administration of that foundation. The administrative work that accompanied this office, together with the further duties which he undertook, made it difficult for him to continue his scientific studies, and he shortly had to give them up altogether. The bitter quarrel between the Catholic Church and the state, which was just being instigated by Bismarck, eventually affected Mendel's health and capacity for work. In 1872 the German Liberal Constitutional Party, of which Mendel was a member, passed a bill in the Austrian Reichsrat which imposed a high tax on monasteries. Mendel challenged this bill, but the consequences of this action for his monastery—the most important of which was the sequestration of one of its estates—made Mendel a lonely and bitter man. After having been bedridden for a long time, Mendel died on January 6, 1884, of Bright's disease, at the age of sixty-two.

Křiženecký's recent discoveries in the Mendel Archives of the Moravian Museum led J. Sajner (1963) to reinvestigate the course of Mendel's illness and the cause of his death. He came to the conclusion that "Mendel's principal affliction was nephritis with a renal syndrome associated with generalized edema. Accompanying hypertension caused encephalopathy and heart failure."

Of the numerous investigations undertaken by Mendel only little has been published: *Versuche über Pflanzenhybriden* (1865, published in English in 1924 under the title *Experiments in Plant Hybridization*), *Über einige aus künstlicher Befruchtung gewonnene Hieracium-Bastarde* (1870), *Die Windhose vom 13. Oktober 1870* (1871). The studies in hybridization were reissued after the turn of the century by E. Tschermak in *Ostwald's Klassiker*, the first essay by Gübel in an issue of *Flora*, and all three works in Volume 49 of the *Proceedings of the Brünn Society for the Study of Natural Science*. Mendel's letters to C. von Nägeli, published by Correns (1905), contain further material to complete the picture of Mendel's personality.

Many of the facts set forth by Mendel with great clarity and precision in his classic paper had already been discovered by his predecessors. It seems quite clear that Mendel, with his exceptional gifts for research, owes his success to the careful choice of his experimental material. The pea—a self-fertilizing plant with constant varieties that was easy to cultivate, to isolate, and to cross, a plant, moreover, whose hybrids were fertile—was particularly suitable for Mendel's purpose (see also Thomas A. Knight). Moreover, he eliminated every source of error by crossing closely related varieties and not species; he simplified the experimental conditions by restricting his investigations to the study of the hereditary processes of individual characters that were quite distinct from one another.

In describing Mendel's basic experiments in detail we shall not be able to mention all the particulars and analyses of the crosses in which two or more pairs of characters are combined, or in which the hybrids are crossed with their parental forms, particularly since Mendel's formal designation of the different combinations differs slightly from that usually employed today.

From thirty-four varieties of peas procured from a number of different seed dealers, Mendel chose twenty-two whose differentiating characters were constant. He used the following pairs of characters in his experiments:

1. Round seeds–wrinkled and irregularly angular seeds
2. Yellow cotyledons[3]–green cotyledons
3. White seed coat–gray seed coat
4. Simply inflated pod–constricted pod
5. Green unripe pod–yellow unripe pod
6. Axial position of the flowers–terminal position of the flowers
7. Long stem–short stem.

Each two of the differentiating characters enumerated were united by cross-fertilization. All the crosses were made reciprocally. The first-generation hybrids revealed that each of the seven characters of the hybrids completely, or almost completely, resembled one or the other of the two parental characters. The characters that remained unaltered or almost unaltered in the hybrids were termed dominant, while those that were latent in the hybrids were called recessive. No differences were observed in the reciprocal crosses. The following differentiating characters were found to be dominant: round seeds, yellow cotyledons, gray seed coat, simply inflated pod, green unripe pod, axial position of the flowers, and long stem.

In addition to the dominant characters, pure recessive characters also appeared in the first hybrid generation (F_2)[4] in a ratio of 3 to 1. Transitional forms were not observed in any experiment. The actual numbers obtained were as follows:

Experiment 1:

5474 round seeds, 1850 wrinkled seeds. Ratio 2.96:1.

[3]Mendel uses the term "albumen."

[4]By first-hybrid generation Mendel means the first-generation descendants of the hybrids produced by the original cross; we now designate this as the second filial generation or F_2.

Experiment 2:
6022 seeds with yellow cotyledons, 2001 seeds with green cotyledons. Ratio 3.01:1.
Experiment 3:
705 seeds with gray seed coats, 224 with white seed coats. Ratio 3.15:1.
Experiment 4:
882 plants with simply inflated pods, 299 with constricted pods. Ratio 2.95:1.
Experiment 5:
428 plants with green pods, 152 with yellow pods. Ratio 2.82:1.
Experiment 6:
651 plants with flowers in axial position, 207 with flowers in terminal position. Ratio 3.14:1.
Experiment 7:
787 plants with a long stem, 277 with a short stem. Ratio 2.84:1.

In the second hybrid generation (F_3) it was observed that all forms having recessive characters did not "segregate"—that is, their character had not varied further; they remain constant. Two-thirds of the forms with dominant characters produced offspring which again "segregated" (that is, displayed both dominant and recessive characters) in the ratio of 3 to 1; the remaining third of the forms with dominant characters bred true.

The individual experiments gave the following results:

1. Of 565 plants raised from round seeds, 193 bred true and 372 segregated in the ratio of 3:1. Ratio of hybrids to constants: 1.93:1.
2. Of 519 plants raised from seeds with yellow cotyledons, 166 bred true and 353 segregated in the ratio of 3:1. Ratio of hybrids to constants: 2.13:1.
3. Of 100 plants raised from seeds with a gray seed coat, 36 bred true and 64 segregated in the ratio of 3:1. Ratio of hybrids to constants: 1.77:1.
4. Of 100 plants raised from the seeds of plants having a simply inflated pod, 29 bred true and 71 segregated in the ratio of 3:1. Ratio of hybrids to constants: 2.45:1.
5. Of 100 plants raised from the seeds of plants with green pods, 40 bred true and 60 segregated in the ratio of 3:1. (When the experiment was repeated the figures were 35 and 65.) Ratio of hybrids to constants: 1.50:1 (1.85:1).
6. Of 100 plants raised from the seeds of plants whose flowers were in

axial position, 33 bred true and 67 segregated in the ratio of 3:1. Ratio of hybrids to constants: 2.03:1.

7. Of 100 plants raised from the seeds of plants with a long stem, 28 bred true and 72 segregated in the ratio of 3:1. Ratio of hybrids to constants: 2.57:1.

These experiments establish an average ratio of 2:1. This proves that two-thirds of the forms having the dominant character in the first generation (F_2) also have the hybrid character, while one-third breeds true.

As Mendel says: "Since the members of the first generation come directly from the seed of the hybrids, it is now clear that the hybrids having one or the other of the two differentiating characters form seeds; of these seeds, one-half develops again the hybrid form while the other half yields plants that breed true and receive both dominant and recessive characters in equal proportions."

If A denotes the dominant character, a the recessive, and Aa the hybrid form, the expression A + 2Aa + a gives the developmental series for the progeny of the hybrids with two differentiating characters. This accounts for the observation, made by Gärtner, Kölreuter, and others, that hybrids tend to revert to the original parental forms. For, in the course of generations, the number of hybrids decreases while the number of true-breeding forms increases. Assuming that all forms have the same degree of fertility and that each plant yields four seeds per generation, we have the numerical relations listed in Table 1. Of the 2048 plants in the tenth generation, 1023 would therefore have the constant dominant character, 1023 the constant recessive character, and only 2 would be hybrids.

If several different characters are united in the hybrid by cross-fertilization, this law of development is equally valid for each pair of differentiating characters.

Experiment I:

AB	seed parents	ab	pollen parents
A	round seeds	a	wrinkled seeds
B	yellow cotyledons	b	green cotyledons

The hybrid seeds were round and yellow. In the first hybrid generation (F_2), 556 seeds were obtained from 15 plants. Of these seeds, 315 were round and yellow, 101 were wrinkled and yellow, 108 were round and green, and 32 were wrinkled and green.

Table 1 Numerical Relations per Generation

Generation	A	Aa	a	Ratios A:Aa:a
1	1	2	1	1:2:1
2	6	4	6	3:2:3
3	28	8	28	7:2:7
4	120	16	120	15:2:15
5	496	32	496	31:2:31
n				2^n-1:2:2^n-1

Seeds gathered from all the plants were sown the following year; the numbers of different characters were carefully counted and arranged in groups. As a result of this analysis the characters under consideration appeared in the form of a combination series "in which the two expressions for the characters A and a, and B and b are combined. The full number of the terms of the series is obtained by combining the expressions A + 2Aa + a and B + 2Bb + b."

Experiment II (using three differentiating characters):

ABC	seed parents	abc	pollen parents
A	round seeds	a	wrinkled seeds
B	yellow cotyledons	b	green cotyledons
C	grayish-brown seed coats	c	white seed coats

This experiment was undertaken with the same techniques as in Experiment I. Like that first experiment it revealed a combination series "in which the expressions for the characters A and a, B and b, and C and c are united. The expressions A + 2Aa + a, B + 2Bb + b, and C + 2Cc + c give all the terms of this series. The constant combinations that occur therein correspond to all the combinations which are possible between the characters A, B, C, a, b, and c; two such combinations, ABC and abc, resemble the two original parent plants."

From these results Mendel derived the following valid principle:

The offspring of hybrids in which several essentially different characters are combined exhibit the terms of a combination series in which the expressions for each pair of differentiating characters are united. This also proves that the relation of each pair of differentiating characters in hybrid union is independent of all the other differences in the two original parent plants.

If n denotes the number of differentiating characters in the two original parent plants, then 3^n gives the number of terms in the combination series, 4^n the number of individuals which belong to the series, and 2^n the number of unions which remain constant. If the original plants differ with respect to 4 characters, then the series contains $3^4 = 81$ terms, $4^4 = 256$ individuals, and $2^4 = 16$ constant forms–i.e., among each 256 offspring of the hybrids there are 81 different unions, 16 of which are constant.

Mendel obtained all the possible constant unions, i.e., $2^7 = 128$, from his crosses with the 7 differentiating characters enumerated. This proves "that constant characters which occur in the different forms of a plant may be obtained–by means of repeated artificial fertilization–in all the associations that are possible according to the laws of combination." The results of the experiments provided further information about the composition of the germ and pollen cells of hybrids.[5] Experience showed that constant progeny occur only when the germ cell and the pollen both contain the same factor. Now since constant forms can be produced by a single plant or even by a single flower, it follows that there are as many kinds of sex cells formed in the ovaries and anthers of the hybrids as there are possible constant combination forms. The development of the hybrids in the separate generations can be explained by assuming that, on the average, the different kinds of germ and pollen cells are formed in equal numbers.

To prove this assumption, Mendel produced the hybrid AB × ab and crossed it with the pollen of AB and also with the pollen of ab. He also made the corresponding reciprocal crosses, namely, AB with the pollen of the hybrid, and ab with the pollen of the hybrid. Germ and pollen cells of the forms AB, Ab, aB, and ab would then be developed in the hybrids, and the following combinations would be possible:

1. Germ cells AB, Ab, aB, and ab with pollen cells AB;
2. Germ cells AB, Ab, aB, and ab with pollen cells ab;
3. Germ cells AB with pollen cells AB, Ab, aB, and ab;
4. Germ cells ab with pollen cells AB, Ab, aB, and ab.

Mendel's usual thorough analysis of these experiments showed that almost equal numbers of all the expected forms appeared, as postulated by his theory. Which of the pollen cells will, in fact, be united with a given germ cell therefore remains a matter of chance: in the hybrid Aa, equal numbers of pollen cells A and a together with equal numbers of germ cells A and a are formed; these cells will unite according to the laws of probability as shown in the following diagram:

[5]Mendel always talks of germ and pollen cells. We should now refer to these as the egg cell and the sperm cell.

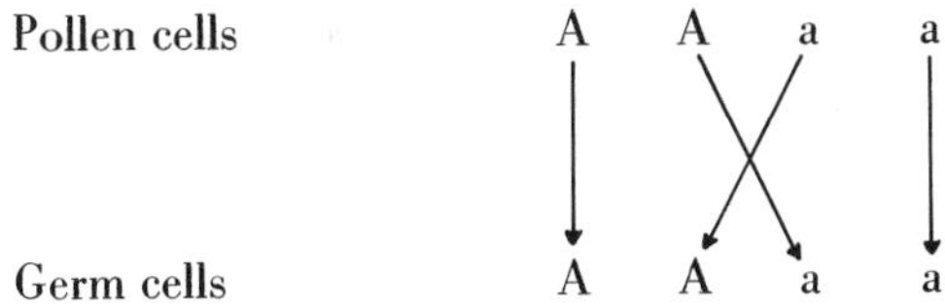

For various reasons, the separate values are subject to fluctuations, and the true ratios may be given only by finding the average. The greater the number of different separate values that are established, the more certain it is that the effects of chance are eliminated.

Mendel concludes his account with these words:

> The law of combination of differentiating characters which governs the development of hybrids is therefore based on, and can be explained by, the principle we have shown to be true, that is, that the equal numbers of the germ cells and pollen cells produced by the hybrids correspond to all the constant forms resulting from the combinations of the characters united by fertilization.

We can summarize the essential results of Mendel's *Pisum* studies as follows:

1. With regard to their appearance, hybrids resemble those parental characters involved in the cross which are called dominant while the corresponding recessive characters disappear. Hybrids are uniform in appearance.
2. The characteristics that distinguish the hybrid's parents from one another are mutually independent. The factors that make up these characteristics, though united in the hybrid, do not blend together; they segregate in the germ cells and in the pollen cells and recombine according to the laws of probability.
3. The different germ cells are formed in equal numbers in the hybrids.

Apart from his work on peas, Mendel also studied a number of other plants to see whether the developmental law that applied to *Pisum* was valid for other genera. His experiments with *Phaseolus* gave results which agreed with what he had discovered in *Pisum.* Crosses using *Phaseolus* species that differ widely from one another (*Phaseolus nanus L.* × *Ph. multiflorus W.*) were only partially successful, especially with regard to the inheritance of flower colors. According to Mendel, however, the difficulties involved in the discovery of hereditary regularities in such cases stem from the fact that the majority of our cultivated plants are members of various hybrid series whose further regular development is greatly varied and interrupted by frequent crosses *inter se.*

Mendel was also an apiarist; he wanted to see whether the laws he had discovered to be true of peas were also applicable to bees (E. von Tschermak-Seysenegg 1956). He obtained queen bees of breeds with various different characters and possessed some fifty hives containing German heath bees, Italian bees, Carniolan bees, along with Egyptian and American breeds; he investigated their coloring, the characteristics of their flight, their behavior, and their inclination to sting. It seems certain (Father Dzierzon 1854) that Mendel knew about the parthenogenetic origin of drones. Unfortunately, his notes on these experiments have been lost.

In his choice of *Hieracium* species for hybridization purposes, Mendel (1870) was less fortunate, for here he did not observe any segregation in subsequent generations. (See Figure 26, Mendel's letter to Nägeli, written July 3, 1870.)* He was unable to account for this phenomenon which was later discovered by Ostenfeld and Raunkiaer in 1903 to be a case of apomixis. The pseudogamous *Fragaria* "hybrids" discovered by Millardet (1894)—he called them "false hybrids"—could not be explained at the time of their discovery.

In view of the clear and unequivocal nature of the results of Mendel's hybridizing experiments on peas, it is natural to ask why his predecessors failed to obtain the same results. This very question was raised by Mendel himself. He refers to Kölreuter, Gärtner, Herbert, Lecoq, and Wichura in his famous essay, though he does not mention Naudin's work. We have already had occasion to note Mendel's critical remarks on the work of Gärtner.

The most important difference between Mendel and his predecessors is that he strove to classify the multiplicity of forms—the *variation désordonnée* of Naudin—which appear in the first hybrid generation (F_2). Proceeding from the simplest kind of experiment (one pair of differentiating characters) to more complex ones (two or three pairs of differentiating characters), he succeeded in discovering the frequency of the occurrence of the characters and combinations thereof and was able to establish a statistical law. Regarding the external appearance of hybrids, he confirmed the results of the authors to whom he referred; for the hybrids either assumed an intermediate condition with respect to their parents or else they resembled one or the other of the parental types to a greater or lesser degree. We could hardly expect the experiments of Kölreuter, Gärtner, and others to contain a statement of

*A translation of this letter is given in the Appendix, p. 291.

the numerical ratios in which individual characters are inherited, since these botanists were studying what were considered to be good species; such species were distinguished by a large number of characters whose differences were not always well marked. Moreover the experimental material was of limited proportions and not likely to reveal segregation ratios with any clarity. This also accounts for Gärtner's statement that he had never been able to observe the reappearance of parental forms in hybrid generations but had only seen forms that closely approximated the parents. According to Mendel the experiments of Kölreuter, Gärtner, and others on the transformation of one species into another, requiring a longer or shorter period of time according to the species, have a very simple explanation. The fact that the hybrid contains as many kinds of germ cells as are permitted by the possible constant combinations of the characters within it means that one of these must always be of the same kind as that of the fertilizing pollen cells. It follows that it is theoretically possible for the second fertilization AB × B to produce a constant form that is identical to the pollinating parent. This possibility depends on the number of experimental plants and on that of the differentiating characters present in the cross. The smaller the number of experimental plants and the larger the number of differentiating characters, the more slowly the transformation will take place. When a plant with a majority of dominant characters is used to continue the crosses, this will also affect the amount of time required to achieve the transformation.

It speaks well for Mendel's spirit of inquiry that he did not draw general conclusions from his discovery of the laws governing the inheritance of individual characters in *Pisum* but insisted on their confirmation with respect to other genera before the establishment of a general law could be permitted. However, the statistical law which he discovered was not restricted to the description of what happened to individual factors, or predispositions in the various subsequent hybrid generations; it effectively analyzed the total form of an organism into its constituent elements by consistently associating one element with each different character. Mendel's procedures of analysis and synthesis provided a definitive explanation of what many of his predecessors had only dimly perceived. In contrast to the descriptions of hybrids—often unclear and too general—of earlier investigators in this field, Mendel's discoveries had the precision which only a statistical analysis confers, but at the time they were neither noticed nor understood.

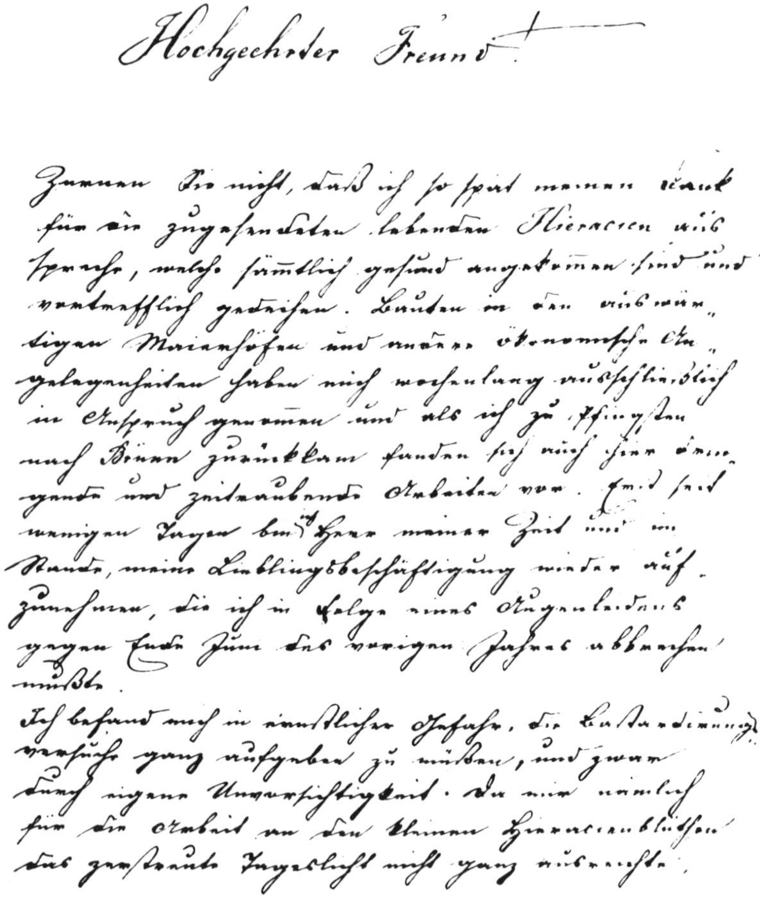

Hochgeehrter Freund!

Zürnen Sie nicht, daß ich so spät meinen Dank
für die zugesendeten lebenden Hieracien aus-
spreche, welche sämtlich gesund angekommen sind und
vortrefflich gedeihen. Bauten in den auswär-
tigen Meierhöfen und andere ökonomische An-
gelegenheiten haben mich wochenlang ausschließlich
in Anspruch genommen und als ich zu Pfingsten
nach Brünn zurückkam fanden sich auch hier drin-
gende und zeitraubende Arbeiten vor. Erst seit
wenigen Tagen bin ich Herr meiner Zeit und im
Stande, meine Lieblingsbeschäftigung wieder auf-
zunehmen, die ich in Folge eines Augenleidens
gegen Ende Juni des vorigen Jahres abbrechen
mußte.
Ich befand mich in ernstlicher Gefahr, die Bastardirungs-
versuche ganz aufgeben zu müssen, und zwar
durch eigene Unvorsichtigkeit. Da mir nämlich
für die Arbeit an den kleinen Hieracienblüthen
das gestreute Tageslicht nicht ganz ausreichte.

26 Gregor Mendel's letter to Carl von Nägeli, written July 3, 1870

nahm ich einen Beleuchtungsapparat (Spiegel mit Sammellinse) zu Hilfe, ohne zu ahnen, welches Unglück ich damit hätte anstellen können. Nachdem ich mich im Mai und Juni viel mit H. Auricula und praealtum beschäftigt hatte, stellte sich eine eigenthümliche Ermüdung und Abspannung in den Augen ein, welche trotz aller sorglich angewendeten Schonung einen bedenklichen Grad erreichte und mich bis in den Winter hinein für jedwede Anstrengung unfähig machte. Seitdem hat sich glücklicherweise das Übel fast vollständig verloren, so daß ich wieder anhaltend lesen und auch die Befruchtungsversuche an Hieracien, so gut es eben ohne künstliche Beleuchtung möglich ist, vornehmen kann.

Mit dem heutigen Schreiben sende ich zugleich einige lebende Hieracium-Bastarde, denen auch, wo es nothwendig ist, die Stammformen beiliegen.

Das bisherige Ergebnis der Versuche ist noch zu gering zu nennen und zu mangelhaft, als daß sich daraus schon endgiltige Folgerungen ziehen ließen. Einige Erfahrungen sind indessen doch schon gewonnen und ich erlaube mir dasjenige davon kurz anzuführen, was mir von einiger Wichtigkeit zu sein scheint.

Zuvorderst will ich erwähnen, daß es mir bei einigen Piloselloidformen, trotz vielfacher Versuche noch nicht gelungen ist, durch Bestäubung mit fremdem Pollen auch nur einen einzigen Bastard zu erhalten. Das gilt z. B. von H. aurantiacum. Bei dieser Art war ich bis jetzt nicht im Stande, die Anwendung des eigenen Pollens aufzuheben. Auch H. Pilosella und H. cymosum machen Schwierigkeiten. Bei anderen z. B. bei den Varietäten von H. praealtum gelingt bei ganz gleicher Behandlung die Befruchtung mit fremdem Pollen schon leichter, und H. Auricula ist, wie ich mich nun mehrfach überzeugt habe, bei einiger Vorsicht eine vollkommen verlässliche Versuchspflanze. Ich habe mehr als 100 Köpfchen dieser Art im vorigen Jahre mit dem Pollen von H. Pilosella, cymosum und aurantiacum befruchtet; etwa die Hälfte davon ist zwar in Folge erlittener Verletzungen eingetrocknet und aus den übrigen wurden nur je 2–6 Samen erhalten, allein die aus denselben gezogenen Pflanzen sind ohne Ausnahme Bastarde. Die noch ganz kleinen Pflänzchen von H. Auricula + H. Pilosella und H. Auricula + H. cymosum wurden mir leider im Warmhause von Schnecken bis auf einige wenige Exemplare abgeweidet, jene von H. Auricula + H. aurantiacum blieben jedoch erhalten und es sind davon 98 im Garten ausgesetzt. Sie dürften im nächsten Monate zur Blüthe kommen.

Für Versuche sehr geeignet scheint auch eine andere Form zu sein, die ich indessen mit der Bezeichnung

N° XII. der Sendung beischließe, da ich dieselbe nicht einzureihen und zu benennen weiß. Ich habe sie in Holzschlägen im großen Gesenke gefunden. Der einzige Befruchtungs Versuch, den ich an denselben im vorigen Jahre mit dem Pollen von H. Pilosella vorgenommen habe, ist vollständig gelungen, da die erhaltenen 29 Pflanzen sämtlich Bastarde sind.

Ich erlaube mir hier zu bemerken, daß von mir bisher nur eine einzige Form von H. Pilosella zu Befruchtungen verwendet wurde. Da ich jedoch nicht ganz sicher bin, ob nicht bei der vorjährigen Sendung eine Verwechslung stattgefunden hat, weil ich später zur Zeit der Blüthe wahrnahm, daß eine andere behaarte Pilosellaform in das Gebiet meiner Versuchspflanze eingedrungen war, lege ich dieselbe noch einmal bei und zwar mit der Bezeichnung H. Pilosella (Brünn). Ob diese Form zu H. echioides in irgend einer Beziehung stehe, darüber wage ich keine Meinung auszusprechen, will nur anmerken, daß dieselbe hier häufig vorkommt und der nächste bekannte Standort des H. echioides etwa 5 Meilen entfernt ist. Dort wurde im vorigen Jahre auch die von mir unter der Bezeichnung H. praealtum (?) eingesendete Pflanze mit H. echioides und H. praealtum zugleich wachsend angetroffen und damit wohl außer Zweifel gestellt, daß die Vermuthung, dieselbe gehöre in die Reihe H echioides – praealtum, richtig war. Ein Vergleich mit den Bastarden zeigt sogar, daß dieselbe dem

H. echioides näher steht.

Ich würde sehr dankbar sein, wenn Sie verehrter Freund mir gelegentlich einmal Ihre Ansicht über das Hieracium No. XII.¹) mittheilen wollten, da diese Pflanze nächst H. Auricula zu den besten Versuchspflanzen gehört, insofern sich davon ziemlich leicht Bastarde in größerer Anzahl gewinnen lassen. Dieser Umstand ist deßhalb von Wichtigkeit, weil es nur bei einer größeren Anzahl aus gleicher Befruchtung stammender Bastard Individuen möglich werden kann, die an denselben vorkommenden Abweichungen einer Beurtheilung zu unterziehen.

In der That kamen bis jetzt in allen Fällen Varianten zum Vorschein, wo aus derselben Befruchtung der Bastard in mehreren Exemplaren erhalten wurde. Die Beobachtung, daß aus der Einwirkung des Pollens der einen Art auf die Ovula einer zweiten von einander abweichende, ja wesentlich verschiedene Formen hervorgehen können, war für mich, ich muß es gestehen, sehr überraschend, und das um so mehr, als ich mich durch Anbauversuche überzeugt hatte, daß die Varianten im Falle der Selbstbefruchtung nur formbeständige Nachkommen liefern. Ich hatte bei Pisum und anderen Pflanzengattungen nur immer uniforme Bastarde gesehen und konnte demnach bei Hieracium dasselbe erwarten. Ich darf es Ihnen, geehrter Freund, verrathen, wie sehr ich

1) H. cinerascens Tausch

mich in dieser Beziehung täuschen ließ. Schon vor zwei Jahren kam bei mir der Bastard H. Auricula + H. aurantiacum in zwei Exemplaren zur Blüthe. Bei dem einen davon war die Vaterschaft des H. aurantiacum auf den ersten Blick herauszufinden; nicht so bei dem andern. Dasselbe wurde vielmehr, weil es andere Blätter und eine ganz verschiedene durchaus gelbe Blüthenfarbe besaß, und es meiner damaligen Meinung nach nur eine Bastardform zwischen den beiden Stammarten geben konnte, als zufällige Einschleppung angesehen und bei Seite gestellt. Ich hatte deshalb meiner vorjährigen Sendung auch nur jenes Exemplar beigelegt, welches in der Blüthenfarbe dem H. aurantiacum sehr nahe kam. Erst als später derselbe Bastard aus der 868ger Befruchtung und der Bastard H. Auricula + H. pratense (var.) beide in 3 Exemplaren und eben so vielen Varianten zur Blüthe kamen, war der richtige Sachverhalt nicht mehr zu verkennen.

Da das eingesendete Exemplar von H. Auricula + H. aurantiacum und eben so die Variant H. Auricula, wie ich aus Ihrem geschätzten Schreiben ersehe, abgestorben sind, ergänze ich dieselben und lege auch den lange verkannten Bastardzwillingsbruder mit der Bezeichnung H. Auric. + H. aurant. 868 b hinzu. Die 3 Exemplare vom vorigen Jahre sind mit 869 c, d, e bezeichnet. Der Variant c ist vollkommen fruchtbar.

Von dem Bastarde H. Auricula + H. pratense (var.) ist ein Variant über den Winter eingegangen, ebenso die Stammart H. pratense. Letzteres war nicht genau das typische H. pratense, da sie einige Sternhaare auf den Blättern brachte. Die mir durch Ihre Güte zugekommenen 2 Exemplare sind gleich im ersten Jahre im Garten zu Grunde gegangen; das eine verkümmerte ohne geblüht zu haben, das andere während der Blüthe. Die neue Art habe ich in der Umgebung noch nicht aufgefunden.

Der Bastard H. ~~praealtum~~ cymigerum No XII + H. Pilosella (Brünn) hat aber abgeblüht. Unter den vorhandenen 29 Exemplaren kommen sehr auffallende Abweichungen vor. Sie stellen zwar sämmtlich Übergangsformen von einer Stammart zu der andern dar, Niemand würde sie jedoch für Geschwister halten, wenn er sie wild wachsend anträfe. Ich werde die ganze Collection einsenden, sobald nur die Ausläufer hinreichend bewurzelt sind, was in wenigen Wochen der Fall sein dürfte. Auch hoffe ich zu jener Zeit schon über den diesjährigen Hauptversuch H. Auricula + H. aurantiacum referiren zu können, von dem ich wegen der grösseren Anzahl der vorhandenen Exemplare einigen Aufschluss erwarten darf.

Das Hieracium cymigerum No XII. wurde heuer mit H. Pilosella vulgare (München) befruchtet und es dürfte im nächsten Jahre der Vergleich der beiden Bastardreihen H. No XII. + H. Pilosella (Brünn) und H No XII. + H. Pilosella vulgare (München) nicht ohne Interesse sein. Auch wurde schon H. Auricula mit H. Pilosella vulgare (M.) und H. Pilosella (Br.) verbunden, und mit H. Pilosella niveum (M) soll es nächstens geschehen. An H. Pilosella incanum habe ich erst eine Blüthe gesehen, andere kommen hoffentlich noch.

Von dem Bastarde H. praealtum (Bauhini?) + H. aurantiacum, von dem ich im vorigen Jahre 2 Exemplare eingesendet habe, werden 25 Pflanzen zur Blüthe kommen. Auch bei diesen sind, so viel bis jetzt ersichtlich ist, Unterschiede vorhanden. Von zwei Exemplaren, die in Töpfen gezogen wurden und längst abgeblüht haben, ist das eine vollkommen fruchtbar, das andere fast steril. Ebenso kamen in der Bastardreihe H. Auricula + H. aurantiacum gänzliche Sterilität neben vollkommener Fruchtbarkeit vor.

Von den Bastarden H. ~~praealtum~~ setigerum (?) + H. aurantiacum und H. praealtum (Bauhini?) + H. aurantiacum ist die zweite Generation zur Blüthe gekommen und ebenso von H. praealtum + H. flagellare die dritte. Auch in diesen Generationen variirten die Nachkommen nicht. Ich kann bei dieser Gelegenheit die Bemerkung nicht unterdrücken, wie sehr es auffallen muß, daß die Bastarde von Hieracium im Vergleich mit jenen von Pisum ein geradezu entgegengesetztes Verhalten beobachten. Wir haben es hier offenbar nur mit einzelnen Erscheinungen zu thun, die der Ausfluß eines höheren allgemeinen Gesetzes sind.

Will man die Nachkommen jener Bastarde, die nur eine theilweise Fruchtbarkeit besitzen, in ihrer Entwicklung verfolgen, so ist es nothwendig, daß die Blüthen sehr sorgfältig vor der Einwirkung fremden Pollens geschützt werden, da durch den Pollen anderer Pflanzen leicht einzelne Ovula befruchtet werden können, welche sonst wegen der großen Theils schlechten Beschaffenheit des eigenen Pollens unbefruchtet geblieben wären. Ich fand zugleich einige Doppelbastarde, die von H. praealtum (Bauhini) + H. aurantiacum, stammen, welche Pflanze ich mitten zwischen H. Pilosella (Brünn), jedoch entfernt von anderen Hieracien abblühen ließ. Diese Bastarde wären demnach als (H. praealtum (Bauh.) + H. aurantiacum) + H. Pilosella (Brünn) zu bezeichnen. Sie sind in mehrfacher Hinsicht sehr interessante Formen.

Werden an den Blüthen theilweise fruchtbarer Bastarde die Narben mit dem Pollen anderer nicht zu weit abstehender Arten belegt, so bilden sie mehr Samen, als wenn sie isolirt gehalten und auf Selbstbefruchtung angewiesen ~~sind~~ werden; was nur von der Mitwirkung des fremden Pollens ~~abhäng~~ herrührt, wie man sich durch den Anbau der Samen leicht überzeugen kann. Bei vollkommen fruchtbaren Bastarden hingegen ist eine ~~At~~ ängstliche Absperrung nicht nöthig. Versuche mit H. praealtum ? + H. aurantiacum haben gezeigt, daß man fremden Pollen, auch jenen der beiden Stammarten, in Menge auf die Narben auftragen kann, ohne daß die Selbstbefruchtung dadurch gehindert wird. Die Samen geben sämmtlich die ursprüngliche Bastardform.

Ich lege der Sendung noch den Bastard H. cymosum (Münch.) + H. Pilosella (Brünn) bei. Es ist der einzige Bastard, den ich bis jetzt von H. cymosum erhalten, obwohl ich diese Art schon vielfach zu befruchten suchte.

Bei den Archhieracien ist es sehr schwierig die Selbstbefruchtung aufzuhalten. Bis jetzt wurden nur zwei Bastarde erhalten. Die Mutterpflanze bei einem ist

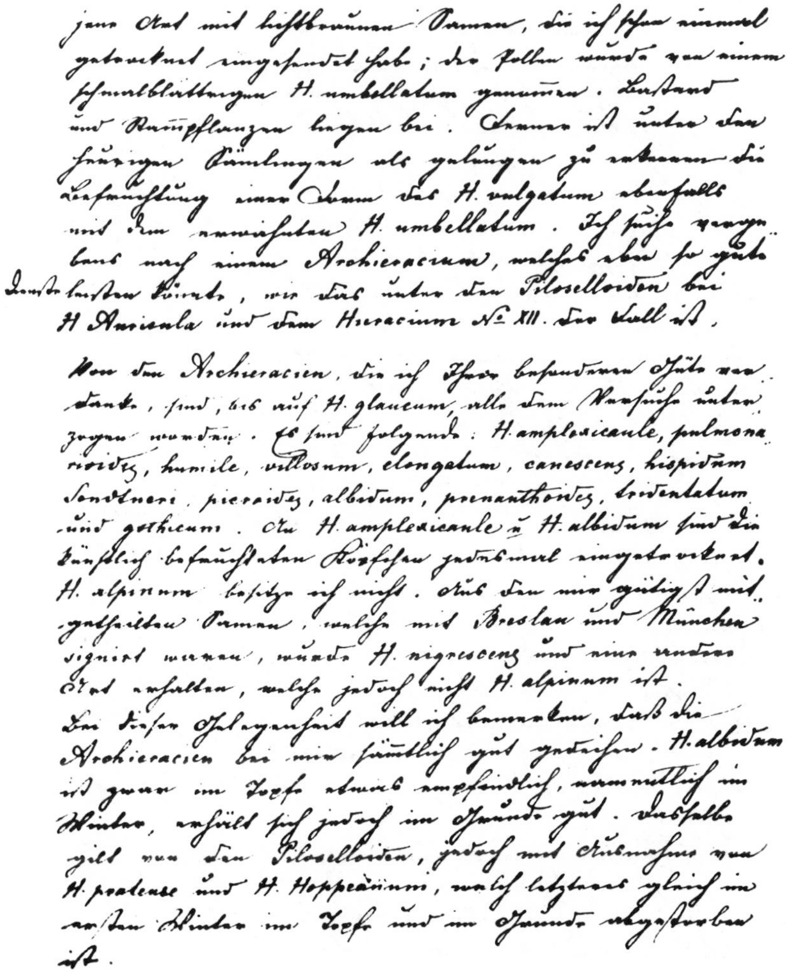
jener Art mit lichtbraunen Samen, die ich schon einmal
getrocknet eingesendet habe; dieselbe wurde von einem
schmalblättrigen H. umbellatum genommen. Bastard
und Stammpflanzen liegen bei. Ferner ist unter den
früheren Sämlingen als gelungen zu erkennen die
Befruchtung einer Form des H. vulgatum ebenfalls
mit dem erwähnten H. umbellatum. Ich suche vergeb-
lich nach einem Archieracium, welches eben so gute
[deutliche] Bastarde liefert, wie das unter den Piloselloiden bei
H. Auricula und dem Hieracium No XII. der Fall ist.

Von den Archieracien, die ich Ihrer besonderen Güte ver-
danke, sind, bis auf H. glaucum, alle dem Versuche unter-
zogen worden. Es sind folgende: H. amplexicaule, pulmonarioides, humile, villosum, elongatum, canescens, hispidum
Sendtneri, picroides, albidum, prenanthoides, bidentatum
und gothicum. An H. amplexicaule u. H. albidum sind die
künstlich befruchteten Köpfchen jedesmal eingetrocknet.
H. alpinum besitze ich nicht. Aus den mir gütigst mit-
getheilten Samen, welche mit Breslau und München
signirt waren, wurde H. nigrescens und eine andere
Art erhalten, welche jedoch nicht H. alpinum ist.
Bei dieser Gelegenheit will ich bemerken, daß die
Archieracien bei mir sämtlich gut gedeihen. H. albidum
ist zwar im Topfe etwas empfindlich, namentlich im
Winter, erhält sich jedoch im Grunde gut. Dasselbe
gilt von den Piloselloiden, jedoch mit Ausnahme von
H. pratense und H. Hoppeanum, welch letzteres gleich im
ersten Winter im Topfe und im Grunde abgestorben
ist.

Anderweitige Bestäubungsversuche konnte ich im vorigen Jahre wegen meines Augenleidens nicht beginnen. Nur ein Experiment schien mir so wichtig, daß ich mich nicht entschließen konnte, dasselbe auf eine spätere Zeit zu verschieben. Es betrifft die Ansicht Naudin's und Darwin's, daß zur genügenden Befruchtung eines Ovulum ein einziges Pollenkorn nicht ausreichend sei. Als Versuchspflanze benützte ich, wie es auch Naudin that, Mirabilis Jalappa; das Resultat meines Versuches ist jedoch ein völlig anderes. Ich erhielt aus der Befruchtung mit einem einzigen Pollenkorne 18 gut entwickelte Samen und daraus eben so viele Pflanzen, von denen bereits 10 in Blüthe stehen. Die Mehrzahl dieser Pflanzen ist aber so üppig ausgebildet, als die aus freier Selbstbefruchtung stammenden. Einige wenige Exemplare sind bis jetzt allerdings im Wachsthume etwas zurückgeblieben, allein nach dem Erfolge, den die übrigen aufweisen, dann die Ursache davon doch nur darin zu suchen sein, daß nicht alle Pollenkörner eine gleiche Befruchtungstüchtigkeit besitzen, und daß ferner bei dem in Rede stehenden Versuch die Mitbewerbung anderer Pollenkörner ausgeschlossen war. Wo mehrere concurriren, da dürfen wir annehmen, daß es immer nur dem kräftigsten gelingen wird, die Befruchtung zu vollziehen.

Ich will übrigens diese Versuche wiederholen, auch dürfte es sich durch ein Experiment direct nachweisen lassen, ob bei Mirabilis zwei oder mehrere Pollenkörner an der Befruchtung eines Eichens theilnehmen können. Nach Naudin sollen wenigstens drei erforderlich sein!

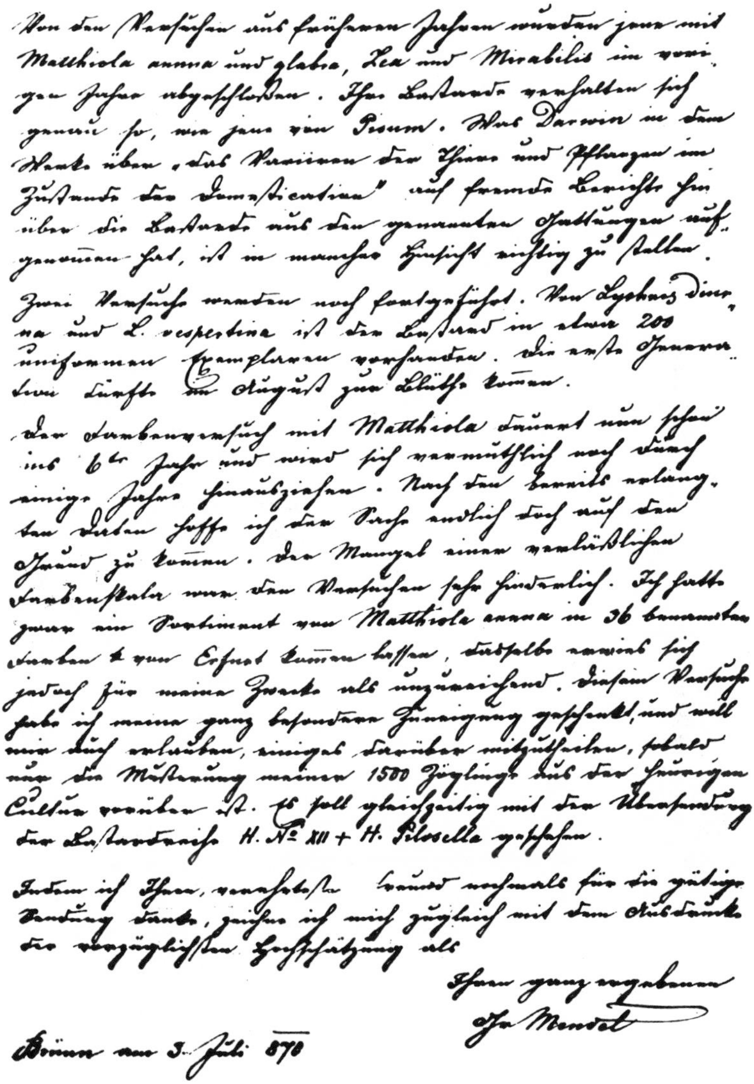
Von den Versuchen aus früheren Jahren wurden jene mit Matthiola annua und glabra, Zea und Mirabilis im vorigen Jahre abgeschlossen. Ihre Bastarde verhalten sich genau so, wie jene von Pisum. Was Darwin in dem Werke über „das Variiren der Thiere und Pflanzen im Zustande der Domestication" aus fremden Berichten hin über die Bastarde aus den genannten Gattungen aufgenommen hat, ist in mancher Hinsicht richtig zu stellen.

Zwei Versuche wurden noch fortgeführt. Von Lychnis diurna und L. vespertina ist der Bastard in etwa 200 uniformen Exemplaren vorhanden. Die erste Generation dürfte im August zur Blüthe kommen.

Der Farbenversuch mit Matthiola dauert nun schon ins 6te Jahr und wird sich vermuthlich noch durch einige Jahre hinausziehen. Nach den bereits erlangten Daten hoffe ich der Sache endlich doch auf den Grund zu kommen. Der Mangel einer verlässlichen Farbenskala war den Versuchen sehr hinderlich. Ich hatte zwar ein Sortiment von Matthiola annua in 36 benannten Farben aus Erfurt kommen lassen, dasselbe erwies sich jedoch für meine Zwecke als unzureichend. Diesem Versuche habe ich meine ganz besondere Zuneigung geschenkt, und will mir auch erlauben, einiges darüber mitzutheilen, sobald nur die Musterung meiner 1500 Zöglinge aus der heurigen Cultur vorüber ist. Es soll gleichzeitig mit der Übersendung der Bastardreihe H. No XII + H. Pilosella geschehen.

Indem ich Ihnen, verehrtester Freund nochmals für die gütige Sendung danke, zeichne ich mich zugleich mit dem Ausdrucke der vorzüglichsten Hochschätzung als

Ihr ganz ergebener
Gr. Mendel

Brünn am 3. Juli 870

C. Zirkle (1951) has raised the question whether earlier publications of other writers might have influenced Mendel to undertake his exact mathematical analysis of the F_2 generation. It hardly seems possible to answer this question. But even if such an influence existed, it would not diminish the value of Mendel's own work. It has also been asked whether his predilection for mathematics, physics, and chemistry may have made it easy for him to think in terms of recombinable units (Dunn 1965). Barthelmess (1952) comments (p. 76):

The whole of Mendel's work is based on the idea of classification. First, he classified the multiplicity of forms contained in his various experimental species in terms of *pairs* of characters so as to be able unequivocally to characterize each pair of parent plants. Gärtner had already worked like this in practice but had not made a principle of his procedure. The "total" units selected by the acuity of Mendel's intuition resolved into differentiating pairs of individual characters which remained constant and enabled him to classify what had been known as the *variation désordonnée* of the progeny. When inspection revealed that the characters were recombined in different ways with a definite frequency of occurrence, combinatory theory was seen to be an adequate method of mathematical analysis. By using a model of the interdependence of hypothesis and experiment, by proceeding methodically from simple to complex relations (1, 2, 3 pairs of characters) Mendel was able to discover the laws governing inheritance—a result which made him famous.

What is interesting—and against the background of other developments in the history of science so significant—is that Mendel proves his law governing the relations between parents and offspring to be *statistical*, and not dynamic, in nature.

Zirkle suggests that Mendel had probably read the papers of the apiarist Johann Dzierzon and perhaps even knew him personally. A parish priest, Dzierzon lived in neighboring Silesia. We may safely assume that Mendel had some kind of contact with him. Mendel, who experimented with numerous breeds of bees, was himself a well-known apiculturist and also a member of the Moravian Apicultural Society, presiding over the meetings from time to time as deputy chairman. According to conference reports contained in Volume I (1867) of the journal of this Society, Dzierzon frequently addressed the Society during the course of its meetings. The seventeenth convention of the Association of German Beekeepers was held at Kiel from September 12 to 14, 1871, and was attended by Dr. Ziwansky, chairman of the Brünn Society, and Mendel. Dzierzon spoke at length. It is thus almost certain that Mendel and Dzierzon knew each other personally (this information is taken from a letter, written June 29, 1966, by Professor Weiling of Bonn).

Dzierzon (1845, 1854) had discovered that workers and queens are produced from fertilized eggs, while drones spring from unfertilized eggs. Crossing German and Italian bees, he found that unfertilized hybrid queens produced German and Italian drones in a clear 1:1 ratio. From this it followed that a hybrid male would also have to produce two sorts of sperm and that, in the absence of a selective fertilization, a ratio of 1:2:1 or 3:1 would result. Zirkle considers it possible that such considerations and results might have shown Mendel the significance of exact segregation ratios. Dzierzon (1854, p. 64) writes:

It must be ascertained that the queen belongs to the pure breed. For if she herself originated from a hybrid brood, she could not possibly produce pure drones but only half Italian and half German drones though, oddly enough, not according to type but according to number, as if it were difficult for nature to fuse both species into an intermediate breed.

But it is also possible that Mendel was familiar with the experiments made by Jean-Antoine Colladon (1821),[6] as suggested by H. Grüneberg (1956) and J. Rostand (1956, 1958). Crossing white and gray mice, Colladon noticed that there was no blending of the colors as a result of the cross, for it was always white and gray mice that appeared in different numbers. It is not known whether he recorded the exact distribution of the differently colored animals because the manuscript describing this investigation has been lost. As we know from the biography of Mendel written by Iltis (1924), Mendel himself worked with mice before he began his experiments with peas. It therefore seems reasonable to conclude that he wanted to repeat and confirm Colladon's experiments and did not begin to experiment with botanical specimens until later.

It is certainly true, however, that a number of authorities in genetics have pointed out that there are now certain questions pertaining to Mendel's classic study which can no longer be answered. Fisher (1936), for example, has asked: What did Mendel discover? How did he discover it? What did he think he had discovered? What did his contemporaries think of this discovery?

To begin with, Fisher examines whether Mendel's studies are to be taken literally. In his experiments with peas, Mendel supposedly used seven varieties, all of which differed from one another only in respect to one gene. Fisher claims that it could also be possible that Mendel

[6] Colladon (1755-1830), one of the founders of the Swiss Society for the Study of Natural Science, studied pharmacy in Berlin and later developed a special interest in botany.

studied varieties that differed from one another with regard to several genes and that he only studied the segregation of a principal gene and neglected the others. Alternatively, if polyhybrid segregation did in fact occur, Mendel might have recorded the segregation of each gene separately—as happens in linkage studies—and then considered this segregation as a single experiment. Nonetheless, it seems from Mendel's account that from the twenty-two cultivated varieties, he did in fact select seven that differed from one another solely in respect to one gene. A mathematical analysis of the numbers of monohybrid, dihybrid, and trihybrid segregations shows that the segregation ratios approach those that one would expect on theoretical grounds to a surprising extent, considering the limited amount of experimental material. Mendel contented himself with these numbers in his essay without subjecting them to a mathematical analysis; he did not mention any repeat experiments to prove the certainty of the segregation ratios; moreover, he did not consider it necessary to test the segregation ratios of reciprocal crosses. He must therefore have been convinced of the equivalence of the parental gametes. According to Fisher, it is also curious that Mendel did not make use of linked genes in his experiments; this would have complicated the results considerably. When Mendel realized that the ratio of 3:1 was actually a ratio of 1:2:1, he quite cheerfully claimed that it was two-thirds of three-fourths of the total number of parent plants which again segregated and did not quote exact segregation ratios in support of this claim. On the strength of these facts, Fisher concludes that Mendel's essay is indeed to be taken literally as far as the framing of the individual experiments is concerned, but that the figures he gives were not objectively discovered; rather, they were approximated to a segregation ratio already known to him. It therefore seems that Mendel either knew what was to be expected because of theoretical considerations and only then conducted his experiments, contenting himself with relatively small numbers, or else he himself did not know what was expected but had it told him by someone else. According to Fisher, either possibility is supported by the fact that Mendel selected genes that segregated independently of one another from the large number of his varieties, and by the fact that he rejected a priori the possibility that the parental gametes were not equivalent. Thus his experiments did not really constitute any new discovery but rather a demonstration of laws that were perhaps already known to him. For Fisher, Mendel's merit lies mainly in the fact that he was well aware of the results of his work, an awareness clearly lacking in his contemporaries.

De Beer (1964) commented on the statistical analysis of Mendel's results undertaken by Fisher. Like Zirkle (1964), de Beer concludes that the excessive statistical exactitude of Mendel's figures may derive from the fact that he disregarded cases of segregation having unusual numerical ratios.

Weiling (1965, 1966) has recently evaluated Mendel's experiments on heredity from a biometric standpoint. He concluded that Mendel intuitively employed patterns of inference based on modern stochastic models of phenomena, e.g., that branch of stochastics that might be called "statistics of critical uncertainties." Having obtained surprisingly good agreement with the expected ratios by subjecting Mendel's data to the χ^2 test, he observed that this test requires material in normal distribution, whereas Mendel's data were in binomial distribution, and thus he explains what would otherwise seem to be too good to be true. Like Platt (1959), Weiling emphasizes once again that it was precisely this marriage of biology and statistics—at a time when this was not the usual practice—that made what Mendel had to say so difficult for his fellow naturalists to understand.

The strange fate that befell Mendel's published work is probably due either to his contemporaries' failure to understand him or to their conscious rejection of his ideas. Some thought that the journal publisher was in part responsible for Mendel's work being neglected for more than thirty years. We learn, however, from Mendel's biography (Iltis 1924) that the *Proceedings of the Brünn Society for the Study of Natural Science* were exchanged with the publications of at least 120 other associations and societies and were therefore available in many other libraries.

Correns (1922), who published Mendel's letters to Nägeli (1905), suggests that Mendel's form of presentation, which was quite new at the time, might have helped cause his essay to sink into oblivion. To Mendel's letter of December 31, 1866, Nägeli replied that he assumed Mendel's work to be merely a preliminary study that would eventually be expanded to include all the details pertaining to the experiments, that Mendel's formulas were merely empirical (and not rational in the sense that they constituted a universal scientific theory), and that the constant forms needed to be further tested. Mendel could have usefully expanded his account to include the similar results he had obtained in experiments with stock, *Mirabilis*, and maize. One reason for his silence may have been that he doubted the general applicability of the discoveries he had made in peas because of the *Hieracium* hybrids; he also found it difficult to obtain exact segregation ratios with other material, especially with the inheritance of the color of the flower in stock. Moreover, he was also occupied with other problems, particularly those of meteorology and astronomy. His wearisome struggle against the government may have finally made him abandon his scientific work.

Until recently the literature of our subject had recorded that four botanists who were contemporaries of Mendel were familiar with his work: A. Kerner von Marilaun, C. von Nägeli, H. Hoffmann, and W. O. Focke. This view is now generally known to be incorrect. On the occasion of the one hundredth anniversary of the publication of Mendel's classic study, Gaisinovich (1965) and Orel (1965) again pointed out that the Russian botanist Ivan Fyodorovich Schmalhausen (1849-1894) mentions Mendel's essay in a footnote to his dissertation, *On Plant Hybrids—Observations on the Flora of St. Petersburg.*

Schmalhausen (Figure 27) was born April 15, 1849, in St. Petersburg where his father was a university librarian. Starting in 1867, he studied botany under Professor A. N. Beketov at the University of St. Petersburg. While he was still a student, he investigated the flora of the province of St. Petersburg and discovered many new species. He was particularly interested in natural hybrids and their role in the origin of intraspecific variability; this was also the subject of his master's thesis, *On Plant Hybrids* (1874). After he had successfully defended his thesis, a two years' stay abroad led him to Zürich, where he conducted paleobotanical investigations under Heer. He then went to Strasbourg and worked under de Bary and Schimper. He studied the flora of Switzerland as well as that of the Tyrol, Italy, and France. After he had returned from abroad, he became a curator at the Botanical Gardens of St. Petersburg. In 1877 he defended his doctoral dissertation on the development of lactiferous tubes in plants and was a visiting lecturer in paleobotany at the University of St. Petersburg. Appointed Professor of Botany at the University of Kiev in 1878, he studied recent and fossil flora of southern Russia. He was the author of a flora of Southwest Russia (Kiev, 1886) and of a flora of Central and South Russia, the Crimea, and the North Caucasus (2 vols.; Kiev, 1895, 1897). Schmalhausen died after a brief illness on April 19, 1894.[7]

[7] I am indebted to A. E. Gaisinovich of Moscow for these biographical details on I. F. Schmalhausen and for his picture.

27 I. F. Schmalhausen (1849-1894)

The attitudes adopted by these five contemporaries of Mendel to his work are of considerable interest for the history of science, and we shall therefore deal with them in detail.

To all appearances, A. Kerner (1831-1878), who was appointed professor at Innsbruck in 1860 and director of the Botanical Garden in Vienna in 1878, did not understand the significance of Mendel's work. He was a follower of Darwin and convinced that natural species hybrids were the cause of variability and the formation of new species. As Correns (1922) states, the results obtained by Mendel from cultivated varieties perhaps seemed to Kerner to be too weak for this role. From a letter which Mendel wrote to Kerner on January 1, 1867 (Figure 28)[8] we know that he had sent him a copy of his essay.

Kerner answered on March 5, 1867, but his reply has been lost. Tschermak (in a letter to the present author), however, considers it improbable that Kerner was familiar with Mendel's essay, for Dr. Ginzberger, Kerner's research assistant, reports that a copy of the essay was found uncut in Kerner's library.

Nägeli's assessment of Mendel's essay is of greater importance. Carl Wilhelm von Nägeli (Figure 29) was born March 26, 1817, at Kilchberg near Zürich. In 1836 he began his studies in medicine at Zürich where he attended the lectures of Lorenz Oken on botany and zoology. He wrote his dissertation under A. P. de Candolle in Geneva and received his degree at Zürich in 1840. He then spent eighteen months in Schleiden's laboratory in Jena. He qualified as a lecturer in 1843 at Zürich University and six years later was appointed Associate Professor there. In 1852 he accepted the offer of the chair of botany at Freiburg. He became Professor of Botanical Studies at the Federal Institute of Technology in Zürich in 1856 and the following year went to the University of Munich where he continued to produce a great deal of varied and important work until his death on May 10, 1891.

Nägeli was one of the outstanding botanists of the nineteenth century. His thorough investigations advanced our knowledge of anatomy, physiology, taxonomy, and evolutionary theory. He was a great expert on hawkweeds and concerned himself with the problems of species hybridization from the standpoint of systematics and phylogeny.

Of the five botanists under review, Nägeli was best acquainted with Mendel's work because of the letters Mendel had written him. Mendel had sent him pea seeds for sowing and had repeatedly asked him for advice about the *Hieracium* hybrids. These letters form the most important supplementary material to Mendel's published works and reveal the great profusion of species with which he worked. In them he mentions that the hybrids produced in the experiments that had meanwhile been completed on *Matthiola*, *Zea*, and *Mirabilis* behave exactly like those of *Pisum*. The letters also contain the first statements about the inheritance of sex in *Lychnis*. Nägeli's replies to Mendel are relatively short and are mainly limited to sending experimental material. It is quite clear from his critical remarks about Mendel's essay

[8]I am indebted to Professor D. von Wettstein for the photocopy of Mendel's letter. The original is kept in the Dörfler collection of Uppsala University Library.

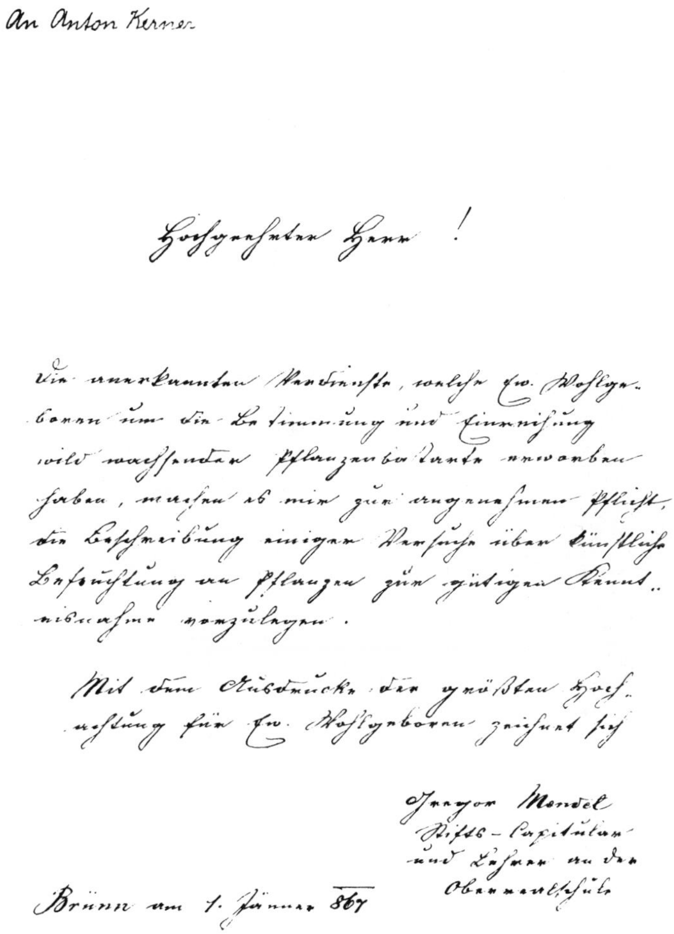
An Anton Kerner

Hochgeehrter Herr!

Die anerkannten Verdienste, welche Ew. Wohlgeboren um die Bestimmung und Einreihung wild wachsender Pflanzenbastarte erworben haben, machen es mir zur angenehmen Pflicht, die Beschreibung einiger Versuche über künstliche Befruchtung an Pflanzen zur gütigen Kenntnisnahme vorzulegen.

Mit dem Ausdrucke der größten Hochachtung für Ew. Wohlgeboren zeichnet sich

Gregor Mendel
Stifts-Capitular
und Lehrer an der
Oberrealschule

Brünn am 1. Jänner 867

28 Gregor Mendel's letter to A. Kerner von Marilaun

that Nägeli did no appreciate its significance.[9] Orel (1966) has published part of a letter which Nägeli wrote to Mendel on February 25, 1867. It reads:

[9]In a sensitive appreciation of Nägeli's role in the development of botanical knowledge, O. Renner (1959) has pointed to the tragic nature of the contact

29 Carl von Nägeli (1817-1891)

It seems to me that your experiments with *Pisum* are far from being completed and are indeed only just beginning. A shortcoming of all the more recent experimenters is that they have shown so much less perseverance than Kölreuter and Gärtner. I am pleased to note that you

between Nägeli and Mendel. He suggests that Mendel wanted to please Nägeli by hybridizing specimens of the *Hieracium* genus. Working with the sexually anomalous hawkweed, however, Mendel came to doubt the general validity of the laws he had discovered; Renner concludes: "For Mendel, the encounter with Nägeli was disastrous" (p. 263).

are not making this mistake and that you are following in the footsteps of your celebrated predecessors. However, you should try to excel them, and in my view this will only be possible—and this is the only way to make advances in the theory of hybridization—if exhaustive experiments are conducted upon a single object in every possible direction. Such a complete series of experiments giving irrefutable proofs of the most important conclusions has by no means yet been undertaken. Should you possess some extra seeds used in your hybrid fertilizations which you do not yourself intend to plant, I should be pleased to cultivate them in our garden so as to see whether they remain constant under different conditions. Those I should particularly wish to receive are A, a (offspring of Aa), AB, ab, Ab, aB (offspring of AaBb). If you agree to this, kindly send me the seeds as soon as possible together with an exact description of the origin of each type. Of course, I leave the choice of specimens to you, though I must observe that I do not have much time or space at my disposal. I shall not remark on any other points in your communication, for without a detailed knowledge of the experiments on which they are based I could only make conjectural comments. Your plan to include other plants in your experiments is excellent, and I am convinced that you will obtain quite different results (with regard to the inherited characters) from these other varieties. I should think it particularly valuable if you were able to undertake hybrid fertilizations in *Hieracium*.

At Nägeli's request Mendel sent him 140 carefully labeled packets of pea seeds with explicit directions for handling and for recognizing their hereditary factors—this despite Nägeli's contemptuous and unjustified criticism. Now Nägeli himself was interested in the problem of the formation of varieties (1865, 1866) and knew that a variety arises owing to inner causes and that new characters occur and become constant as it develops. But he did not conduct any control experiments, nor did he refer to the work with peas in subsequent letters.

The scientific thought of his time could not grasp the originality of the thesis contained in Mendel's essay: that it is not the overall appearance of an individual but individual characters that are inherited. It is nonetheless incomprehensible that such an eminent botanist as Nägeli should not remember the correspondence he had had with Mendel over a period of more than seven years and thus fail to use Mendel's results in the formulation of his own theory of inheritance. Twenty years later, in the year of Mendel's death (1884), Nägeli proposed his mechanistic-physiological theory of descent which accounts for inheritance in terms of rudiments for individual features and characters contained in the idioplasm. Nägeli's neglect of Mendel's work seems all the more incomprehensible when we recall that by this time the fundamentals of karyokinesis and of fertilization had been discovered, along with the part played by the cell nucleus in inheritance.

Gaisinovich (1935) and Weiling (1965) have pointed out that Nägeli and Peter (1885) cite Mendel's essay in their monograph on hawkweed which appeared one year after the chemicophysiological theory of descent. But the fact that Nägeli merely mentions the essay shows even more clearly that he did not understand the importance of that work.

In his theory, Nägeli introduced the concept of the idioplasm which contained every perceptible character as a predisposition or hereditary factor. An organism transmitted all of its characters through the idioplasm during the process of reproduction. The composition of the idioplasm was determined, according to Nägeli, by its molecular structure—especially by the arrangement of its smallest particles, which he called micellae. These micellae were combined to give units of a higher order, which in turn contained the predispositions, or hereditary factors, for systems of cells, tissues, and organs. Moreover, the fact that each micella can have a different chemical composition adds to the complexity of the constitution of the idioplasm. Nägeli estimated the size of a micella to be 1.2 trillionths of a cubic centimeter. He held that a micella was composed of crystalline groups of molecules and derived the growth of the idioplasm from the intussusception of micellae into clearly partitioned idioplasmic structures. He did not assume that the idioplasm in the cell was exclusively located in the nucleus although, as A. Weismann (1892) subsequently emphasized, it was already suspected that the hereditary substance was contained in the cell nucleus. Nägeli held that the idioplasm extended in reticular fashion in the form of the most delicate parallel threads or cords, occasionally grouped together to form bundles, and thus permeated the whole cell. Because of the idioplasmic links between one cell and another, the idioplasm pervaded the whole body. The constancy of the idioplasm is guaranteed provided the micellar longitudinal rows of cords develop further in strict parallelism. However, should the cords fuse or divide—that is, if the cross-section of the idioplasm is altered—hereditary variations occur in the individual in question. According to Nägeli, such mutations occur all the time, but a sudden alteration of a character is effected only when the small and continuous breaks in the parallelism of the idioplasm have accumulated to a certain degree.

Nägeli's micellar hypothesis, which goes far beyond the purely genetic problem of determining the material basis of heredity, received only passing approval and was in fact rejected by many of his contemporaries. J. Wiesner (1892) subjected it to a rather detailed critique.

Since Nägeli was much more familiar with Mendel's work than were any of his other contemporaries, the extremely important notion of idioplasm would probably have been formulated more clearly and divested of its speculative overtones, and its structure and location in the cell would have been more correctly determined, had he paid greater attention to Mendel's contribution.

We must comment in similar vein on the study by H. Hoffmann (1869). A professor of botany at Giessen, he attempted to refute Darwin's theory of evolution with a wealth of experiments of his own and of others, for he had been unsuccessful in fixing numerous variations and therefore doubted the importance of variations as a basis for the formation of new species. He cites Mendel five times (pp. 52, 86, 112, 119, 136), but three of these citations refer only to the species hybrids obtained by Gärtner from *Aquilegia*, *Geum*, and *Lavatera*, which are mentioned by Mendel.

Concerning Mendel's experiments with beans, Hoffmann writes: "According to Mendel, hybrids can be raised from *P. vulgaris* × *nanus*, and from *multiflorus*♂ × *nanus*♀. The hybrids from *mult.* × *nanus* had a low degree of fertility or were even sterile. The color of the flowers and seeds was usually more like that of *multifl.* In the second generation, several hybrids appeared resembling *nanus.*" (*Proc. of Brünn Soc. for Study of Natural Science* 1865, Vol. IV, p. 633).

On Mendel's experiments with *Pisum*, Hoffmann can only say (p. 136): "*Pisum.* Observations during a period of 6 years by G. Mendel (*ibid.*)," and "*Pisum sativum* hybrids and such—produced from forms that were constant from seed." Hoffmann then gives some trifling details about the techniques used in the crossings and continues: "Hybrids have a tendency to revert to their parental forms in subsequent generations."

It is not necessary to comment further on Hoffmann's study. He completely failed to recognize the significance of Mendel's work, for had he done so the form and content of his book—which appeared four years after the publication of Mendel's essay—would certainly have been considerably modified. Hoffmann's quotations, however, probably had the merit of drawing W. O. Focke's attention to Mendel's essay.

A Bremen physician and a great authority on plant hybrids, Wilhelm Olbers Focke (1834-1922) (Figure 30) wrote his well-known book *Die*

30 Wilhelm Olbers Focke (1834-1922)

Pflanzen-Mischlinge in 1881. It contained the findings that had been amassed during the course of the preceding 120 years about hybridization in the vegetable kingdom. Focke had relatively limited personal experience with species hybridization in the genera *Raphanus*, *Melandrium*, *Rubus*, *Anagallis*, *Digitalis*, *Nicotiana*, *Berberis*, *Cochlearia*, *Rosa*, *Begonia*, *Primula*, and *Galeopsis*. He describes the luxuriant development of such hybrids, indicates which species crosses can be more easily effected in the various genera, and points to reciprocal differences in the production of hybrids. In his systematic inventory of vegetable hybrids, Focke mentions Mendel a number of times in connection with his descriptions of hybrids produced from *Pisum*,

Phaseolus, and *Hieracium*. But he too failed to appreciate the importance of Mendel's experiments. Commenting on *Pisum* hybrids, for example, Focke writes: "Mendel's numerous crossings yielded results that were quite similar to those of Knight, but Mendel believed that he had found constant numerical relations between the different types of hybrids" (p. 110).

The many hybrids described by Focke and his interpretations of them reveal that he did not believe in uniform laws governing the production of hybrids and their descendants. His high standing in this field may have discouraged many of his contemporaries from a close study of Mendel's own essay. Elsewhere in the book the experiments with peas are no longer mentioned (p. 444):

Of recent scientific experiments in hybridization, those undertaken by Robert Caspary on *Nympheaceae*, those of G. Mendel on *Phaseolus* and *Hieracium*, and those of D. A. Godron on *Datura*, *Aegilops*, *Triticum*, and *Papaver* deserve to be called particularly instructive. The series of experiments conducted by Godron on *Datura* hybrids is to be regarded as a most distinguished accomplishment. These experiments proved that fertile hybrids, although extremely variable in their progeny, can produce in the course of generations hybrid varieties that remain constant when grown from seed. We are more indebted to Godron than to any other nineteenth-century experimenter for new factual knowledge about the characters of hybrids.

For Focke, Mendel was clearly only one of many who were worthy of mention for having produced vegetable hybrids. It may be the case that Mendel the outsider was not taken seriously by professional botanists, particularly since he worked mainly with cultivated plants.

Ivan Fyodorovich Schmalhausen was the only contemporary of Mendel who has the distinction of having correctly understood the significance of Mendel's work.

He did not become acquainted with Mendel's essay until after he had completed his dissertation, but he then appended a footnote in which he wrote:

I did not know of Mendel's study, *Experiments in Plant Hybridization*, until this work had gone to press. I nevertheless consider it necessary to refer to Mendel's study, for his methods, particularly his way of expressing his results in formulas, merit our attention and deserve to be further elaborated (with completely fertile hybrids). The author's task was to determine with mathematical precision the number of forms obtained from pollinating the hybrids and the numerical relations holding between individuals of each form. For the crossings he selected plants with constant and easily distinguishable characters, whose hybrids remained completely fertile in subsequent generations. Varieties of the pea met these conditions perfectly. . . .

Schmalhausen then goes on to give an exact account of the segregation ratios that occur in monohybridism and dihybridism, together with the conclusions that Mendel drew from them. A translation of his dissertation (according to Gaisinovich 1965) was published in *Flora* in 1875, but without the historical chapter; it therefore lacked the footnote mentioning Mendel's essay. In Russian literature, J. A. Filipchenko (1925) was the first to point out Schmalhausen's reference to Mendel. In addition, Vavilov (according to Orel 1965) mentions Schmalhausen's work in his introduction to the sixth edition of the Russian translation of Mendel's classic essay (1935), as does Gaisinovich (1935) in his chapter "Gregor Mendel und seine Vorgänger" (Gregor Mendel and His Predecessors) in *O. Sageret, Charles Naudin, G. Mendel, Selected Works on Vegetable Hybrids.*

Gaisinovich (1965) was able to say of Schmalhausen:

> He emphasized all the merits of Mendel's work: the successful choice of experimental material, the constancy of the "easily distinguishable characters," and the exact numerical assessment of all the forms in the offspring. He correctly sets forth the numerical relations that were observed in the crossings and concludes by proving that characters "are not blended" in the hybrid. Schmalhausen finally states that Mendel's conclusions "essentially agree with the theoretical reflections of Naudin."

Schmalhausen's observations on Mendel's essay in 1874 unfortunately remained unheeded in Russia in the nineteenth century and were of course unknown outside the frontier of that country.

In addition to Kerner von Marilaun, Nägeli, Hoffmann, Focke, and Schmalhausen, we must mention two more scholars who were familiar with Mendel's essay, or at least knew his name. In his lecture, *Cross-Breeding and Hybridization*, delivered in 1891 and published in 1892, L. H. Bailey mentioned Mendel's essay in the bibliography. According to Roberts (1929), Bailey himself was not acquainted with Mendel's essay, and had taken the reference from Focke's book. In 1896 Bailey incorporated this lecture into a book, *Plant Breeding*, which originally appeared without a bibliography. A bibliography was added in the second edition (1902), following a suggestion by de Vries.

We learn from a study by Stomps (1954) that Professor Beijerinck of Delft knew of Mendel's essay several years before de Vries, to whom (according to Stomps) it was sent by Beijerinck in 1900. But the dates given by Roberts (who quotes de Vries's own words) and Stomps do not agree (see p. 274).

Finally, we must add that Edwardson (1962), Zirkle (1964), and Weiling (1965) report that Mendel's essay, *Experiments in Plant*

Hybridization, was mentioned in the *Royal Society Catalogue of Scientific Papers* (1864-1873) and in the ninth edition of the *Encyclopaedia Britannica* (1881-1895) in an article on hybridization by Romanes.

These names are the last of those who knew or referred to Mendel's essay before the turn of the century. Whatever else was written about genetics until that time was therefore accomplished without any knowledge of Mendel's work.

6.4 Genetic Problems in the Work of Charles Darwin and the Provisional Hypothesis of Pangenesis

Much more decisive for the fate of Mendel's works, however, may have been the fact that the interest of the scientific world was completely riveted on the question of evolution because *The Origin of Species* (1859) by Charles Darwin had appeared seven years before the publication of Mendel's essay. Supported by a wealth of his own observations made during the course of many years and by his extensive knowledge of the literature, Darwin therewith presented the problem of evolution to the public.[10]

The son of a physician, Darwin (Figure 31) was born February 12, 1809, in Shrewsbury, and died April 19, 1882, at Downe. He is buried in Westminster Abbey. Darwin was a grandson of the English doctor, naturalist, and poet Erasmus Darwin (1731-1802), who wrote a four-volume treatise, *Zoonomia* (1794-1796), containing his ideas on the change and development of species, on adaptation, on rudimentary organs, and so on. Such notions were later expanded and fully substantiated by his grandson Charles. Erasmus Darwin also described the differences between vegetative and sexual propagation, though his grandson did not recognize these because of the appearance of "bud sports." In *Zoonomia* Erasmus wrote of the offspring of buds, tubers, and bulbs that they "exactly resemble their parents, as is observable in grafting fruit trees, and in propagating flower roots; whereas the seminal offspring of plants, being supplied with nutriment by the mother, is liable to perpetual variation."

Charles Darwin first studied medicine at Edinburgh, and then divinity at Cambridge. At the invitation of Captain Robert FitzRoy, Darwin traveled around the world on a voyage of exploration aboard

[10]In this historical treatise we must confine ourselves to those aspects of Darwin's work that pertain to specific problems in genetics: hybridism, the provisional hypothesis of pangenesis, and problems associated with the varied observations on the formation of breeds and varieties recorded by Darwin in his works.

31 Charles Darwin (1809-1882)

the survey ship H.M.S. *Beagle* and thus became a naturalist. In the course of this voyage, Darwin gathered together extensive geological, zoological, and geographical material. After his return, he worked in London for a while, and then from 1842 as an independent researcher in Downe, Kent. This far-ranging material, which he investigated thoroughly, enabled him to present a compelling proof of the natural development of organic life. The first scientific work to result from his world voyage, a treatise on the structure and distribution of coral reefs, was published in 1842. His second work, a monograph on cirripeds, was completed in 1854. In 1859 he published his celebrated study, *The Origin of Species by Means of Natural Selection*, which marked a turning point in the history of biology.

This first great work of Darwin was rapidly followed by two others: *The Variation of Animals and Plants under Domestication* (1868) and *The Descent of Man* (1871). His revolutionary theories were vociferously accepted or rejected by science, politics, and the Church during the last four decades of the nineteenth century. There is hardly any doubt that a number of lines of biological research were neglected or even completely forgotten amid this ferment. We must therefore ask how far Darwin's ideas on heredity approached those of Mendel or whether they actually confirmed them.

Mendel is nowhere mentioned in Darwin's works. We can therefore assume that Mendel's essay remained unknown to Darwin despite his wide interest in hybridization and his extensive knowledge of the literature. And so we must conclude with O. Richter (1932), who has made a detailed study of the journeys undertaken by Mendel, that a personal meeting between Darwin and Mendel never took place. It might have been possible for this to have happened in 1862 when Mendel visited London. Although he had not yet published his essay by that time, he most certainly possessed enough material to present to Darwin. In that year, however, Darwin and his family were very ill; they lived in seclusion and received no guests. We do not know whether Mendel intended to go and see Darwin. The latter probably never did hear of Mendel, even though he did not die until 1882, that is, seventeen years after the publication of Mendel's fundamental essay.

Punnett (1925) has pointed out that in *Cross- and Self-Fertilization*, Darwin discusses the study of Hoffmann (1869) in which a reference to Mendel's work on *Phaseolus* and *Pisum* appears. Darwin had therefore studied a work mentioning Mendel only four years after the publication of the essay.

In his critical study *Darwin's Place in History*, C. D. Darlington (1959) has asked what would have happened had Darwin known of Mendel and his work. He answers this question by quoting Darwin's opinion, as expressed to his friend Hooker (September 13, 1864), of Naudin's experiments: "I cannot think it will hold. The tendency of hybrids to revert to either parent is part of a wider law . . . that crossing races as well as species tends to bring back characters which existed in progenitors hundreds and thousands of generations ago. Why this should be so, God knows."

Darlington's judgment is devastating. Even Darwin, whose thought encompassed dimensions quite different from those of Mendel, would probably not have recognized the importance of Mendel's essay.

I consider it idle to pose such a question. Mendel's essay differs from the work of Naudin in its extraordinarily exact analysis of the inheritance of individual characters, in its statement of numerical ratios leading to the establishment of statistical laws, whereas Naudin merely observed a *variation désordonnée* in F_2. Nobody will ever know whether Mendel's essay would have made a deeper impression on Darwin than the admittedly correct, but all too general, statements of Naudin and whether some of the ambiguity in Darwin's work might thereby have been removed.

We must now consider the related question whether Mendel was acquainted with one or more of Darwin's great works. This can be answered in the affirmative. The Mendel section of the Moravian Museum in Brünn contains the copy of the German edition of *The Origin of Species*, published in 1863, which belonged to Mendel and was underlined by him in places. De Beer (1964) pointed out the similarity between certain passages in Darwin's book and Mendel's essay. For example, in the first chapter of *The Origin of Species* we read:

> It seems clear that organic beings must be exposed during several generations to new conditions to cause any great amount of variation; and that, when the organization has once begun to vary, it generally continues varying for many generations. No case is on record of a variable organism ceasing to vary under cultivation. Our oldest cultivated plants, such as wheat, still yield new varieties: our oldest domesticated animals are still capable of rapid improvement or modification.

In his own essay, Mendel writes:

> It is readily granted that the origin of new varieties is favored by cultivation and that many divergent varieties that would perish under natural conditions can be preserved by the intervention of man; but nothing justifies the assumption that the tendency to form varieties is thereby furthered to such a degree that species quickly lose their stability, with their offspring diverging into an endless series of extremely variable forms. Were change in the conditions of vegetation the unique cause of variability, we should expect to find that those cultivated plants which have been grown for centuries under almost identical conditions would have reacquired stability. As is well known, however, that is not the case; for they are the very plants which have not only the most varied, but also the most variable forms.

The extent to which Mendel agreed with Darwin's ideas is difficult to ascertain. Mendel never mentions Darwin; this is probably due to the fact that he did not wish to become publicly involved in the conflict over Darwin's revolutionary theories. We can assume that Mendel concurred with Darwin on many points but that he had other views on the causes of hereditary variability since he held the view, on experimental grounds, that it was segregation and recombination of the genes that was decisive, rather than change in environmental conditions. With regard to the demarcation of species and varieties, however, both Darwin and Mendel thought that there were no fundamental differences between these two categories.

In this connection we must also mention that as early as 1819 the English anthropologist and surgeon W. Lawrence (1783-1867) had argued vigorously for the importance of hybridization in the origin of new forms in animals and in man. In so doing he had run contrary to all the disciples of the prevailing theory, Lamarckism, which sought to explain the origin of hereditary variability exclusively in terms of changes in environmental influences.

Darwin wrote on the problem of species hybridization in *The Origin of Species* and dealt, in his first treatment of the subject, with the fertility or sterility of hybrids. He held that the ease with which a cross could be effected was not necessarily related to the degree of resemblance between the two parent species with which the cross was to be made. He says that "the facility of making a first cross between any two species is not always governed by their systematic affinity or degree of resemblance to each other"(1872, chapter 9, p. 271).

He knew that the possibility of obtaining reciprocal crosses can vary (as Kölreuter had observed with *Mirabilis*); species that were difficult to cross can sometimes yield fertile hybrids and those that were easy to cross frequently produce sterile hybrids; and cross-pollination usually produced better seed setting than is the case with self-pollination. Darwin rejected the widely debated claim that the fertility or sterility of a pair of parent plants provides a criterion for classifying them as varieties or species: "It can thus be shown that neither sterility nor fertility affords any certain distinction between species and varieties. The evidence from this source graduates away, and is doubtful in the same degree as is the evidence derived from other constitutional and structural differences" (*ibid.*, p. 263).

Like many of his predecessors, Darwin observed the luxuriant development of hybrids, and he also recognized the debilitating influence of inbreeding: "I have made so many experiments, and collected so many facts, showing on the one hand that an occasional cross with a distinct individual or variety increases the vigor and fertility of the offspring, and on the other hand that very close interbreeding lessens their vigor and fertility, that I cannot doubt the correctness of this conclusion" (*ibid.*, p. 263).

This conclusion was not merely based on external impressions, however, for Darwin often employed the techniques of weighing and measurement to secure an exact verification of what was visible to the eye.

According to him, the causes of the varying degrees of fertility in hybrids and of their luxuriant development did not reside in some mysterious force; they resulted from the fact that individuals of this kind had been exposed to different conditions in the course of earlier generations or, alternatively, varied spontaneously ("or their having varied in a manner commonly called spontaneous"), thus enabling their sexual elements to be differentiated in every case. From this it follows that "after plants have been propagated by self-fertilization for several generations, a single cross with a fresh stock restores their pristine vigor and we have a strictly analogous result with our domestic animals" (*The Effects of Cross- and Self-Fertilization in the Vegetable Kingdom*).

For Darwin (*Variation of Animals and Plants . . .*), internal and external conditions are also the causes of variability in other features and characters. If living creatures are placed in unnatural conditions over a period of several generations, they have an extreme tendency to vary. He maintained that this fact is analogous to what takes place in hybrids whose offspring tends to vary a great deal in later generations. For the changed conditions in the hybrids exert an influence on the reproductive cells, producing a kind of stimulation which has a beneficial effect on the progeny. With regard to external conditions, he thought that food supplies form a very important cause of variability. However, the sudden appearance of bud variations led him to remark: ". . . but I cannot imagine a class of facts better adapted to force on our minds the conviction that what we call the external conditions of life are in many cases quite insignificant in relation to any particular variation, in comparison with the organization or constitution of the being which varies" (1893, Vol. 2, p. 279).

It is sometimes said of Darwin that he thought the causes of variability consisted solely of environmental conditions and the changes that occurred in them. However, this view is based on an incomplete knowledge of Darwin's writings. The occurrence of bud variations in particular—which he termed "sports" like all other sudden variations—led him to conclude that the decisive cause of variability is to be found in the genetic constitution of an organism. Describing a red magnum bonum plum which appeared on a forty-year-old yellow magnum bonum tree, he writes (*ibid.*, p. 282):

> When we reflect on these facts we become deeply impressed with the conviction that in such cases the nature of the variation depends but little on the conditions to which the plant has been exposed, and not in any special manner on its individual character, but much more on the inherited nature or constitution of the whole group of allied beings to which the plant in question belongs. We are thus driven to conclude that in most cases the conditions of life play a subordinate part in causing any particular modification.

Darwin's work contains numerous descriptions of the behavior of individual characters in hybrids. With few exceptions, he observed an intermediate condition in hybrids, though he was also familiar with dominant characters which appeared unaltered in the hybrid, as was the case, for example, with crosses between white and gray mice or white and black rabbits. Similar results were observed when crossing Ancon sheep with common sheep; also in the patterns of inheritance of animals in which the tail or horns were absent. Like many of his predecessors, Darwin knew that the generations following the first cross reverted back to their original parent types, though the time it took for this to be accomplished could vary: "As a general rule, crossed offspring in the first generation are nearly intermediate between their parents, but the grandchildren and succeeding generations continually revert, in a greater or lesser degree, to one or both of their progenitors" (*ibid.*, p. 23).

He explained the recurrence of parental characters in terms of the separation of elements in the hybrid, just as Naudin had done" (*ibid.*).

> If pollen which included the elements of one species happened to unite with ovules including the elements of the other species, the intermediate or hybrid state would still be retained, and there would be no reversion. But it would, as I suspect, be more correct to say that the elements of both parent species exist in every hybrid in a double state, namely, blended together and completely separate.

Crossing zygomorphous and peloric *Antirrhinum majus*, Darwin noticed that the flowers of all the hybrids were zygomorphous. When these hybrids were inbred, the next generation consisted of eighty-eight plants with zygomorphous flowers and thirty-seven plants whose flowers were peloric. He had therefore found a regular case of a 3:1 segregation ratio in the first hybrid generation, as Mendel would have called it—that is, in F_2. This case is reported in *Variation of Animals and Plants under Domestication*, which appeared two years after the publication of Mendel's essay, but Darwin was unable to give the correct explanation of this phenomenon (1868, Vol. 2, p. 70):

Here we have a good instance of the wide difference between the inheritance of a character and the power of transmitting it to crossed offspring. The crossed plants, which perfectly resembled the common snapdragon, were allowed to sow themselves, and, out of a hundred and twenty-seven seedlings, eighty-eight proved to be common snapdragons, two were in an intermediate condition between the peloric and normal state, and thirty-seven were perfectly peloric, having reverted to the structure of their one grandparent. This case seems at first sight to offer an exception to the rule formerly given, namely, that a character which is present in one form and latent in the other is generally transmitted with prepotent force when the two forms are crossed. For in all the *Scrophulariaceae*, and especially in the genera *Antirrhinum* and *Linaria*, there is, as was shown in the last chapter, a strong latent tendency to become peloric; and there is also, as we have just seen, a still stronger tendency in all peloric plants to reacquire their normal irregular structure. So that we have two opposed latent tendencies in the same plants. Now, with the crossed *Antirrhinums* the tendency to produce normal or irregular flowers, like those of the common Snapdragon, prevailed in the first generation; whilst the tendency to pelorism, appearing to gain strength by the intermission of a generation, prevailed to a large extent in the second set of seedlings. How it is possible for a character to gain strength by the intermission of a generation, will be considered in the chapter on pangenesis.

This example is one of many that illustrate how Darwin's work touches on the statistical laws discovered by Mendel, though Darwin did not realize that such cases were examples of laws of heredity. He explained the phenomenon of segregation in a completely different way, saying that plants with peloric flowers have a strong tendency to produce zygomorphous flowers, and vice versa. Darwin talked in terms of two opposed latent tendencies in the same plants. In the crossed *Antirrhinum*, the tendency to produce zygomorphous flowers prevailed in the first generation while the tendency to form peloric flowers appeared in the second generation.

This explanation agrees precisely with the ideas Darwin developed in his provisional hypothesis of pangenesis for, expounding that hypothesis, he writes (1893, Vol. 2, p. 395):

And when two hybrids pair, the combination of pure gemmules derived from the one hybrid with the pure gemmules of the same parts derived from the other would necessarily lead to complete reversion of character; and it is, perhaps, not too bold a supposition that unmodified and undeteriorated gemmules of the same nature would be especially apt to combine.

By "reversion," Darwin simply means the product of the reunion of identical hereditary predispositions, or factors, that results in the formation of one of the parent types. He continues: "Pure gemmules in combination with hybridized gemmules would lead to partial reversion. And lastly, hybridized gemmules derived from both parent-hybrids would simply reproduce the original hybrid form. All these cases and degrees of reversion incessantly occur."

In Mendelian terminology, the above quotation says that back-crossing the hybrid with the recessive parent type yields offspring, some of which (50 percent) resemble that recessive parent. Crossing the hybrids *inter se* would not, however, as Darwin assumed, produce the hybrid form alone since it would appear only in one-half of the progeny.

In the phenomena he observed and in the account he gave of them, Darwin evidently came close to making the same observations as Mendel, except that it did not occur to him to ascertain exactly the numerical ratios of the hybrid offspring.

Darwin was essentially an observer and not an experimenter. That Mendel's crucial essay was not read by the man who, familiar with all the old theories of reproduction and heredity contained in the literature, sifted the findings of earlier investigators and assimilated them into his work, is one of the incomprehensible facts about the history of our science.

We cannot conclude this review of Darwin's abundant observations and ideas on genetics without discussing the basic principles of the provisional hypothesis of pangenesis which Darwin expounds at the end of his treatise *The Variation of Animals and Plants under Domestication.* This hypothesis constitutes yet a further development of the doctrine of pangenesis which had appeared, under different guises, in almost every century during a period of more than 2000 years. For example, we have already discussed the work of Maupertuis, an eighteenth-century exponent of this doctrine; Galton, Nägeli, and de Vries, at the close of the nineteenth century, were the last to espouse a variant of this theory.

In this connection we must note that Darwin had a nineteenth-century predecessor in Herbert Spencer, who developed a theory of the material basis of heredity in 1864.

The fact that every vegetable cell was able to form a new and complete organism and that the sperm and egg cell could transmit all the characters of the organism led Spencer to postulate the existence of "physiological units," each one of which represented the total species character. For Spencer, these "units" were not simple molecules but complex aggregates of molecules capable of growth and reproduction. Because of their complex structure they were easily modified by external conditions that could bring about structural changes in the units or disturb their equilibrium. According to Spencer, the way the units were grouped was the cause of the differences between bodily parts, while their individual composition was responsible for the variability of individuals within a given species and for the diversity of species. These units therefore possessed physiologically variable dimensions and had a determinate structure. This view comes close to that held at about the same time by E. Brücke (1862), who had postulated on the basis of considerable factual material the existence of minute organic particles. In other words, he had proposed a theory of organized cell structure.

A few years later (1868) Darwin developed his "Provisional Hypothesis of Pangenesis." He proceeded from the assumption that each character able to vary independently of another was linked to a material bearer. He claimed that every cell of the developing organism was capable of producing countless tiny granules, all different from one another, which were responsible for the formation of individual characters and organs. According to Darwin, these granules, or gemmules as he called them, can reproduce independently of one another and in varying degrees of strength. Whenever cell division occurs, they enter into the daughter cells and can freely circulate within the body, passing into the reproductive cells which ensure their transmission to subsequent generations. If gemmules of a given kind diminish in number as compared with the others, the characteristic they determine is weakly developed; if they are reduced still further, it eventually becomes latent. This latent character can reoccur as a case of atavism, however, if the number of gemmules determining it becomes subsequently increased.

According to the hypothesis of pangenesis, variability depends on two different processes in the gemmules. First, a change in the relative proportions of existent gemmules, their incapacitation or renewed strength, can produce wide fluctuations in the way individual characters are marked. Second, the nature of the gemmules itself can change, since new ones can be produced from the old. For Darwin, the gemmules were not single molecules but aggregates composed of several molecules. They were able to assimilate, grow, and divide, with the result that two identical gemmules usually came from a common ancestor. Darwin's account of where precisely the gemmules were situated in the cell is unclear. He discussed the possibility, however, that they did not always occur singly but in groups of varying magnitudes, as is suggested by phenomena of correlative variability.

W. K. Brooks (1883) sought to amend Darwin's hypothesis as follows: He held that the male reproductive cells were able to exert an especially strong attraction on the gemmules circulating in the blood, with the result that they accumulated in larger quantities in the male cells. Brooks's modification stems from the fact that he thought the male reproductive cells played a much more prominent part in causing variation than the egg cells.

6.5 Francis Galton and Some Nineteenth-Century Views on the Inheritance of Diseases and Deformities in Man

The first person to comment on Darwin's provisional hypothesis of pangenesis was his cousin Francis Galton (1822-1911), who did so in 1876. To prove that Darwin's gemmules were circulating in the blood, he performed blood transfusions on a given breed of animal using blood taken from a different breed, thereby hoping to influence the character of the offspring. His experiments on rabbits had a negative result. He nonetheless assumed the existence of gemmules, and characterized them collectively as the "stirp" (stirpes = root or stock). But he rejected the idea that they circulated freely in the blood and thus the claim that they reaccumulated in the reproductive cells. On the other hand, he assumed that all the gemmules which are used to construct the cells are thereby consumed. The remaining gemmules—those that did not contribute to development—were contained in, and transmitted by, the reproductive cells. It has been frequently pointed out that this view may contain the first, albeit superficial, reference to the notion of the continuity of the germ plasm, subsequently developed by A. Weismann.

As is seen from an exchange of letters with Darwin concerning the hypothesis of pangenesis, Galton derived a 1:2:1 ratio as recorded in a letter written December 18, 1875, in which he explains to Darwin his view of heredity in terms of a finite number of existent particles (Pearson 1914-1930, Olby 1965, 1966). He assumed that the hereditary units (gemmules) do not blend in the hybrid, and there is only a certain number of them contained within each cell. Given only two gemmules per cell, one white and one black, it will frequently occur that the offspring are one-quarter white, one-quarter black, and one-half gray.

32 Francis Galton (1822-1911)

The son of a banker, Francis Galton (Figure 32) was born in Birmingham in 1822, the same year as Mendel. Like Charles Darwin, he was a grandson of Erasmus Darwin. He studied medicine at Birmingham, then at King's College in London, at Cambridge, and finally at St. George's Hospital. Although he did not complete his studies in medicine, he did obtain a thorough grounding in mathematics at Cambridge. Like Darwin, he suffered from a certain degree of nervous instability. He undertook long journeys to Malta, Egypt, and South Africa, returning to England in 1852 where he devoted his extremely varied talents to writing and research in a number of different areas of study. He experimented with musical instruments, conducted research in physics and meteorology, and was the first person to recognize the importance of fingerprints as a means of personal identification. He died January 17, 1911, at Grayshott House in Haslemere.

Galton may be regarded as the originator of empirical research in medical genetics. He was also the first person to recognize the importance of studying twins for work in human genetics (1876). Taking his cue from the successes of plant and animal breeders who were able to mold the physical constitution of future generations according to their will, he undertook detailed research on the inheritance of mental qualities and on the methods of its control (1865, 1869).

The first of two short essays on the subject, entitled *Hereditary Talent and Character* and published in 1865, begins as follows:

> The power of man over animal life, in producing whatever varieties of form he pleases, is enormously great. It would seem as though the physical structure of future generations was almost as plastic as clay, under the control of the breeder's will. It is my desire to show, more pointedly than—so far as I am aware—has been attempted before, that mental qualities are equally under control.
>
> A remarkable misapprehension appears to be current as to the fact of the transmission of talent by inheritance. It is commonly asserted that the children of eminent men are stupid; that, where great power of intellect seems to have been inherited, it has descended through the mother's side; and that one son commonly runs away with the talent of a whole family. My own inquiries have led me to a diametrically opposite conclusion. I find that talent is transmitted by inheritance in a very remarkable degree; that the mother has by no means the monopoly of its transmission; and that whole families of persons of talent are more common than those in which one member only is possessed of it.

Galton perused biographies to determine the frequency and degrees of kinship among the people named in them for his studies in the transmission of inheritance of intellectual abilities. He thus examined the kinship relations among groups of distinguished persons—the justices of England from 1660 to 1868, the politicians in the reign of George III, and the prime ministers of the nineteenth century—in order to obtain a general understanding of the laws governing the inheritance of mental qualities. He also examined the family relationships of the most famous generals, authors, mathematicians, scientists, poets, painters, musicians, theologians, and linguists; he wrote a work comparing these findings to the transmission by inheritance of physical factors, obtaining his data from the family relationships of various groups of oarsmen and boxers.

From his observations Galton concluded that intellectual ability was inherited in an extremely high degree. He proposed the utopian ideal that the quality of mankind be improved by the deliberate application of the techniques of breeding. He coined the term "eugenics" as the name of the science which would study the conditions under which hereditary characters could be perfected in future generations. Galton's fundamental work in biostatistics had a profound influence on theoretical and applied genetics well into the present century, but it seems that he was not aware of Mendel's essay.

The introduction of the methods of mathematical statistics into genetics derives from the work of the Belgian anthropologist and mathematician Adolphe Quètelet (1796-1874),[11] two of whose works were widely read. In *Sur l'homme* (1835), he established numerous frequency distributions for growth, life-span, and such, calculating the mean and emphasizing various characteristic values. He showed how the Gaussian Law of Errors could be applied to the measurement of living organisms. His book *Anthropométrie* (1871) gives an introductory account of the essentials of mathematical methods in genetics.

The mathematical treatment of variability resulted in the founding of the English biometric school. It is clear that Galton would have thought Mendel's essay extremely important if he had had the occasion to read it. Galton (1889) introduced the quantitative measurement of populations into genetics, by means of which random differences in the properties of single individuals were to be discovered in the variability

[11] There is, however, another sense of "biometry" according to which that concept is older and may be traced back to some of the seventeenth- and eighteenth-century mathematicians and statisticians (Heinisch 1966). In this sense biometry is to be understood as the theory of the application of mathematical methods to the study of the multiplicity of organic life in the service of exact summary description. Using the techniques of probability theory, the mathematician de Moivre (1667-1754) had tried to clarify whether the ratio of the sexes, that had been determined in given populations as 1.06♂:1.00♀, was a random deviation from 1:1 or whether this difference was genuinely significant. The Berlin theologian and scientist Johann Peter Süssmilch (1707-1767) was also a precursor of the modern school of biometry. His well-known book *Die göttliche Ordnung in den Veränderungen des menschlichen Geschlechts* . . . published in 1741, made an important contribution to demography and to early studies in biometry. Süssmilch was already aware of the importance of the size of samples, of the representative selection of numerical data, and of the probalistic nature of biometric conclusions.

of characters (Galtonian normal curve of frequency distribution, Galton, 1877.) He treated populations as units and established a series of laws, the most important of which is the Law of Regression. According to this law, selection causes offspring to show the same deviation from the mean type of a given population as their parents, though the deviation is usually not so highly marked. Selection therefore produces a shift in the formation of characters in the direction of the selection in question.

In his day, Galton's formulation of the Law of Regression was an important accomplishment; it was the first attempt to treat genetic problems in populations with mathematical methods. However, Galton's Law of Regression and his Law of Ancestral Inheritance contributed little to the understanding of inheritance of individual characters or to the discovery of the material basis of hereditary substance. Karl Pearson (1857-1936), a pupil of Galton, developed the Law of Regression further. In Denmark, W. Johannsen (1857-1927) later subjected it to a detailed study with his analysis of populations in terms of pure lines.

Pearson, a student of philosophy, jurisprudence, physics, and biology, was the first person to be appointed to the chair of eugenics which had been endowed by Galton. In 1884 after studying at the universities in Berlin and Heidelberg, he was appointed Professor of Applied Mathematics and Mechanics at the University of London. From 1890 he taught geometry, mathematical statistics, and probability theory at Greshaw College, where he also worked on graphic methods for representing frequency distributions. He was interested in the problems of fitting curves to observational data, and developed the χ^2 test as a measure for goodness of fit. However, Pearson's philosophical treatises, together with his works on genetics and evolution, were not particularly well received. In 1901, together with Galton and Weldon, Pearson founded the journal *Biometrika* which played an important part in introducing biometric methods into biology.

The nineteenth century witnessed a rekindling of the interest taken by students of medicine in congenital defects and diseases. As we have shown, observations on the hereditary transmission of disease were made throughout history (J. W. Ballantyne, 1895, on antenatal disease and deformity). J. Labus (1929) has reviewed the most important literature of the eighteenth century; she gives an abstract of the thirty-one works she studied and discusses the main problems they contain. We shall here mention only a few of the numerous nineteenth-century studies concerned with the hereditary transmission of pathological features. J. F. Meckel (1812) gave illustrations of a number of hereditary deformities in the first volume of his *Handbuch der Pathologischen Anatomie.* Particularly noteworthy is the contribution to medical genetics of the man who (according to A. G. Motulsky, 1958) must be regarded as the forgotten founder of that branch of study, Josef Adams (1756-1818). In 1814, Adams published his book entitled *A Treatise on the Supposed Hereditary Properties of Diseases Based on Clinical Observation.* In contains the following observations:

1. Clear distinctions between congenital, familial ("recessive"), and hereditary ("dominant") conditions and their patterns of inheritance.
2. Some of the characteristics of familial ("recessive") disease were noted; for instance, mating between relatives would increase the frequency of disease.
3. Hereditary diseases may become manifest at different stages in life, not necessarily at birth.
4. Hereditary susceptibilities exist which need external precipitating factors to make illness apparent. In terms of risk for offspring, it is immaterial whether clinical disease is present or not.
5. Genetic illness can be treated, especially if the precipitating factors can be removed.
6. Intrafamilial correlations between the ages of onset of hereditary diseases exist and are clinically useful in genetic prognosis.
7. Diseases that are clinically identical may be genetically heterogeneous.
8. The higher incidence of hereditary disease in isolated areas may be caused by inbreeding.
9. The early death of the affected would lead to the disappearance of genetic disease, were it not for the resurgence of the condition in offspring of parents who were free from it. He therefore logically predicted mutations.

C. F. Nasse (1820) gave an account, supplemented by his own observations, of the then known cases of hereditary disposition to bleed to death (hemophilia). This disease was described for the first time at the close of the eighteenth century. Everyone who had written on this topic had observed that hemophiliacs were always males and that women, though married to men from nonafflicted families, nonetheless transmit their fathers' tendency to bleed. In such cases, all male children were usually hemophiliacs, though exceptions to this rule were not unknown. The causes and subsequent occurrence of the hemorrhages, concomitant symptoms of the disease, its fatal consequences, and methods of stanching the bleeding were aspects of hemophilia that received detailed description.

The transmission by inheritance of mental traits and aberrations was described as well. In 1847 the French physician Prosper Lucas published a two-volume work on the hereditary nature of mental characters, though some of the material it contains is derived from imaginative fancy rather than from observation. In 1857 a French psychiatrist, B. A. Morel, wrote a treatise on the heritability of cases of mental derangement. Writing in 1865, C. von Seidlitz noted many instances of the hereditary transmission of mental and physical characteristics; further examples are contained in an essay on heredity in historical perspective by E. Roth (1885) and in a study by E. Ziegler (1886). Although these works are referred to, it must be understood that they give some faulty explanations along with correct ones.

6.6 Ernst Haeckel

The year in which Mendel published his *Experiments in Plant Hybridization* also saw the appearance of a work containing a philosophical account of theories of inheritance, the *Generelle Morphologie*, written by Ernst Haeckel (1866).

Haeckel (Figure 33) was born on February 16, 1834, in Potsdam and died in Jena on August 9, 1919. He studied medicine, was a pupil of the great physiologist Johannes Müller, and worked under Virchow as a research assistant. We are indebted to Haeckel for important monographs on Radiolaria, Medusae, and sponges. In morphology he was a disciple of Goethe and became a passionate admirer of Darwin, for whose ideas he tried to gain acceptance with his "evolutionary trees" drawn according to the principles of morphology. Haeckel sought to achieve a synthesis of the great scientific currents of his age, comprised of elements of morphology, physiology, and theory of descent. The influence of the theories of Lamarck and Darwin, the great discoveries in cytology (that the cell was the basis of all organic life, Virchow's doctrine "*omnis cellula e cellula*," the realization that the cell was composed of nucleus and cytoplasm), together with the fact derived from experience that morphological and physiological characters of parents are transmitted to their children—all this led him to develop a number of new ideas on the nature and importance of heredity.

33 Ernst Haeckel (1834-1919)

Thus the year 1866 witnessed the publication of two important works in genetics, independently conceived. One of these, which comprised two weighty tomes, attempted to introduce order into the multiplicity of phenomena without personal knowledge on the part of the author of any experiments in genetics. He incorporated that order into his own world view, but he did not formulate any strikingly new conclusions. The other, a modest essay, provided an answer to a specific, clearly formulated question by using the procedures of induction. The author of the first work uncritically adopted a profusion of unsubstantiated ideas, incorporating them into the structures of his *Weltanschauung*, while the other, constantly criticizing his own experiments and those of others with logical consistency, made discoveries about the inheritance of individual characters which are still valid today.

We may wonder why Haeckel and Mendel had never heard of one another or why they did not attempt to establish contact. There are probably several reasons. Haeckel's thought was far removed from experimental work in genetics. Important advances in this field were chiefly made by botanists with whose work Haeckel was probably not acquainted, for in the last decades of his life he became involved in political and religious strife. It is uncertain how much Mendel knew of Haeckel's work. He had been able to study Darwin's theories with care, but it is possible that he was unable to obtain the works of Haeckel, a great adversary of Catholicism.

These are the most important theses on heredity developed by Haeckel:

1. The inherent power to transmit certain characters, termed heredity (atavism), and the effected transmission, known as inheritance (hereditas), are general physiological functions of an organism that are intimately associated with reproduction and are indeed a partial manifestation of that process. The basic law of heredity is that every organic individual directly or indirectly reproduces a creature resembling itself. The causes of heredity are to be found in the direct transfer of material parts of the parental organism to the filial organism.
2. The material continuity between parents and children is secured by identical albuminous compounds of a specific constitution. It is difficult to understand how the egg and sperm can transmit the complicated morphological and physiological characters and features of the parental organism to the child. This can only be explained in terms of the complexity of the molecular structure and anatomical constitution of the cytoplasm, as yet unknown to us; and by the fact that individual development is a continuous chain of phenomena of molecular motion in the active plasm:

The impetus for this specific motion is conveyed together with the material substratum itself from the parental to the filial organism during the reproductive act; and the direct continuity of these infinitely diverse and complicated developmental movements is the effective cause of the infinitely diverse and complicated manifestations of inheritance (Vol. 2, pp. 174-175).

3. The degree of heredity can vary; that is to say that the measure of agreement in form and function between parent and offspring is determined by the ratio between the size of the parent and that of the superfluous product of growth which it releases. The degree of heredity is therefore greater in gemmation than in germ cell formation and sexual reproduction. Moreover, the longer the material connection holds between the parents and the child, the greater the degree of heredity. This is seen to be the case with grafted varieties of fruit: they transmit their characters in asexual reproduction, whereas in sexual reproduction they produce offspring that differ widely from their parents. Anomalous varieties—e.g., copper beech, or horse chestnut with double flowers—can be propagated only by asexual means, while it often happens that monstrosities are produced by sexual reproduction. These considerations lead to the following thesis:

Every manifestation of inheritance in organisms is determined by the material continuity between the parental and filial organism; the degree of inheritance [i.e., the degree of morphological and physiological resemblance between the parental and filial organism] is directly proportional to the time for which this continuous connection between parent and offspring holds, and in inverse ratio to the difference in size between them (Vol. 2, p. 176).

4. There are two main kinds of inheritance: The transmission of hereditary characters and that of acquired characters. The first is known as conservative inheritance, the second as progressive inheritance. The acknowledgment of conservative inheritance leads to the mistaken view that all the members of a species are held together by a certain number of immutable characters and represent a natural unit. Such a view imposes strict limits on variation. It is progressive inheritance alone which allows us to recognize the unlimited variation of organic forms and the free transmutation of species, and thus to explain all the facts of organic morphology.

According to the law of conservative inheritance, every organism transmits to its offspring the same morphological and physiological characters which it inherited from its parents and ancestors: that is to say, all the characters which an organism inherits from its parents, and only those, are transmitted to its offspring. It follows that each generation of a given species is, in all important respects, identical and so the species is constant. In their daily work, breeders and gardeners have occasion to disprove this view; in artificial breeding, conservative inheritance is not exclusively operative, progressive inheritance plays a part as well.

According to the law of progressive inheritance, every organism transmits not only those morphological and physiological characters which it has inherited but also those it has itself acquired by adaptation during the course of its life. Unlimited species variation follows from this law:

For the perpetual fluctuation between conservation and modification, constancy and transmutation, present in all species of animals and plants is simply explained by the fact that the inheritance of characters is never exclusively conservative, but always progressive as well (Vol. 2, p. 179).

5. In most organisms the generations which directly follow one another are almost identical or at least very similar in respect to all their morphological and physiological characters. Exceptions occur, however, when certain characteristics reappear after a lapse of several generations. It frequently happens that children resemble their grandparents more than their parents. It follows that there are latent characters which reappear at intervals during the course of generations, to be characterized as cases of reversion to type.

Unisexual organisms transmit primary and secondary sex characters unilaterally: i.e., with regard to the overall sum of secondary sex characters male offspring have a greater resemblance to their father, while female offspring have a greater resemblance to their mother. Nonsexual characters, however, are inherited in mixed proportions: that is, male offspring have a greater resemblance to their father with respect to most characters, especially the most important ones, though in others they resemble their mother to a greater extent; the reverse is true of female offspring. This law of blended inheritance is particularly important in hybridization:

The laws of hybridism, which cannot be clearly stated at the present time, will be largely derived from this law. . . . It is precisely this blending in the offspring of the characters from both sexes which is the principal cause of the infinite diversity of individual characters (Vol. 2, p. 184).

As individual members are omitted with the passing of time, the series of different characters successively transmitted in a specific order during ontogenesis is condensed. This means that the greater the speed with which embryos or larvae pass through their developmental stages, the more likely they are to survive in the struggle for existence.

6. "Under favorable circumstances, an organism can transmit to its offspring all the characters it acquires through adaptation during its own lifetime, even though these characters were not present in its ancestors." The probability for the transmission of such characters is determined by the number of generations which were influenced by the causal conditions of adaptation and by whether this influence is continuously sustained. The work of gardeners and farmers has repeatedly shown that this is true.

The specific changes in a bodily part which were not present in the ancestors but acquired by adaptation are transmitted in the same form to the same bodily part of the offspring. Similarly, the changes acquired by adaptation at a given time during the life of an organism can be transmitted at exactly that same stage in life to the offspring. The perplexing nature of these facts becomes apparent when we reflect that the egg, cell, and sperm are therefore responsible for identity in respect of place and time in the inheritance of such characters (homotopic and homochronous inheritance).

7. The continuous reciprocal action of the physiological functions of inheritance and adaptation is the cause of the infinite diversity in organisms:

> All the properties or characters of organisms are the product of the reciprocal action of two formative physiological functions: the internal formative power of inheritance which is determined by the material composition of the organism and conveyed by means of reproduction; and the external formative power of adaptation, which is determined by the reaction of the organism to the external world and conveyed by means of nutrition (Vol. 2, p. 224).

The hereditary characters are termed homologous characters or characters of inheritance, while the adapted characters are known as analogous characters or characters of adaptation. One of the principal tasks of morphology presupposes the acknowledgment of this distinction.

From our review of Haeckel's theses on heredity we can easily see that what they expound had previously been said, and sometimes repeated, in one way or another in the earlier history of genetics.[12] While Mendel was at pains to discover numerical ratios for the

[12]We must emphasize the fact that this criticism of Haeckel refers solely to his theses on heredity. There is no doubt whatever that he played an extremely important part in the development and popularization of Darwin's theory of natural selection.

inheritance of individual characters, Haeckel contented himself with data and explanatory remarks of a quite general nature. A confirmed disciple of Charles Darwin, he took up the search for the causes of variation in external and internal conditions, though unlike Darwin he did not maintain an attitude of constant doubt in a genuine endeavor to solve the problem. Haeckel spoke of the inheritance of acquired characters as an established fact equivalent to the inheritance of hereditary characters, though he did not produce any really convincing reasons for doing so. Indeed, the examples of the inheritance of acquired monstrous characters quoted by Haeckel, such as hexadactylism, nicely illustrate how he adopted and interpreted the unproved facts of earlier times.

Haeckel's unwavering conviction that acquired characters were inherited is also expressed in a letter to his good friend August Weismann, dated January 15, 1894 (Uschmann and Hassenstein 1965). He writes:

> In any case, I consider the problem of progressive inheritance to be one of the most important in the whole of biology. I am so firmly convinced of the "inheritance of acquired characters" that it seems to me that were we to proceed without it—and thus without the "phyletic adaptation" which it entails—the theory of descent would lose its value as a causal explanation.

However, it would be a mistake to judge Haeckel's importance in the history of genetics exclusively in terms of the theses we enumerate here. It was probably Haeckel who was the first to discover the role of the nucleus and cytoplasm in the cell and to define clearly their different functions. Thus in the first volume of his *General Morphology* he writes (p. 287):

> In the cells in which the nucleus operates as active material in addition to the cytoplasm, we must view the nucleus and cytoplasm jointly as the formative substance. It is true that the nucleus, since it is the primary source, is to be considered as the original differentiating factor of the plasm, with the reservation, however, that cytoplasm and nucleus subsequently become coordinated parts, as if they were different organs of equal status, and fulfill different functions.

According to Haeckel, since every organism is the product of hereditary characters and of characters acquired by adaptation, this law should be applied when considering the elementary particles of the cell: the nucleus attends to the transmission of hereditary characters while the cytoplasm takes care of the adaptation or accommodation to the conditions of the external world. Haeckel (1866) was therefore the first to declare that the cell nucleus acts as the material basis for heredity, a claim that was later made, after thoroughgoing investigations, by O. Hertwig and E. Strasburger in 1884.

In spite of the approval Haeckel lavished on Darwin's theory of natural selection, he opposed the provisional hypothesis of pangenesis, advocating instead his own hypothesis of the perigenesis of plastidules (1876), or the "wave production of vital particles." Though this hypothesis has long since fallen into oblivion, we must consider it here in order to show the muddled, extravagant ideas that were still possible at a time when exact research had already established the fundamentals of heredity.

In formulating his theory of perigenesis, Haeckel assumed that cells are physiologically and morphologically autonomous organisms which, through a division of labor, are differently developed but dependent on one another. Moreover, the protoplasm of the cell is considered to be the bearer of life and the active factor in cell vitality. At the lowest level of organic development, cell matter is not differentiated into protoplasm and nucleus, but forms homogeneous, albuminous particles of a mucilaginous texture, known as monera, each molecule of which can perform everything that is necessary for the functioning of life. This albuminous substance is known as vital or formative matter, plasson or bioplasson. Anucleate cells that consist only of plasson, like monera, are called cytods. Haeckel assumed that every higher organism also passed through such a stage at the beginning of its development. This stage occurs after fertilization when the egg cell and the sperm lose their nucleus thus producing an embryonic form which he called a "monerula." It is only later that the plasson of the monerula is differentiated into nucleus and protoplasm, giving rise to the first cell from the first cytod. In accordance with the fundamental biogenetic law, higher organisms therefore pass through the primitive moneron stage. "The cytod is the first and primitive form of the unit of life, the cell the second, more advanced form" (p. 31). Both are already characterized in the *General Morphology* as formative components or plastids whose functions are derived from the elementary chemical activity of the plasson. In accordance with Haeckel's monistic viewpoint, the molecules of the plasson, or plastidules, are equipped with a capacity for reproduction or for memory, which becomes active at every stage of development, especially when organisms reproduce. An individual organism is said to "reproduce" when it grows beyond a size which is characteristic of it as an individual. Every time a cell divides, the two daughter cells inherit the nature of their mother cell, for they are identical parts of it, and the molecular motion of the plastidules is essentially the same in both of them. "Inheritance is the transmission of plastidular motion, the propagation of the individual molecular motion of the plastidules from the mother to the daughter plastid" (p. 45). The different conditions under which the daughter cells continue to exist induce partial modifications in the original plastidular motion. Such adaptation means that the plastid acquires new characters. Thus the acquisition of new characters proceeds from cell generation to cell generation, and changes in plastidular motion therefore occur partly as a result of the inheritance of earlier plastidular movements and partly from the acquisition of new plastidular movements by means of adaptation. Changes in the plastidules depend on the rearrangement of their atoms. By way of analogy with Darwin's theory of natural selection, the plastidules which predominate when the plastids reproduce are those that best adapt themselves to the external conditions of existence by assimilating new foodstuffs so as to accomplish the rearrangement of their atoms in the most efficient way. The division of labor among the cells in the differentiated organism depends on that among the plastidules that in varying degrees undergo a process of separation or differentiation to produce different kinds of plasson. The difference between the sexes and, above all, sexual reproduction are understood as the concrescence or fusion of two plastids that have developed differently as a result of the extensive division of labor among their plastidules. After fertilization, an intimate blending of the plastidules takes place, and a perfect union between the different molecular movements of both plastids is established. This process of concrescence or fusion occurs after the decomposition of the nuclei. The "monerula" is produced and gives rise to the first cell through the formation of a new nucleus. The plastidular motion of this first cell is the product of both the different plastidular movements of the egg plastid and the sperm plastid. This explains why the child inherits characters from both parents.

As the author of the fundamental biogenetic law, Haeckel held that the effective cause of the biogenetic process resided in plastidular motion. In his hypothesis of perigenesis he considered plastidular motion as the branchings of a wave motion which, at the level of phylogeny, are responsible for differentiation in the organic world and are recapitulated, at the level of ontogeny, in the embryonic development of every living creature. Perigenesis is Haeckel's name for this ramified wave motion of the plastidules. Perigenesis depends on the reproductive power of the plastidules, synonymous with their capacity for memory, which is responsible for the propagation of the plastids: "The memory of the plastidules enables the plasson to transmit by inheritance through a continuous periodic motion its characteristic properties from generation to generation, and to add to them the new qualities which the plastidules acquire by adaptation during the course of development" (p. 68). He adds: "Heredity is the capacity for memory, variation the mental power inherent in the plastidules" (p. 69).

Darwin's gemmules were considered as aggregates of molecules. The plastidules of Haeckel's theory of perigenesis are single molecules that transmit their individual motion to the adjacent plastidules and form new plastidules of the same kind by a process of assimilation. Their atomic structure, and thus their motion, is expressed as a result of external influences. Darwin assumed that every cell deposits gemmules in each part of the body and that the reproductive cells contain gemmules that come from all the cells of the organism and its ancestors. Haeckel's theory of perigenesis, however, stems from the mechanical principle of the transmission of motion. Plasson is composed of plastidules. The molecular motion of the plastidules, which as the plastids reproduce is transmitted to those that are newly formed, is a ramified wave motion that is modified when the plastids are required to adapt under the pressure of external circumstances.

The concept of the inheritance of adaptations with which we became familiar in our study of Haeckel's "laws" of heredity is a recurrent feature in his hypothesis of perigenesis. We can be sure, however, that Haeckel's postulation and unwavering acceptance of the inheritance of acquired characters in no way helped to clarify objectively this important topic of heredity and development.

7 Nineteenth-Century Observations on Sudden Variations in Animals and Plants; the Mutation Theory of Hugo de Vries

We are indebted to Charles Darwin for an extremely rich and varied account of his own observations and of the well-attested observations of famous breeders and gardeners on the sudden appearance of new, true breeding forms. Darwin's greatest interest lay in questions concerning the causes and nature of such changes. Out of the wealth of material available, we shall confine ourselves to those variations whose hereditary nature is guaranteed, or at least made highly probable, by Darwin's own observations or by those of reliable contemporaries.[1]

Darwin (1868) is disposed to draw a conceptual distinction between a variety and a monstrosity, though he grants that a variety can sometimes change into a monstrosity. Generally speaking, he characterizes all the modifications within a given species which has similar counterparts in other genera and families as variations or varieties, while the modifications that lack such parallel formations he designated as monstrosities. Godron (1872) also reports on the sudden appearance of unfamiliar, abnormal varieties.

In his theory of heterogenesis, S. Korschinsky (1901) discussed the extent to which sudden changes should be viewed as monstrosities belonging to the study of teratology. He held that they are legitimately categorized in this way and quoted the well-known example of an anomaly in the common poppy. The anomaly lies in the fact that all the stamens in the flowers are converted into small ovaries supported on tall stems. There are, nonetheless, countless cases of sudden variations that exhibit scarcely any unusual characteristics, such as the change from spiny to glabrous fruits, from divided to simple leaves and so on. In such cases it would be impossible to decide which of the two varieties was the normal one and which the unnatural one. The only important thing to note is that many different kinds of modifications to typical structures may occur suddenly, some of them accompanied by a functional organic disturbance (monstrosities), while in other cases the vital functions of the organism are not impaired (varieties or races).

A third way of construing these new forms was to consider them as cases of atavism or monstrosities, to view them as reversions to

[1] With the exception of quotations taken from authors listed separately in the bibliography, all excerpts in this chapter derive mainly from Darwin (1868) and Korschinsky (1901).

phylogenetically older forms, or to characterize them as teratological forms. Whereas the concepts of species, race, and variety embody an idea of the range of the "leap" or bound which distinguishes the new form from its ancestral one, this notion is usually missing in the theory of reversions and monstrosities.

Darwin distinguished between several different kinds of reversion: reversion within a pure race to characters previously typical of that race which had subsequently been lost; reversion to characters that have long since disappeared following a cross effected many generations ago; reversions that appear as "bud variations" in somatic tissue; and—a subcase of the third kind—instances of reversion where individual segments of, say, a flower are affected by the change.

Darwin's study of sudden variations with unknown causes centered primarily on domesticated animals and cultivated plants. He held that these were much more likely to vary than were wild varieties and breeds (an idea that has been held ever since). On this topic Darwin writes as follows:

> Yet domesticated animals and plants can hardly have been exposed to greater changes in their conditions than have many natural species during the incessant geological, geographical, and climatal changes of the whole world. The former will, however, commonly have been exposed to more sudden changes and to less continuously uniform conditions. As man has domesticated so many animals and plants belonging to widely different classes, and as he certainly did not with prophetic instinct choose those species which would vary most, we may infer that all natural species, if subjected to analogous conditions, would, on an average, vary to the same degree.

Darwin's writings contain numerous references to the hereditary causes of different characteristics of various races and breeds, and he also sought to fix the date of their origin. He was interested in the frequency of such changes and also in their immediate causes. We cannot always tell whether the breeds of domesticated animals and varieties of cultivated plants he mentions are the result of a single variation or of a series of changes comprising several steps, but we can safely say that the occurrence of sudden changes constituted the beginnings of the formation of a new variety or breed.

Darwin remarks that different varieties of the domesticated dog, such as the hound and the greyhound, are mentioned in what are probably the earliest historical records, for they are found on Egyptian monuments dating from the fourth to the twelfth dynasties. The dachshund and lapdogs appear shortly after. These in turn are quickly followed by mastiffs, pug-dogs, and poodles, all of which materialize without passing through any transitional stages. Darwin comments explicitly on the way in which these races originated and on the abnormal nature of many of their characters. Anderson mentions a bitch with a defective leg giving birth to several pups with the same deficiency. With regard to cats, in which because of indiscriminate crossing it is more difficult to observe hereditary variations, Darwin's interest was captured by the tailless Manx cats, by a toothless cat from the West Indies, and by families of six-toed cats. While there are numerous references to the formation of monstrous varieties of dogs and cats, there are very few that pertain to the horse. Changes in the shape and number of bones and differences in dentition were the chief variations that were discovered. Waterson tells of a mare that produced three foals successively without tails. Horses with frizzled, negroid hair or with hair having variable stripes had been observed frequently. The strong tendency for different breeds of horses to be formed is evidence of a high degree of variation within a given species. Of the ass, Darwin mentions the existence of four different breeds in Syria. The variability of coloring and the transverse stripes on the legs are important. The formation of different races of pigs had already been more closely studied (H. von Nathusius 1860, Rütimeyer 1861). The *Sus scropha* group includes breeds that differ from one another, as do the breeds of the *S. indicus* group, in several characters pertaining to the shape of the skull and the form of the body. Solid-hoofed pigs had been known since the time of Aristotle; it is probable that this peculiarity appeared at various times in different countries (A. D. Buchanan Smith, O. J. Robinson, and D. M. Bryant 1936). A further hereditary anomaly, appendages attached to the corners of the jaw, is described by Eudes-Deslongchamps. Hallam tells of a race of pigs whose hind extremities were completely missing. This deficiency was transmitted through three generations.

Many different races of cattle had long since been identified. In all the *Bos indicus* groups, as well as in those of *B. taurus*, numerous differences in the shape of the body and of the ears, curvature of the horns, and coloring of the hide, and so forth had been observed; for example, in the Falkland Islands, herds of conspicuously different colors had been noted on hunting expeditions. And when hunting such cattle, as Darwin relates, a sharp watch was always kept for white spots on the distant hills in one district, and for dark spots in another. As with all domesticated breeds, it is clear that many domesticated breeds of cattle are the result of interbreeding. But there is no doubt whatever that other breeds were known to arise solely as the result of sudden variation, without any contribution from foreign species or breeds.

Roulin described the South American breed of "pelones" with extremely thin and fine hair and the "calongos" which were completely hairless. Castelnau described other Brazilian races of this kind, one of which had remarkable horns. Thanks to d'Azara (1801), we know of the Paraguayan breed of "chivos" with straight, vertical horns which are conical and very large at the base. In Corrientes the same author discovered a dwarf race with short legs and an unusually large body. Cattle without horns and other breeds with reversed hair had also been discovered in Paraguay. On the northern bank of the Plata River Darwin himself observed two small herds of another peculiar breed of cattle. Their forehead was very short and broad (*Primigenius* type) and the nasal end of the skull, together with the whole plane of the upper molar teeth, was curved upward. The lower jaw projected beyond the upper jaw and had a corresponding upward curvature. This breed was therefore related to other breeds of cattle in the same way as pug-dogs are to other breeds of dogs. The first published note on this race was given by d'Azara around 1790, but we know that such cattle were kept as curiosities near Buenos Aires around 1760. And, although the date of their origin is unknown, they must have originated after 1552 when cattle were introduced, supposedly, among the Indians south of the Plata.

As with cattle, the formation of different races among sheep is considerable. The fat-tailed breeds of the Orient are particularly distinguishable. An increase in the number of horns occurs quite frequently, though it also happens that in one or both sexes they are completely missing. A tendency for different breeds to adjust or adapt to various localities is very pronounced among sheep, and spontaneous generation of new races was frequently observed.

An example of this adaptation ability was the sudden appearance of the Mauchamp sheep (1828). Of the many new varieties of Merino sheep that had been observed, the Mauchamp had received the closest description. Although the first specimen of this breed was weak and had no horns, it did have long, smooth, and silky wool. Its owner crossed it with other animals, and the qualities of the wool turned out to be hereditary. Since this kind of wool was needed in manufacturing Indian scarves, the cultivation of the new breed with silky wool was expected to result in large sales. As the scarves became unfashionable, however, the cultivation of the Mauchamp breed declined.

Historically speaking, nothing was known about the sudden appearance of new breeds, or of defective varieties, among goats. But hereditary variations in the shape of the mammae and the horns, in hair coloring, and other characters were numerous. The great diversity among different breeds of rabbits is partially due to the very early domestication of that animal. It appears that species hybridization did not influence the development of the rabbit, and so the numerous varieties of this animal must have been derived solely from the natural evolution of different breeds and from interbreeding. Only one species of wild rabbit was known with certainty to exist in Europe, of which melanic and albino forms together with dwarf races had been known for a long time. The most important characters differentiating the various tame breeds—size, shape of the ears, body structure, color and composition of the coat—were probably developed gradually during the process of domestication, and then propagated on economic or sentimental grounds. In 1794 Anderson reported the sudden appearance of an unusual breed of rabbits. In a litter he found an animal having only a single ear; it gave rise to a new one-eared race. Hairless specimens of the common mouse were described by G. Gordon (1850).

Darwin gave a meticulous account of variation and development of races among pigeons, since the descent of all the domesticated European races from a single wild form could safely be assumed. He describes in detail the various races and subraces of pouters, carriers, runts, barbs, fantails, turbits, trumpeters, and tumblers, and gives notes on the dates of the origin of the different races. Pouters were already known about 1600; the first reference to the existence of fantails also dates back to 1600 when this race appeared in India; jacobins were known earlier; turbits had been described ever since 1677; tumblers were present in India before 1600; runts were probably already known at the time of Pliny; four subraces of barbs existed about 1600; carriers appeared at the close of the seventeenth century. Little is known about the dates of the appearance of certain sports, because breeders tended to keep valuable varieties a secret and to discard defective varieties that were unsuitable for breeding. From the tactics adopted by pigeon breeders we learn that a chain of several cumulative variations can sometimes lead to the formation of a new race. Similar remarks apply to fowl. We shall not enumerate the different races of fowl known at the time of Darwin, since the origin of the domesticated hen is a controversial issue, and, moreover, crosses between different varieties may have contributed much to the development of that animal. Darwin stresses that we must be careful not to overestimate the effects of crossing and that we must consider the probability of the occasional hatching during the course of centuries, of birds with abnormal and hereditary peculiarities, such as the possession of supernumerary toes.

In his chapter on the formation of races in ducks Darwin notes that the eggs may vary in color. He records the case, observed by Hansell, of a common duck which always laid eggs whose yolk had a dark brown color, like melted glue. Appearing suddenly in a particular breed, this character trait was also seen to be hereditary: the ducks hatched from such eggs also laid eggs with a dark brown yolk, which meant that the breed had to be abandoned.

Darwin describes in detail the frequent appearance in England of so-called "japanned" or "black-shouldered" peacocks. They always appeared unexpectedly in flocks of common peacocks and bred true. Darwin records seven well-authenticated cases of the sudden appearance of black-shouldered peacocks during his lifetime. It seems, however, that this character had also occurred in earlier times, for it can be seen in ancient paintings. Birds of this type differ from the common peacock in the coloring of their secondary wing feathers, scapulars, wing coverts, and thighs. They are smaller than the common variety, and the females are paler in color.

Darwin described a number of characters that appeared spontaneously in canaries, including the top knot, feathered feet, and black wing and tail feathers.

With regard to goldfish, Darwin mentions varieties of different colors with modifications to the structure of the fins and a triple tail. A further variety had a hump on its back near its head. L. Jenyns described a variety in which the fleshy part of the tail was almost cut away; the caudal fin was situated a little behind the dorsal and directly above the anal fin. The anal and caudal fins were double; the anal fin was vertically attached to the body; the eyes were large and protuberant.

Very few new varieties of bees were known at the time of Darwin. Woodbury stated that he had often seen variations in the color of queens and drones. Lowe discovered a form that differed from the common bee in having a far greater number of lightly colored hairs on the head and thorax.

By way of contrast with the bee, a high degree of variation in the silk moth, *Bombyx mori*, had been observed throughout the ages. Black caterpillars sometimes appear among those of the usual kind. The silk of the cocoons can be white or yellow and the degree of fineness can vary considerably; but these characters are not always inherited. This is also true of the reduction of the wings in imagoes.

Like domesticated animals, human beings were also subject on occasion to sudden variations whose hereditary nature could be observed for generations. Thus Darwin reports that of the six children and two grandsons of Lambert, the remarkable "porcupine man" whose skin was thickly covered with callouses, all showed similar changes. In a Siamese family the face and body were covered with long hair and dentition was defective over a period of three generations. A woman with a completely hairy face was exhibited in London in 1663. According to a report by Hodgkin, a white lock of hair was inherited in an English family for several generations. The different forms of polydactylism were frequently described. Wilder gave a tabular summary of a number of cases; he held that supernumerary fingers are more common than supernumerary toes and that men exhibit such abnormalities more frequently than women. In every case, the hereditary nature of these deformities was well authenticated. The transmission by inheritance of other sudden changes, like deficiency of the phalanges, thickened wrists, crooked fingers, and so on was also observed. Thus Ogle described a case of defective phalanges over a period of four generations. Sproule confirmed the transmission, during a century, of harelip accompanied by cleft palate in his own family. The hereditary nature of a disposition within a certain family to contract specific diseases was also noted. Holland mentions one such case of diabetes. In Darwin's lifetime it was already known that albinism and an associated constitutional frailty ran in certain families.

That hereditary abnormalities in man were also observed by the peoples of ancient civilizations is seen from a remark made by W. H. Prescott (1844) in his *History of the Conquest of Mexico.* He writes:

> I must not omit to notice a strange collection of human monsters, dwarfs, and other unfortunate persons, in whose organization Nature had capriciously deviated from her regular laws. Such hideous anomalies were regarded by the Aztecs as a suitable appendage of state. It is even said, they were in some cases the result of artificial means, employed by unnatural parents desirous to secure a provision for their offspring by thus qualifying them for a place in the royal museum.

Darwin has little to say about the origin of new forms among feral animals. In 1828, Hofacker reported on a one-horned stag that was said to have been seen in a forest in 1781. In 1788 two stags with this abnormality were discovered in the same forest, and several of them were subsequently observed. Godron (1872) gave a detailed treatment of the distribution of color variations, albinism, melanism, and erythrism. He held that they were primarily due to environmental influences like nutrition, soil, and climate and did not consider the transmission by inheritance of these variations. General remarks on the origin and nature of new characters in wild plants may also be found in the same work.

C. E. Keeler and S. Fuji (1937) report on ancient variations in mice; some of them were known 200 years ago in China, and details of them were portrayed in objets d'art.

There were countless observations made in the nineteenth century on new color variations in birds; some of the more notable were the white swallow; albino varieties of *Hirundo rustica*; the yellow hammer, *Emberiza citrinella*; the blackbird, and pheasants, a white twite, a cream-colored rook, and a color variation in blackbirds in which the male was silvery white without any spots, and the female was a dirty white with brown spots.

New forms in the vegetable kingdom were often treated with skepticism by Darwin. In his opinion, the botanists' general neglect of cultivated plants had delayed knowledge in this field. Since we have only very vague descriptions of the origin of many cultivated plants, it is impossible to decide whether a new form is to be regarded as the product of a species crossing, or as a variety or monstrosity suddenly produced for unknown reasons.

Justified though these doubts may be, a large number of hereditary variations in cultivated plants—of which Darwin mentioned several—were already known in the nineteenth century. With regard to variability in cereals he pays more attention to the parallel variations which, for example, are seen in every species of wheat: the seed is pubescent or glabrous, the flowers aristate or nonaristate, the colors of the testae vary in a uniform way, and so on. Many varieties of wheat were already known at this time: Dalbret had 150 to 600, Le Couteur 150, and Phillipar 322 varieties. Already at that time many of these new forms were found as isolated specimens in pure crops and were then propagated. Many of these sports were developed into new varieties by the great cultivators of the nineteenth century.

Other groups of plants had already been thoroughly investigated as regards their variability. One of the oldest accounts is a book by A. Munting (1672) *Waare Oeffeninge der Planten*, in which numerous varieties of trees, shrubs, flowers, and herbaceous plants are mentioned. A comprehensive description of all the varieties of conifers known by 1867 was summarized in a work of that year by E. A. Carrière. Variations in cucurbitaceous plants were described in detail by Naudin (1856), who emphasized the extensive variability of the flowers and of the shape and number of the carpels.

In 1864 B. Verlot gave an account of the production and conservation of varieties of ornamental plants. In this work he gives a valuable tabulated survey of all the possible variations in plants, but he displays a tendency not to distinguish between hereditary and nonhereditary variations.

Darwin points out the high degree of constancy of the flowering organs in different species of cabbage and contrasts it with the numerous variations in the shape and color of the leaves and in growth. In his catalogue of 1851, Vilmorin described ten varieties that were considered to be merely ornamental plants. In his own garden Darwin observed forty-one varieties of English and French peas which differed from one another mainly in height, maturation time, and leaf size. The pods and seeds also displayed significant differences in size, shape, and color.

Loudon described at least twenty-nine varieties of the hawthorn which were based on differences in the shape and color of the blossoms, in the shape and color of the leaf, and in the shape and firmness of the berries.

The number of varieties found in fruit trees knows no bounds. The London Horticultural Society listed 80 varieties of cherries in 1842. In the catalogue for the same year, 897 varieties of apple were named, the number of pears known at that time was no smaller, and 149 varieties of gooseberry were mentioned. The various forms differed in the manner of their growth, in the time they took to produce foliage and flowers, in the size, color, and lobation of their leaves, in the composition of the outer fruit skins, in the size and color of their fruits, and in the armature of their branches.

A. Gautier (1886, 1901) observed numerous races of the genus *Vitis*, and drew far-reaching conclusions about the origin of these forms on the basis of numerous chemical analyses. He was of the opinion that variation and speciation depend on the chemical structure of the plasm which could be altered by hybridization, grafting, or parasitic influences.

The number of observations on sudden new variations in crop and ornamental trees was already extremely large at the time of Darwin. A catalogue by Lawson in Edinburgh contained, for example, twenty-one varieties of the common ash (*Fraxinus excelsior*).

The weeping or pendulous growth was known in willows, ashes, elms, oaks, in the marshmallow, and in other trees (Korschinsky 1901). Various forms of this type of growth were found throughout Europe in the eighteenth and nineteenth centuries, but nothing is known about the heredity of many of these forms. Carrière (1859) relates, however, that thirty-two of the offspring of one generation of the weeping peach, *Persica vulgaris* var. *pendula*, had inherited the pendulous manner of growth. The seedlings of the weeping oak also bred true. According to MacNab, seedlings of the weeping birch, *Betula alba pendula*, in the Botanical Gardens in Edinburgh grew straight up for fifteen years but then developed pendulous branches. The transmission of this character was also observed in the weeping ash and in the weeping variety of arbor vitae.

A further abnormal form of growth, pyramidal growth, was discovered in the blackthorn, juniper, and ash. Trees of this kind have relatively short branches, symmetrically distributed, thereby giving the crown a narrow, attenuate shape. Carrière tells of a columnar variety of *Cupressus fastigiata*. It was given the name of var. *cereiformis* and was obtained from seeds of the common cypress by Ferrand in 1838. This character was strictly inherited. Another form which is also constant from seed, *Quercus pedunculata* var. *fastigiata*, occasionally appeared in the Pyrenees and southern France.

By way of contrast, *Robinia pseudacacia* var. *pyramidalis*, established in France in 1833, was seen to have seeds that produced variable offspring. In such cases, however, we must remember that complete isolation of the flowers was not always achieved. It also seems that the character we are considering was not strictly inherited in *Taxus baccata* var. *fastigiata*—a consequence of the fact that the first specimen was wholly pistillate; the seeds were invariably produced from normal pollen. Among the pyramidal varieties found growing in the natural state are *Abies pectinata pyramidalis*; the hornbeam, *Carpinus betulus* var. *pyramidalis*, which originated near Kassel; and the pyramidal oak, *Quercus pyramidalis*, which differs from *Quercus fastigiata*. It is probable that all the pyramidal oaks of northern and central Germany are descended from this latter tree which was discovered in a forest near Babenhausen on the Württemberg border.

Still another peculiar trait of some members of the vegetable kingdom is nanism (dwarfishness), seemingly a hereditary characteristic. In 1825 Billiard discovered *Lonicera tatarica nana* in Fontenay-aux-Roses. Carrière (1865) reports that the *nana* variety may often be found among typical specimens of *Cedrus libani*. According to the same author, *Biota orientalis nana* was also constant from seed.

Korschinsky stresses that there is no doubt that cases of the horticultural phenomena known as nanism and giantism, those that were hereditary at least, arose suddenly and sporadically among perfectly normal siblings and then remained constant. The varieties described as *nana, pumila,* or *compacta* are thought to have originated in this way; for example, *Ageratum coeruleum* var. *nanum, Scabiosa atropurpurea* var. *nana, Coreopsis tinctoria* var. *pumila, Tagetes patula* var. *nana*, dwarf varieties of the balsamine, of the aster, or the cineraria, *Tropaeolum*, and so on.

Verlot (1864) reports on two similar cases discovered by Vilmorin. In a bed of *Tagetes signata* cultivated by Vilmorin a particularly robust and bushy specimen appeared in 1860; this form was inherited by the offspring of that plant. An equally vigorous specimen was also discovered among normal specimens of *Saponaria calabrica* about the same time; most of the offspring of the next generation were of the same character. Verlot discovered a variety of *Lablab vulgaris* (in which the tendrils can be up to three meters long) which grew to be only eighty centimeters tall. The same author holds that nanism is generally more common among annuals than among perennials. He points out that such new varieties can in turn give rise to further new varieties and mentions the following examples: *Scabiosa atropurpurea* var. *nana purpurea*; a *pumila* variety of *Calliopsis tinctoria* which produced *Calliopsis tinctoria pumila purpurea*; *Tagetes patula nana* gave rise to a new form with yellow flowers, and so on.

The appearance of a fasciated stem also counts as a change in growth form. According to Korschinsky, in *Sambucus nigra* this phenomenon led to the formation of two new varieties, *S. nigra monstrosa* and *S. nigra monstrosa compacta.*

In 1811 a variety of strawberry having no stolons, named "Fraisier de Gaillon" or "Fraisier des Alpes," appeared in Normandy. The first specimen of this new variety had appeared among numerous plants of the usual kind in a crop of ever-flowering wood strawberry (*Fragaria semperflorens*). Raised from seed, the new form was constant. Its history was described by P. L. Lévêque de Vilmorin (1886). A similar variety had been discovered in the outskirts of Laval in 1748, and probably other, similar forms appeared at different times in different places.

A large number of hereditary variations in the color of leaves was also found. There were descriptions of dark red leaves in the hazelnut, the barberry, the ash, the oak, the maple, and the beech. *Fagus silvatica* var. *purpurea* Aiton is particularly well known, for it is often found in public parks. This form was first discovered in the forests of Thuringia in the middle of the eighteenth century. According to Loudon, seedlings of the copper beech usually have the appearance of their female parent, though seedlings of an intermediate, greenish character and some that are completely green in color also occur. Seeds of the copper beech were often planted under experimental conditions, for example, by Cappe in Périers in 1840; all the seedlings produced were red. According to Pépin (1853), seeds from these latter plants gave rise to more seedlings, of which two-thirds were also red. In subsequent experiments the seeds did not always produce the copper-colored beech, though this was probably due to cross-pollination. J. Jäggi (1893) emphasizes that the copper beech found in the forests of Thuringia (Germany) was not the sole original specimen of its type. Two old copper beeches discovered in Buch am Irchel in the canton of Zürich (Switzerland) are particularly well known. According to Hausmann (*Flora von Tyrol*), a similar form is also found in the mountains near Rovereto (Italy).

Bertin obtained *Berberis vulgaris* var. *atropurpurea* from seeds of the common barberry in Versailles. The seeds, obtained in 1839, were all of a deep red color. Korschinsky claims that Carrière's *Prunus Pissardi* is actually a red-leaved variety of *Prunus cerasifera* which was imported from Persia in the late 1870s and spread throughout Europe. A variety having red leaves with green blotches subsequently appeared in Germany. A form very similar to *Prunus Pissardi* was discovered in Späth's plantation. A dark red variety of the peach first appeared in America in about 1871, and later also in France. A further example of a sudden, isolated variation in color among a large number of normal seedlings is an elm found by Angebault in Fossay in 1828, which had a yellow bark and variegated leaves.

There were relatively few observations of variegated leaves in trees. Pépin described a specimen of *Sophora japonica fol. variegatis* in Versailles in which most of the seedlings had the appearance of their female parent. He also mentioned a specimen of *Celtis australis* with variegated leaves which was constant when raised from seed. According to Beissner, *Acer pseudoplatanus fol. variegatis* was also grown from seed. In a large number of cases, however, the transmission by inheritance of variegation of the leaves was found erratic. This prompted the writers of the nineteenth century to point out that the mode of inheritance of such variations was probably different from that which operated in varieties which—during individual development at least—remained constant.

We must briefly note some of the variegated varieties of horticultural and ornamental plants. Scarlet dahlias are said to have been produced from a single specimen of that type which originated in Ghent and was discovered by Louis van Houtte. Leaves streaked with white were also known (de Vries 1901). Verlot reports that crimson leaves were known in *Ocimum basilicum*, *Oxalis corniculata*, and in *Atriplex hortensis*; raised from seed, these forms were completely constant. He also describes a number of constant forms with variegated leaves. Other variegated varieties, however, did not turn out to be constant; such varieties are known in *Plantago*, *Humulus*, *Helianthus*, and in celery and cabbage. Like the flowers whose leaves were streaked with white, they are probably largely due to factors that were unknown at that time. We must also remember that variegated forms occur independently of genetic control as the result of virus action (for example, *Aucuba japonica*) and that the variegated character is nonetheless transmitted from generation to generation.

In trees other variations of a morphological and physiological nature were frequently observed. Thomas Meehan (1884) discovered a seedling of *Halesia tetraptera* which had broad, oval leaves like those of an apple tree, dark green in color, with a rough surface and prominent veins. The flowers also differed from the common kind and were sterile for a few years. The seeds, formed later than was customarily the case, were small though they had relatively large wings. Meehan concludes his observations with the comment that new characters appear suddenly and are usually inherited. Darwin mentions that Bose described three varieties of the elm: the broad-leaved, the lime-leaved, and the twisted elm. Darwin's own observations led him to distinguish two subvarieties of the ash with simple instead of pinnated leaves, forms which were generally constant from seed.

Korschinsky remarks on *Robinia pseudacacia* var. *monophylla* (or *unifoliata*) in which the pinnules are reduced, except for the terminal leaflet which can become greatly enlarged. Further characters pertaining to armature, to the inflorescences, and to the color of the leaves were also affected. The form in question suddenly appeared among numerous seedlings of the usual type in a plot owned by the horticulturist Deniau, in Brain-sur-l'Authion in 1855. In the following generation, about only one-quarter of the seeds produced *monophylla*. A similar case is that of *Fraxinus excelsior* var. *monophylla* which has undivided, lanceolate leaves with serrated edges. Although its origin is unknown, it seems highly probable that it originated in England in the late eighteenth century, and Korschinsky is convinced that it was produced by heterogenesis. But experiments designed to investigate the transmission by inheritance of the monophyllous character of this form showed conflicting results: the new variety was by some described as being absolutely constant, while it was said by others that it reverted back to an earlier type.

A simple-leaved variety of the walnut, *Juglans regia* var. *monophylla*, was also known, although nothing is known regarding its origin. It is first mentioned in *Arboretum Muscaviense* in 1864. Carrière (1865) describes a slightly different variety which he calls var. *mono-heterophylla* because it has simple leaves of different shapes. The first specimen of this type was discovered growing alone in a woodland marsh and probably arose spontaneously from seeds of the common kind.

Zabel discovered *Acer platanoides* var. *integrilobium* around 1870; it differed from the usual maple in having entire, undivided lobes. In 1820 among plants of the tree nursery owned by Fennessy and Son, *Quercus pedunculata* var. *Fennessii* Hort. was found; it had narrow attenuate leaves with few lobes, short in length. Similar forms possessing, in part at least, entire leaves seem to have been common in England. In 1879 Magnus discovered near Berlin a specimen of the common oak which had undivided leaves.

But the opposite step—from simple and entire leaves to laciniate leaves whose divisions were more deeply incised and more numerous—is more common than the move from the complex to the simple. There are, however, relatively few statements about the origin of these laciniate forms. In 1858 a *filicifolia* variation appeared in a common oak, *Quercus pedunculata*, in the nursery of forest trees at Muskau. The new trees had been produced from seed of the trees which grew in the vicinity of Muskau where such a form had never before been observed. Another variety, *Quercus pedunculata Doumeti*, appeared in France. It has narrow, lanceolate lobes on the leaves. The elm variety *Ulmus pedunculata* var. *urticaefolia*, described by Jacques in 1830, provides a further example of this kind. The name refers to the leaves which are shaped like a nettle. *Ulmus glutinosa* produced numerous laciniate varieties. It appears that individual trees of this type were found growing wild in Normandy. A similar form, named var. *imperialis*, appeared in 1855 and was put on the market in 1858 by the Desfossé-Thullier company. W. Hofmeister (1868) reports that lacerated leaves appeared on certain branches in specimens of *Carpinus betulus* growing in Heidelberg and in the Botanical Gardens in Leipzig. Horticulturalists from France to Scandinavia reported cases of plants with laciniate leaves. Among the discoveries there was variety to the degree of laceration, and scholars' opinions differed as to whether the characteristic of serrated edges was hereditary; one writer held that 50 percent of the offspring had divided leaves.

The appearance of plants lacking spines was frequently observed. In 1823 Caumzet discovered a specimen of *Gleditschia sinensis* which lacked spines among numerous seedlings of the usual kind; *G. triacanthos* L. also produced a variety of this type. Varieties of *Robinia pseudacacia* which were destitute of spines are common: Utterhart (Ysabeau 1859), for example, discovered the *Utterharti* variation in 1833. The *umbraculifera* variation which lacks spines but possesses a globose corolla is likewise said to have originated from seeds of the common *Robinia.* Several specimens of the variety *inermis Rehderi* were discovered in a bed of seedlings at Muskau. Nothing is known about the inheritability of spinelessness in these varieties.

In 1847 Trochu discovered six spineless specimens of *Ulex europaeus* among a large number of seedlings of the normal type. Since this plant was agriculturally very important in northwestern France, the new variety could have become extremely valuable for purposes of cultivation. Trochu and Vilmorin tried without success to propagate the new form; Verlot (1864) and Naudin (1861) attributed this failure to imperfect isolation.

In 1860 Billiard obtained from seed the first gooseberry that was bereft of prickles. Propagated by grafting, it was placed on the market under the name of "groseillier Billiard." From 1884 onward E. Lefort sowed seeds of this type at Meaux and obtained a number of new varieties, all of which lacked prickles. Carriere (1892) described the four best varieties.

On the other hand, in 1861 he had described a variety of *Crataegus monogyna* which was named var. *horrida* because of its spines; they are much more prickly than those of the common type. In the new variety most of the spines were fasciculate, but nothing is known regarding the heredity of this character.

A large number of hereditary variations in cryptogams was observed during the course of the nineteenth century. Thomas Moore wrote a comprehensive monograph (1859-1860) on the species and varieties of ferns found in Britain. He noted considerable variation in the following genera: *Athyrium*, *Asplenium*, *Blechnum*, *Polypodium*, *Lastrea*, and *Polystichum* and described as many as 155 varieties of *Scolopendrium vulgare* growing in England. In this species the leaves are known to display countless variations. The margin of the leaf can be undulated, dentate, lobed, or incised, while the blade of the leaf can be rounded at the top or bilobed. Narrow pinnatipartite leaves and the formation of round, bud-like leaves on the leaf base were frequently observed. In England most varieties were found growing in the natural state, though some were obtained by sowing the spores of wild plants. Characters are not always strictly inherited, though Bridgman holds that they are invariably transmitted when spores are taken from a portion of the leaf that has changed. In several of the varieties it is only a part of the leaf that changes; this is what causes reversion to the original stock. A further and equally common type of variation in ferns is found in variegation of the leaves. Variegation—doubtlessly inherited—is a characteristic of the East Indian fern, *Pteris quandriaurita* Retz. var. *argyraea*, and *P. aspericaulis* Wal. var. *tricolor*.

Remarkable physiological changes are manifest in plants which vary from the parent types in respect to the length of the flowering period and number of the flowers. In 1862 a single specimen of the *semperflorens* variation was discovered among 1000 other seedlings of *Robinia pseudacacia* in France. In the first flowering period of the fourth year of its life the peculiarity of the new variety became evident. It was put on the market in 1875. In the cherry a similar variety, *Prunus acida* Koch var. *semperflorens*, was also identified, though nothing is known with certainty about its origin. Further, there exists an ever-flowering variety of the strawberry, *Fragaria semperflorens*, which was cultivated after having been found growing wild in the Alps. In the Jardin d'Acclimatisation the *semperflorens* variation appeared among numerous other specimens of *Sambucus nigra*. A *Carlierie* variety of *Cytisus nigricans* exists; after the fall of the blossom its racemes form a new shoot, complete with leaves, which bears flowers for a second time. This variety has been cultivated since about 1850; when grown from seed it is constant.

We must also mention a variety of *Gloxinia*, found in France in 1897, which begins to flower for a second time directly after the end of the first flowering period. The new form is similar to a tall variety of lily of the valley, *Convallaria majalis* var. *prolificans*, which always develops a second inflorescence but has atrophied ovaries.

Hereditary differences in the shedding of leaves were also recorded. Whathely grafted an early-shedding blackthorn onto a late-one and conversely; both grafts kept to their proper periods. Darwin knew of a Cornish variety of the elm which was almost an evergreen. The varieties of the Turkish oak (*Quercus cerris*) may be classified as deciduous, semievergreen, and evergreen.

Hereditary variations in the shape and color of flowers were found mainly in annuals, less frequently in perennials.

Sabine reported in 1793 on the sudden appearance of new varieties of roses. From wild Scotch roses (*R. spinosissima*) he obtained plants with semimonstrous, semidouble flowers. Continued selection yielded twenty-six well-marked varieties, and as many as three hundred different varieties were counted near Glasgow in 1841.

Darwin mentions that G. Mackenzie (1845) described a Peruvian variety of potato that always produced two kinds of flowers: the first double and sterile, the second single and fertile.

Darwin made a comprehensive study of the origin of new variations in the flowers of horticultural and ornamental plants. He reports on the different forms of doubling, calycanthemy in primulas, the conversion of stamens into pistils in the poppy, the formation of peloric forms in *Gloxinia spinosa* and *Antirrhinum majus*, and so forth. Verlot (1864) also mentioned several of these variations. Darwin considered the case of *Begonia frigida*, brought to his attention by Professor W. H. Harvey, to be of singular importance. This plant usually forms male and female flowers on the same inflorescences. In the female flowers the perianth is superior. In England, however, a plant appeared whose flowers became hermaphrodite and the perianth in each was inferior. Darwin mentions that Harvey had said that any botanist "would probably have considered (this plant) as the type of a new natural order."

Darwin did not categorize the new form as a monstrosity on the ground that analogous structures occur naturally in *Saxifraga* and in the Aristolochiaceae. Seedlings obtained from normal flowers of the new variety produced plants that had mostly hermaphrodite flowers with inferior perianths like the parent plant. Fertilized with their own pollen, the hermaphrodite flowers were sterile.

Darwin makes the following quite general remark on changes in flowers: "Anyone who will habitually examine highly cultivated flowers in gardens and greenhouses will observe numerous deviations in structure; but most of these must be ranked as mere monstrosities, and are only so far interesting as showing how plastic the organisation becomes under high cultivation."

We are indebted to Vilmorin (1894) for a good example of the sudden appearance of a new variety of *Salpiglossis sinuata*. It differs from the typical form in that the corolla is wholly absent. This character was constant, and the new variety was named *S. sinuata* var. *corolla nulla*. Körnicke tells of a peculiar specimen of *Hyoscyamus niger* which has a fasciated stem with abnormally developed flowers. Calyx, corolla parts, and stamens were greatly enlarged. The corolla was unilaterally divided, the ovary was multilocular, and the fruit cap had dentate edges; all characters were hereditary.

According to de Vries (1901), G. Vrolik was the first to describe *Lilium candidum plenum*. Using the methods of vegetative propagation, this plant retained its characters for generations.

Korschinsky, too, studied variations in the structure of flowers in detail. He emphasized that flower doubling depends without exception on an unexpected departure from the normal form, which usually first occurred in a single individual. The reappearance of flower doubling within a given species hardly ever took place. In 1843 Vilmorin observed the first double-flowered specimen of *Ipomoea purpurea*, a plant that had been cultivated since the seventeenth century. In 1895 the second double plant, white in color, was found in the same nursery. A specimen with double flowers also appeared among seedlings of the common kind in France in 1883. About 1850 L. van Houtte discovered a double specimen of *Cyclamen persicum* at Ghent. In 1875 a semidouble form of the same species was found in Warsaw; the offspring of this plant were wholly double. This form was nonetheless fertile, and crossing gave rise to several double varieties that were put on the market in 1880. In the commercial nursery of Haage and Schmidt at Erfurt a double form of *Sanvitalia procumbens* appeared among normal seedlings in 1864. With few exceptions it bred true right from the time of its discovery and acquired appreciable market value in view of certain other characters it possessed.

Fertility in double varieties is not always entirely reduced. It could therefore be shown that, when raised from seed, the poppy, balsamines, carnations, dahlias, chrysanthemums, double-flowered peaches, the double variety of the apple (*Malus spectabilis*) and of the hawthorn retained the character of having double flowers. But other forms, the double variety of the sloe for example, produced plants having single flowers exclusively. Numerous double varieties of lilac and petunia were produced by crossing the original variety that was still fertile in one sex with single varieties. Further double varieties, like Lambotte's petunias (letter to Carrière), fuchsias, *Primula*, and *Matthiola* were developed by continued selection from semidouble forms. It was also observed that double stocks did not breed true.

Double flowers were frequently discovered in wild plants as well. Korschinsky mentions that a specimen of *Anemone alpina* was found in the Vosges Mountains and that double specimens of *Convolvulus arvensis* grew near Toulon and Toulouse. A double variety of *Linaria vulgaris* appeared in England. Near Astrakhan in Russia Korschinsky observed a number of double specimens of *Ranunculus repens.* According to Komarov (1892), there exists in the valley of the Jagnoba a large number of double *Rosa lutea.* Double flowers were also found in *Rubus, Ranunculus, Cardamine,* and *Lychnis.* In 1864 Seemann compiled a list of all the plants known to have double flowers. In the Compositae a different type of double flower is known whereby the disk florets are converted into ligulate florets. De Vries (1901) described in detail the specimen of *Chrysanthemum segetum plenum* which appeared among his cultures. J. Sabine (1824) may also be consulted for further descriptions of new varieties of *Chrysanthemum.* The converse variation, in which all the flowers become greatly elongated or tubular in shape is said to occur less frequently, though it appears, for example, in asters and in the daisy. A third kind of doubling occurs when the calyx is converted into a corolla, resulting in two corollas, one encased within the other. Varieties of this type are known in *Primula elatior, Mimulus luteus, Campanula medium,* and *Gloxinia speciosa.* A double variety of *Gloxinia speciosa* was put on the market in 1873 (Veitch Company). Ten years later the same variety originated independently in France. It seems probable that the *calycanthema* variation appeared in the harebell, *Campanula medium,* in England around 1870. In 1874 it was already being marketed as a constant variety. Drawing on a statement made by R. W. Burbidge, I. Lynch (1900) also notes the existence of this variety. A form of *Campanula persicifolia albiflora* having a double corolla has also been identified. In 1897 Regel obtained a similar variety among numerous individuals of the common kind in St. Petersburg.

K. Goebel (1886) gave a detailed account of doubling in flowers from an evolutionary standpoint. He considered, in turn, the different forms of doubling as they occur in various families of plants.

De Vries (1901) reports that a number of plants became sterile when the flowers turned green: this phenomenon was encountered in dahlias, roses, *Dianthus barbatus, Pelargonium zonale,* and in *Dahlia variabilis viridiflora,* a variety which originated in Holland around 1850. A further type of sterility was termed petalomania. The sterile varieties of sugarcane, banana, and the spruce originated suddenly without passing through any transitional stages. The sterile varieties of rye (Rimpau 1899) and *Nigella syncarpa,* described by A. Ernst (1901), probably belong to the same category.

Laciniate petals constitute another fairly well-known variation in flowers. The first variety of petunia with this character probably appeared in the United States (Rochester, N.Y.). The *fimbriata* variation of *Rosa rugosa* originated from a cross between two roses of the common variety in France. In 1895 an abnormal form of *Begonia erecta cristata* was also found in France; it possessed excrescent structures on the upper surface of the petals. There exists a similar variegy of *Cyclamen latifolium* which has a number of feathery, ramose outgrowths on the petals. This form originated in England in 1893, though it probably also appeared on several other occasions.

The *fistulosa* varieties observed by de Vries (1901) constitute a similar kind of change. He found them in dahlias and in other Compositae, as for example in *Chrysanthemum segetum*, *Coreopsis tinctoria*, and *Dahlia variabilis*.

W. T. Thiselton Dyer (1897) described a number of spontaneous variations in *Cyclamen latifolium*, including a form with diageotropic petals, another in which the petals were laciniate, and a third having pilose structures set within the flowers. The first two varieties arose on several occasions.

We are indebted to J. Murr (1896) for his account of the appearance of nonstellate flowers in the Compositae.

De Vries made a thorough study of variations in the flowers of the Compositae. According to Moquin-Tandon, the *Discoidea* varieties are peloric forms of the Compositae. The terms *Discoidea* and *Radiata* are used interchangeably. These varieties were observed to breed true without exception in *Senecio Jacobaea*. De Vries reports that *Matricaria Chamomilla discoidea* was similarly constant.

Pelorism in different species was mentioned long before Darwin and was, in fact, a fairly well-known phenomenon. Besides the peloric form of *Corydalis*, that of *Linaria vulgaris*, found by Zioberg near Uppsala in Sweden in 1742 and described by Rudberg (1744) in Linnaeus's *Amoenitates academicae*, is most frequently named. J. H. Wakker (1891) reports on the peloric forms of *Linaria vulgaris* and *Cytisus laburnum*. In their study on variations in floral symmetry, W. and A. Bateson (1891) described peloric forms of *Linaria*, *Veronica*, and *Streptocarpus*. I. Lynch (1900) also mentions a peloric form of *Streptocarpus Rexii*. The brief historical study by M. J. Sirks (1915) gives the dates of the discovery and description of the first peloric *Linaria* and of some other peloric forms. From the time of its first discovery the peloric variant has been found wherever *Linaria vulgaris* was growing in large quantities. In some cases the entire plant was peloric, while in others only individual flowers were affected. We cannot say with any certainty whether the peloric varieties are constant from seed. An unusual modification to the peloric form having five spurs is seen in the unspurred, tubular flower of *Linaria vulgaris* found in 1857 in France. According to Verlot, however, the spurred variety was formed from old rootstocks over a period of several years. De Vries reports that unspurred peloric forms were found in *Antirrhinum*, *Viola*, and *Tropaeolum*. According to Viollet, the peloric form of *Antirrhinum majus* has been known since 1857. A single specimen of this type originated among numerous plants of the common kind. In 1865 Helye discovered several peloric specimens of *Antirrhinum majus* in a garden in Paris. Peloric forms were also known in *Calceolaria*, *Limosella*, *Plectranthus*, and *Pedicularis* and in the Caprifoliaceae, Orchidaceae, and Papilionaceae. The peloric variety of *Gloxinia speciosa* was distinguished by its upright flower, its regularly formed corolla, and its five identical stamens which were also regularly developed. Because of these considerable differences from the other species of the genus, Lemaire put the peloric variety into a separate genus, *Orthanthe Fyfiana*. This variety originated in England between 1840 and 1850 when a single specimen was discovered by the horticulturist Fyfe. A variety of fuchsia with upright flowers instead of the usual drooping ones also originated in England about 1860 and was named *Fuchsia erecta superba*. Experiments designed to test the hereditary nature of this character produced conflicting results. According to Sirks (1915), the well-known terminal peloria of *Digitalis purpurea* was first described in 1812, then by de Candolle, and finally by Vrolik in 1844. It also appeared in Brest in 1890. Vilmorin obtained a similar variety of *Digitalis gloxinioides*.

The sudden appearance of concrescence of the petals in species where the petals are normally divided was observed by Vilmorin in the poppy (Groenland 1860). From seeds taken from a plant of this type, Decaisne obtained a number of plants with a sympetalous corolla. The converse process, whereby among plants having sympetalous flowers there appear individuals with choripetalous flowers which thus yield a new variety, was also observed, though there seem to be no precise statements about the origin and heredity of such varieties.

De Vries (1901) describes *Stellaria Holostea apetala*, which he discovered in 1889, and also an apetalous variety of *Capsella Bursa pastoris.* He observed ramose spikes in the variable form, *Plantago lanceolata ramosa.*

With regard to variations in the bracts and such, Case, to quote but one example, discovered changes of this type in specimens of *Richardia aethiopica* Kunth. in which each inflorescence had two involucral leaves. This character remained constant when the plants were grafted.

Variations in the color of flowers occur far more frequently, however, than changes in their structure. Korschinsky remarks that such variations do not occur indiscriminately in all directions and claims that certain directions are preferred. He holds that the variation producing white flowers is the most common. Most of our cultivated plants have produced varieties having white flowers, and these varieties generally bred true. Korschinsky holds that the opposite cases, in which varieties with a colored corolla originate from species having white flowers, are much rarer. He quotes the example of *Begonia semperflorens* which produced in France about 1880 a variety whose flowers were deep red or pink; this new variety bred true. Vernon cultivated it for about ten years, at which time it produced yet a further variety having red flowers with some scarlet leaves. The new variety was named *Begonia semperflorens atropurpurea.* It too retained its characters when propagated. Similar changes from white to colored varieties are found in *Robinia pseudacacia.* Thus a variety with pink flowers, named *R. pseudo-acacia* var. *Decaisneana* by Carrière, originated in France in 1862. Differing reports were given regarding the constancy of the new variety when grown from seed. Some record a complete reversion to the normal form, while others, that given by Robillard for example, note a degree of constancy. This variant subsequently produced from seed yet another new form that had flowers of a lighter color, *R. Decaisneana rubra.*

Carrière reports the sudden appearance of a variety of the cherry having pink flowers. In 1842 there appeared in France a seedling of *Chrysanthemum frutescens* having yellow labiate flowers. The leaves of this plant were more finely divided than was usually the case, and it was therefore named var. *tenuifolium.*

Korschinsky developed a theory to explain variations of the coloring of flowers. It is based on an idea of de Candolle who classified the colors that appear in plants in terms of two series: the xanthic and the cyanic. To the first belong the shades ranging from white through yellow and orange to red; to the second belong the colors ranging from white through azure, blue, and violet to red. Thus white and red are common to both series, while the remaining colors appear only in one

or other of the two. It was held that variations in flower color could be understood in terms of this rule in that all the changes that were possible occurred only within one or the other of the two-color series. Yellow and blue, colors that are typical of a given series, were therefore characteristic of whole families of plants. But exceptions to the rule do occur. Korschinsky records the case of the yellow-flowering variety of *Dahlia Merckii* Lehm., discovered by Pépin in 1843, which had apparently sprung from a violet variety and was constant. A variety of *Ageratum mexicanum* with pale yellow flowers was also described.
Using this system of classification, Verlot (1864) gives detailed lists of differently colored varieties of ornamental plants.

The change from uniformly colored flowers to flowers having irregularly spotted or dappled colors was frequently recorded. In 1844 an unusual, large-flowered specimen was found among seedlings of *Erythrina Crista galli* in France: the standard was yellowy white, though carmine at the edges, the carinas, pale red in color, were cuneate, the stamens green, and the saffron-yellow calyx was globose. Verlot mentioned a number of true-breeding varieties having variegated leaves that were cultivated by Vilmorin. Korschinsky does not attach much importance to varieties with spotted or striped flowers since they did not breed true nor did they grow in the wild state. The cases of nontrue-breeding striped flowers described by de Vries (1901), the so-called intermediate varieties which he observed in *Antirrhinum majus striatum*, *Hesperis matronalis*, and *Clarkia pulchella*, also belong to the same category. Large numbers of naturally occurring color variations were known to the florists of the nineteenth century. Korschinsky mentions *Salvia pratensis rubra*, discovered by Briot, as an example of a variety that originated in the natural state and was the cultivated.

The common yellow-flowered form of *Sarothamnus scoparius* Koch produced two varieties with variegated flowers. In the *bicolor* variation, found near Paris, the standard is pure white in color and the wings and carinas are white at the base with yellow tips. Puissant (1886) discovered the *Andreana* variety among common shrubs of Normandy. It has larger flowers, and the velvetlike wings were of a deep red or purple color. These plants are more delicate and have a low degree of resistance to frost, and this variety spread throughout the gardens of Europe and was frequently propagated by grafts on *Cytisus Laburnum*. Experiments designed to show whether this variety bred true yielded conflicting results ranging from partial to complete constancy. A third variety of *Sarothamnus scoparius* originated in Scotland in 1891; its flowers were milky brown in color. There are no precise statements about its further development. De Vries verified the constancy of numerous varieties having white and colored flowers, but we cannot here enumerate all the varieties he investigated.

From the very beginning a great deal of attention had been paid to the hereditary variability of the shape and color of fruits, particularly among cultivated varieties. Darwin mentions a note in "Arbres Fruitiers" (1836) which states that in England a golden variety of the grape had issued from a red variety that had not been crossed. Variability within the orange group yielded several monstrosities together with numerous varieties. The origin of *Citrus aurantium fructu variabili*, described by A. Risso, is unknown but it may also be an example of a variety that originated suddenly and unexpectedly. In this form the young shoots have rounded oval leaves spotted with yellow, borne on petioles with heart-shaped wings. These are succeeded by long, narrow leaves. The young fruits are yellow, longitudinally striated, and sweet, but ripe ones become reddish yellow and bitter. With regard to peaches, numerous variations in the fruits have been observed from earliest times. Carrière records the case of a double-flowered almond which, after producing almonds for several years, suddenly bore spherical, fleshy, peachlike fruits for two years in succession and then reverted to its former state. Numerous North American varieties were known: e.g., the white-blossomed and the yellow-fruited freestone peaches, the blood clingstone, and the lemon clingstone. In his discussion of the peach, Darwin emphasizes the parallel variations between the peach and the nectarine. In both these forms the flesh of the fruit can be white, red, or yellow; both can produce clingstones or freestones, large or small flowers, and leaves that are serrated or nonserrated. Similar variations parallel to the peach and the plum were also observed in the apricot. In fact, the size, shape, quality, and color of the fruits all vary considerably, and of the plum, Darwin describes the curious varieties known as the double or Siamese and the stoneless plum.

Darwin also mentions the flower of the cluster cherry that has twelve pistils of which the majority abort. It produced from two to six cherries borne on a single peduncle. Sageret describes the variety known as "le griottier de la Toussaint" which simultaneously bears flowers and fruit of all degrees of maturity. P. Ascherson and P. Magnus (1891) mention a variety of *Prunus Padus* L. which bears white fruit and grows in considerable quantities in the eastern stretches of the Alps. R. Zdarek designated the new form as a new species, *Prunus Salzeri*, while other authors term the same form *Prunus Padus* var. *leucocarpus*. It appears that varieties bearing red fruits were also observed. Regarding varieties of the apple, Darwin writes at length about the St. Valery apple in which the flower has a double calyx with ten divisions and fourteen styles surmounted by oblique stigmas. Artificial fertilization is always necessary since the flower has no stamens or corolla. The fruit is constricted round the middle and is formed of five seed cells surmounted by nine other cells. The pigeon apple, however, has only four seed cells.

There are relatively few statements about the date or place of origin of these variations. In 1869 Carrière told of a variety of the quince in which the fruit had the appearance of a lemon (*Cydonia citropomma*). *Prunus japonica sphaerica* in which the fruit is large and round originated from seeds of *Prunus japonica* Thunb. Varieties that adapt to different types of soil are also mentioned.

A variety of *Cerasus semperflorens* has several pistils and produces on every fruit stalk several drupes of different sizes. De Candolle had earlier described a similar variety which he called *C. caproniana* var. *polygyna.* Nothing seems to be known about the heredity of this character. From a stone of the pêcher-amandier (peach almond), Carrière (1870) obtained a new variety which he named *Amygdalus monstrosa.* Each flower contained several pistils, and the petals were small and atrophied. United on a single stalk, the drupes are irregularly formed and fall off before the plant is mature. When grafted, the plant almost always bred true in respect to this character. Flowers having several ovaries were also observed in almonds, but there are no statements about the hereditary nature of this character.

The walnut also produced variations on many different occasions. The shape and size of the fruit, the thickness of the husk, and the thinness of the shell can vary. In the grape walnut, grown in France, the nuts grow in bunches of fifteen to twenty. Barthère described a variety in which the leaves were of different shapes and bore elongated (6 to 7 centimeters long), large (3 centimeters broad), thin-shelled nuts. It was constant from seed. A variation of the walnut that was valuable for purposes of cultivation appeared in France in 1830. This new variety, known as *Juglans fertilis*, bears fruit as early as the second or third year. Although the nuts are small, they are of excellent quality. A delicate plant requiring considerable care and attention, it was put on the market in 1837. For the most part the seeds yielded forms that resembled their female parent.

Naudin (1856) reports that the number of varieties of one of the oldest cultivated plants, the melon, knows no bounds. There are varieties in which the fruits are no larger than small plums, while in others they can weigh as much as twenty-six pounds. In another kind the fruit is only one inch in diameter but over a yard in length and twists about in all directions; many other parts of the plant also show a tendency to become elongated. De Vries (1901) mentions fruits of *Papaver somniferum inapertum* and of *Linum usitatissimum* that no longer open up. And H. Count of Solms-Laubach (1900) describes a variation of *Capsella Bursa pastoris* found by Heeger in 1897 and named *Capsella Heegeri*. It differs from the original stock by the shape of its fruits.[2]

It is extremely doubtful whether all the light-fruited varieties of the European Vacciniaceae that were described from the eighteenth to the nineteenth century are the result of infection from *Sclerotinia* species. In their detailed study of the distribution of white-fruited varieties in the genera *Vaccinium* and *Empetrum*, P. Ascherson and P. Magnus (1891) observe that *Sclerotinia* species occur throughout Europe and contrast this fact with the distribution of white-fruited varieties, which is concentrated in some places though quite sporadic in others. This type of distribution suggests the genuine formation of different varieties.

Numerous different characters pertaining to the seed in cereals, especially maize, were also known. Thus the *rugosa* variety (sweet corn), described by Bonafons, has wrinkled seeds, and the *cymosa* variety carries its ears so crowded together that it is called "maïs à bouquet." Varieties in which the seeds contained a great deal of glucose as well as those whose seeds contained only starch were also frequently observed, as were numerous other characters pertaining to the flowers.

For Darwin varieties formed from buds provided a fund of material for the study of sudden and unexpected changes. He points out that some of our most beautiful and useful plants have arisen by bud variation, a term used by Darwin to refer to all those sudden changes in structure or appearance "which occasionally occur in full-grown plants in their flower buds or leaf buds." Darwin stressed two points: First, and contrary to a commonly held view, it is not the case that all variability may be attributed to sexual union, and second, we cannot account in all cases for the appearance through bud variation of new characters in terms of the principle of reversion to long-lost characters. Darwin holds that there are numerous cases of peach trees producing

[2] G. H. Shull (1914), conducting a genetical analysis of *Capsella Heegeri*, concluded that the divergent shape of the fruit is determined not by a single gene but by two genes.

buds that yielded nectarines. In 1741 Collinson recorded the first case of this kind, and in 1766 he added two other examples. At about the same time, D. E. Smith described the case of a tree in Norfolk, England, which usually bore both types of fruit, nectarines and peaches. In 1808 Salisbury added six other examples of the same kind. A tree planted in Devonshire in 1815 formed a sport yielding nectarines and was studied in greater detail. Still further instances of this phenomenon could easily be given. The "grosse mignonne" peach at Montreuil produced by bud variation the "grosse mignonne tardive," a variety in which the fruit ripens a fortnight later than the parent tree. The "early grosse mignonne" is another variant from the same stock. Thomas A. Knight, a contemporary of Darwin, records (1824) that an old tree of the yellow magnum bonum plum produced a branch bearing red magnum bonums. He also states that W. Hofmeister (1868) mentions a forty-year-old egg-shaped plum tree (Dame Aubert Duhamel) which suddenly produced red plums on one of its branches in 1814. Rivers quotes the case of a branch of an early prolific plum tree (a purple variety) that bore bright yellow plums. Knight also discovered several cases of bud variation in the cherry. A branch on a May-Duke cherry tree, for example, produced cherries with a somewhat changed shape which always ripened later. Carrière records the case of a tree that bore three different kinds of fruit. Bud variations were also known in grapes. A stock of the purple Frontignan produced spurs that bore white grapes. Another variety with purple grapes produced a cluster of amber-colored berries. Lindley described a gooseberry bush that bore at the same time four different kinds of berries, hairy and red, smooth small and red, green, and yellow tinged with buff. Similar cases were known in the currant. The Champagne variety, for example, bore red and white berries on separate branches. In France a bush was found that also bore red and white berries as well as some that were striped red and white. De Vries (1901) wrote in detail on atavistic bud variations.

Numerous bud variations are known in flowers. Darwin mentions some that occur in *Camellia myrtifolia*, *Crataegus oxyacantha*, *Azalea indica*, *Hibiscus*, *Althaea rosea*, *Pelargonium*, *Chrysanthemum*, and roses. Thus he reports Rivers' statement that the white moss rose, the scarlet semidouble moss rose, and the sageleaf moss rose originated from the common moss rose in 1788. We could cite several other cases of a similar nature. Darwin also mentions that *Dianthus*, *Antirrhinum*, *Matthiola*, *Cheiranthus*, *Cyclamen*, *Oenothera*, *Gladiolus*, and *Mirabilis* have a tendency to form bud variations from their flower buds. Cases of bud variation in leaves and in shoots were also frequently observed. Mason mentions an ash (*Fraxinus excelsior*) in which one bough was short-jointed and densely covered with foliage; its characters could be propagated by grafts. Trees with cut leaves, the oak-leaved *Laburnum*, the parsley-leaved vine, and the fern-leaved beech, for example, were often seen to revert by buds to the common form. Darwin also refers to bud variation in the cryptogams, for example, in ferns. Cases where bud variation caused changes in the color of the leaves such as blotches, tinges of red, and such were also discovered.

In addition to the bud variations that took place in organs growing above the ground, variations in subterranean buds, in suckers, tubers, and bulbs, were also discovered. A variety of the barberry with seedless fruit originated in this manner. Nineteenth-century plant breeders and gardeners observed and described countless cases of bud variation in potato tubers and in hyacinth bulbs. Darwin comments in *Animals and Plants*: "We must attribute all such cases to actual variability in the buds. The varieties which have thus arisen cannot be distinguished by any external character from seedlings."

Finally, there are a number of borderline cases in which, as Darwin put it, a separation of the parental characters takes place in seminal hybrids as a result of bud variation. Many of the cases observed during the last century are uncertain, while others, as we subsequently learned, depend on fundamentally different processes, and we shall therefore not consider them here.

In the foregoing pages we have tried to illustrate the wealth of observations made during the past two centuries on sudden variations in the various organs of animals and plants. The works of Charles Darwin, who was endowed with unusually sharp powers of observation, are a vast repository of material for the study of such changes. However, his account of the large amount of facts available to him betrays an uncertainty which never quite left him as far as his interpretation of these findings is concerned. Seeing that highly cultivated plants varied far more than plants in the wild state, he concluded at one point that variability must depend solely upon the direct influence of changes in the conditions of life. He developed this thesis of direct influence to the extent of describing a large number of cases in which accidental damage to the structure of the organism had become hereditary. The ultimate development of this idea leads to the assumption that the prolonged use or disuse of the parts of an organism can cause the formation of different races whose characters are inherited. Other observations, however, such as the variation in the blossoms of a single peach tree, the only ones to change among many thousands of others, or the appearance of a certain, particularly conspicuous, character in different places where the environmental conditions were quite dissimilar, led him to doubt the cogency of this view. Instead, he came to the conclusion that changes are not directly related to the conditions of life but depend on unknown laws which affect the organization or constitution of the individual. He finally expressed his uncertainty in these words: "Our ignorance of the laws of variation is profound. Not in one case out of a hundred can we pretend to assign any reason why this or that part has varied."

Despite his vacillation between these views, he realized that the laws of inheritance would have to be the same in both seminal and bud variations. He was also certain that the law of analogous variations would have to be true of varieties which sprang from seeds as well as of those from buds. He held that variability was more pronounced in plants raised from seed than in those produced from buds. The reasons for this fallacy are obvious to the contemporary reader. Darwin, however, also seems to have thought that both these kinds of variation could have different causes, for he says on one occasion that differences among seedlings are usually very small and that sharply marked changes appear only after long intervals of time. As we have seen, he did not always pursue these sharply marked changes with the same degree of interest. Toward the end of his life he considered them to be most unusual, not a part of the normal course of things and thus extremely rare, particularly in feral animals and wild plants. The excerpts from Darwin here quoted might give the impression that he considered the sudden appearance of the new varieties that he observed in domesticated animals and cultivated plants of great general importance. However, he disavows this possibility elsewhere: "The frequency of these cases is likely to lead to the false belief that natural species have often originated in the same abrupt manner. But we have no evidence of the appearance, or at least of the continued procreation, under nature, of abrupt modifications of structure; and various general reasons could be assigned against such a belief."

A contemporary appreciation of the works of Darwin recognizes him as one of the greatest naturalists who understood what are now considered to be the essential facts of the problem of evolution. But even as he grew older and acquired greater experience, he still emphasized what were, with respect to the problem of evolution, secondary forms of variability, namely, variations in the phenotype caused by external influences. He ascribed to such variations a potential for forming new species, made possible through natural selection.

Not all of Darwin's contemporaries shared his indecision. W. Hofmeister writes:

One of the most conspicuous and remarkable features of variation in plants is undoubtedly the sudden and abrupt appearance of far-reaching modifications to the usual form, as happens in the phenomena we have just mentioned, in others analogous to them, and also in the formation of monstrosities in general. A new form is not produced by small differences from the usual development, all tending toward the same direction and accumulating through the course of generations; it appears suddenly, with all of its considerable modifications to the ancestral form already present and complete.

The zoologist William Bateson is probably the most important of the researchers who worked on the problems of the nature, causes, and significance of variability after Darwin. In his work *Materials for the Study of Variation* (1894), Bateson opposed Darwin's claim that variability was primarily due to environmental influences, basing his opposition on a minutely detailed account of numerous cases of variation. The most important feature of his work is the distinction he drew between meristic and substantive variation. He used the term meristic variations to refer to those that comprise differences from the ancestral type in respect of the number or arrangement of organs that were severally present (fingers, toes) and maintained that phenomena of symmetry were particularly important. Substantive variations, on the other hand, pertained to changes in the actual constitution or substance of the parts themselves. By way of illustration, Bateson relates that the flower of a *Narcissus*, usually divided into six parts, may through meristic variation be divided into seven or into four parts, as the number of the petals varies. The variations known in the color of the flowers of *Narcissus corbularia* (dark yellow or sulphur yellow) were substantive variations which caused changes in the substance of the plant without producing any modification in respect of number or symmetry. Bateson supports both kinds of variation with a vast amount of illustrative material primarily designed to show that such phenomena of variation are discontinuous, that is, are not connected with the original stock by transitional steps. At one point he writes, "We have said that the variation is discontinuous, meaning thereby that the change is a large and decided one, but it is more than this; it is not only large, it is complete."

He also commented on the nature of these variations. He maintained that meristic variations—changes in number—were purely mechanical and achieved a kind of mechanical stability. He derived the discontinuity of substantive variations, however, from chemical differences which were determined by a chemical stability. He did not elaborate on the hereditary nature of these variations, but there is no doubt that many of the cases he mentioned belong to the same category as the sports or single variations of Darwin. He notes, for example, the appearance within a family of certain variations in fingers and toes. His conclusion was that the discontinuity of species results from the discontinuity of variation. In his words: "The evidence of variation suggests that this greater stability depends primarily not on a relation between organism and environment, not, that is to say, on adaptation, but on the discontinuity of variation. It suggests in brief that the discontinuity of species results from the discontinuity of variation."

In the work of Korschinsky (1901) the emphasis is also placed on sudden variations. He characterized them collectively as heterogenesis. This concept was introduced by A. von Kölliker in 1864 who opposed Darwin's theory with his own theory of heterogeneous reproduction, in which he proceeded from the assumption that "under the influence of a general law of development, living creatures produce from their seeds others which are different."

This idea is taken up by Korschinsky as is seen from the following passage:

On the contrary, there are facts of a most compelling nature which prove that because of unknown circumstances and contrary to the law of inheritance an organism can be formed from the fertilized egg cell so unlike its parents that we can consider it a distinct species both in respect of the totality of its external characters and of its capacity to transmit them by inheritance.

He defined in clear terms the nature of heterogenesis by stating that among the progeny of completely normal plants certain individuals unexpectedly appear which differ to a varying degree both from the others and from their parents. These differences may be considerable and affect several characters, or else they may be restricted to a few characters or even to a single one. He continues: "It is remarkable, however, that these characters exhibit a high degree of constancy and are transmitted unaltered from one generation to the next. In this way, a distinct race suddenly originates and is just as stable and constant as those which have been in existence from time immemorial."

Korschinsky also realized that the races that originated abruptly as a result of heterogenesis were in no way different from those that had been in existence for a long time. This raised the question how often sudden changes should be expected. Now, this question can only be tackled by investigating the distribution of the different varieties. Here, once again, the observations of breeders and gardeners provide useful material.

While discussing Darwin's observations, we emphasized that the appearance of sports had been known to breeders and gardeners for a long time, and that in many cases sports were viewed as the only means of obtaining new varieties that were constant. Sudden variations are mentioned in breeders' catalogues as early as the first decades of the nineteenth century. Korschinsky therefore made extensive use of this source for the development of his theory. His aim was to identify the kinds of changes that could be ascribed to heterogenesis. He soon discovered that such changes can proceed in all directions, though he emphasizes as the principal directions variations in growth, in the stem, in the corolla, in the fruits, in the shape and color of the leaf, and in the color and structure of the flowers. Varying degrees of resistance to cold or draft, early or late flowering, profuse or meager flower formation, fragrant or nonfragrant flowers, and so on are mentioned as changes of a secondary kind. The first naturalist to investigate this field, Korschinsky held that the sum total of phenomena of heterogenesis was much greater than was usually assumed, even though this process occurred but rarely in a given species. He points out that single variations occur more frequently than other kinds despite the fact that the true identity of two apparently identical variations had never been established. He also says that the frequency of heterogenetical changes depends on the number of specimens that are cultivated within a given species and that valid comparisons could be made only on the basis of extensive numerical data. He finally reminds us that a number of small variations were undoubtedly overlooked and that breeders and gardeners were particularly interested in new variations that were useful for cultivation. According to Korschinsky, a new variation usually appears in a single individual. Wherever it was found that several individuals had changed in the same manner, the variation had in fact already appeared a generation earlier but had been overlooked.

Deeply influenced by men of practical experience, Korschinsky also develops a coherent view of the conditions under which heterogenetical changes most likely occur. On this point, too, he is more positive than Darwin, for he realized that the causes of heterogenesis are not to be found in external conditions but in internal processes, in the changes that take place in the egg cell, though he was unable to formulate the nature of these changes clearly. He also opposes the view that changes in the conditions of life can in part account for the appearance of heterogenetical variations. An old idea of Darwin's—that plants taken over into culture quickly begin to vary—was also rejected. According to that view plants in the wild state never, or seldom, vary. Korschinsky's statement that the difference between wild and cultivated plants is that variations in the former usually die out while those in the latter are recorded and conserved reveals that he knew the earlier view to be mistaken. We may assume that his correction of the old theory is also intended to include the animal kingdom.

We must mention in this connection that Charles T. Druery (1903-1904) also opposed the thesis that cultivated plants are especially apt to vary. During the last twenty years of the nineteenth century, he studied the ferns of Great Britain and observed a profusion of sports appearing in the wild state. He claimed that no other group of plants had been so thoroughly investigated in respect of the occurrence of variations. From his observations he concluded that all plants vary just as much under natural conditions as under conditions of cultivation, that sports do not spring from changes in environmental conditions, and that variability—as Darwin also thought—eventually proceeds in all directions. Druery observed that in a list compiled by Lowe comprising 2090 different varieties, 1360 were found in the wild state and only 730 originated while under cultivation.

We shall conclude this discussion with a note on an important discovery made by Darwin and Korschinsky. In a letter to J. H. Gilbert dated February 16, 1876, Darwin writes that he had worked during the preceding ten years on hybridization and self-pollination in plants. Cultivating plants over several generations in pots under glass and thus exposing them to very similar conditions, he had noticed to his amazement that when they were self-pollinated for several generations the color of the flowers sometimes changed and finally became constant as in a wild species. From this he concluded that substances were absorbed from the soil and that these substances were responsible for variability. In a rather different way Korschinsky came to hold the same view, though he expressed it with a little more clarity. According to him, those who wish to examine the phenomena of heterogenesis more closely should realize that they can be observed only among the offspring of pure species, that is, in species whose characters are established and secure. It is important to understand that heterogenesis cannot occur among hybrids since they do not allow for the most important precondition of heterogenetical variations, that is, a pure form of heredity. The views of both scholars focus on that important point, the purity of the material, which—as we have already seen—is one of the first conditions that must be satisfied in any study of hereditary variations.

After Darwin, many other scholars studied the phenomenon of sudden variations and added their own views on the significance of such changes. It is clear that H. Hoffmann (1881), for example, observed numerous hereditary variations during the course of his experiments on variation undertaken from 1855 to 1880. He gives a survey of the formation of peloric forms, of cases where the petals turn green, of double flowers, and of variations in the colors of flowers and leaves. He was perfectly familiar with the occasional sudden appearance of new hereditary forms, and his observations included cases where crossings had not been effected. His most important examples are those involving variations in the buds. He also doubted whether cultivated plants had a stronger tendency to vary than those in the wild state for, intent on finding some, he had discovered several variations in the color of the flowers of *Papaver Rhoeas* and *Centaurea cyanus.* He claimed that the causes of hereditary and nonhereditary variability were in part to be found in the salt content of the soil, in meteorological conditions, and in parasites. Toward the end of his study, however, he says that the essential cause of variability is an internal one.

Thomas Meehan, who for many years investigated hereditary variations in plants, also wrote on the importance of sudden changes (1875, p. 9): "Not only do strikingly distinct forms come suddenly into existence, but once born they reproduce themselves from seed, and act in every respect as acknowledged species."

Meehan, who championed the existence of sudden changes in evolution, is the subject of a short paper written by R. R. Gates (1915). In his volume of essays entitled *Darwiniana,* Thomas H. Huxley (1893) enthusiastically acknowledges the occurrence and importance of sudden changes (cf. R. R. Gates 1916). He states that Darwin's position would have been clearer and stronger had he not have relied so heavily on the aphorism *Natura non facit saltum.* He continues: "We believe, as we have said above, that Nature does make jumps now and then, and a recognition of the fact is of no small importance in disposing of many minor objections to the doctrine of transmutation" (p. 77).

And elsewhere: "We greatly suspect that she (Natura) does make considerable jumps in the way of variation now and then, and that these saltations give rise to some of the gaps which appear to exist in the series of known forms" (p. 97).

After a detailed discussion of the Ancon sheep, Huxley points out that we do not know why varieties arise and that it is more than probable that the majority of them appear in the same spontaneous manner. According to Huxley, however, the most important thing to note is that since these varieties did exist, they conformed to the fundamental principle of heredity, namely, that like begets like, that their progeny retained the same differences from the original stock.

Darwin, Bateson, and Korschinsky compiled a vast amount of factual material concerning the problems of sudden changes, single variations, sports, and heterogenesis and the associated questions about the origin of wide differences among forms. But it was the Dutch botanist de Vries who unified these facts, supplementing them with an imposing number of his own observations, and welded them into the important theory known as the Theory of Mutation. As he observes in the introduction to his two-volume work, *Die Mutationstheorie*, the theory of mutation hinges on the fact that the characters of organisms are composed of units which differ sharply from one another. These units are combined into groups, but there exist no transitional forms between them, just as there are none between the molecules of chemistry. Whenever a new unit appears, a phenomenon known as mutation, the new form is separated from the species from which it was produced. These new forms, species, and varieties therefore originate suddenly from the original stock and do not pass through any transitional stages. The term "mutation," chosen by de Vries to characterize these saltations, was in no way original. Long before Darwin, the naturalists who shared the belief—rarely formulated in clear terms—that species developed from genera or that even smaller units developed from species were known as transmutationists, and their doctrine as the theory of transmutation or transformation. Darwin himself seldom used the term. In the autobiography he writes: "I had become, in the year 1837 or 1838, convinced that species were mutable productions. . . ." (p. 75)

In the work of Bateson this term is found but rarely, and in Korschinsky's theory of heterogenesis there is no mention of it whatever. Transmutation or transformation, mutability, immutable and mutable, and the verb, to mutate, are concepts, though inexact ones, that existed in biology throughout the nineteenth century. It is therefore surely unwarranted for G. Steinmann to make the following critical remark in his book *Geologische Grundlagen der Abstammungslehre* (1908, p. 18):

Nothing provides a better testimony of the indifference shown by botanists in their historical researches than the distressing fact that the concept of mutation—now elucidated in every textbook on paleontology, and formulated by Waagen in 1867 to refer to the smallest, yet still perceptible changes, as it were to the differential of organic transformation during the course of time—could recently be employed by a botanist to describe what is a quite different phenomenon.

From what has been said, it seems more probable that Waagen took the concept of mutation from biology, defined the term more closely, and introduced it into paleontology. In any case, his use of this term to describe sudden changes, whatever kind they may have been, was perfectly natural.

De Vries believed that the genus *Oenothera*, and the species *Oe. Lamarckiana* in particular, constituted excellent material for an exact study of the process of mutation. In the first decades of the twentieth century, however, it was shown that the majority of the phenomena that de Vries observed and interpreted as mutations required a quite different explanation. We know, however, that his basic ideas were correct since genuine cases of mutation were discovered shortly after the publication of *The Theory of Mutation*, and these helped to make the theory one of the most important in modern genetics.

De Vries began to conduct experiments on *Oe. Lamarckiana* in 1886 in an abandoned potato patch in northern Holland (Hilversum), where he discovered a number of these plants which had been growing there since about 1875. Closer examination revealed that the plants varied in all their organs. In addition to cases of fasciation and ascidia, differences in life-span were also apparent. One year later two new forms appeared, each quite distinctive; they were characterized as elementary species as opposed to the collective species of Linnaeus. There were only a few specimens of each variety, and each group of specimens was found growing on a particular spot and nowhere else in the whole potato patch. One of these varieties was *Oe. brevistylis*, a very short-styled plant which was at first wholly staminate. The other variety, named *Oe. laevifolia*, was conspicuous because of its beautiful, smooth-leaved foliage, particularly so in autumn when the petals become narrower and lose their cordate appearance at the upper edge. Both varieties bred true, and de Vries therefore suspected with good reason that they had originated as a result of mutation. In 1886 two samples were taken from the locality in Hilversum: first, nine large, beautiful rosettes with fleshy roots; second, seeds of a quinquelocular fruit from the middle of the potato patch. In 1887 seeds of *Oe. laevifolia* were gathered. These three families, cultivated in the experimental garden at Amsterdam, provided the whole of the extensive experimental material that was then cultivated year by year. *Oe. brevistylis*, which did not appear in the cultures, was also taken directly from the locality at Hilversum. New elementary species appeared in all three families. From seeds of the first and second cultivated generations of the *Lamarckiana* family that was produced from the nine rosettes there arose three new unknown forms: *Oe. nanella*, *Oe. lata*—several specimens of both varieties—and *Oe. rubrinervis* of which there was only a single specimen. *Oe. gigas* was found in the fifth generation. It appeared as a single specimen and was a particularly robust, broad-leaved, large-flowered variety with short fruits, and bred true forthwith. Varying numbers of *Oe. albida*—a delicate, brittle, pale green, narrow-leaved variety that also bred true—were found in almost every generation. The aforementioned *Oe. rubrinervis* was conspicuous because of red veins on the leaves and broad red streaks on the calyx and fruits. Its flowers were somewhat larger and of a darker yellow color, and the brittleness of the stems and leaves was particularly striking. *Oe. rubrinervis* appeared frequently—in the third, fourth, fifth, and sixth generations—and yielded a total of thirty-two specimens. It, too, bred true in every case. A fourth form, *Oe. oblonga*, appeared in 1895 and 1896; it differed from the others since it was able to produce further different forms. The leaves of *Oe. oblonga*, sharply set off against their stalk, are narrow and macropodous with broad, pale veins. *Oe. nanella*, which originated in the first cultivated generation, is a dwarf form, though in every other respect it exhibited all the typical *Lamarckiana* characters. Since it differed from the original strain in growth alone, de Vries characterized it as a variety, though he emphasizes that it behaved exactly like an elementary species as far as constancy was concerned. One of the earliest of the new forms, *Oe. lata*, was wholly pistillate and so its constancy—about which, however, there seemed to be no doubt—could not be tested in an exact manner. Its leaves are broad and have a wide base and long stalks. *Oe. lata* has low plants with limp stems and drooping tips. In later generations this form appeared regularly, though in fluctuating quantities. One of the rarest forms was *Oe. scintillans* which appeared only on eight occasions in the *Lamarckiana* family. It also behaved abnormally in another way, for it did not breed true but always produced—after having been fertilized with its own pollen—together with further *scintillans* plants, certain percentages of *oblonga* and typical *Lamarckiana*. The previously mentioned forms were the most important ones to originate in the *Lamarckiana* family. A number of additional ones appeared, however, including *Oe. sublinearis*, *subovata*, *leptocarpa*, *elliptica*, *semilata*, and *spathulata*, but they were not so closely studied. This series of events was paralleled in the *laevifolia* family: in the course of seven generations, forty-one mutants were found and without exception they were already known from the *Lamarckiana* family. Other families, besides the *Lamarckiana* and the *laevifolia*, were also studied, and these investigations yielded in every case a similar picture of events: new mutants appeared, though in fluctuating quantities, in every crop of any size. On the basis of these exact and extensive investigations, de Vries proposed a number of theses—the Laws of Mutation—concerning the way in which new elementary species originate. These theses constitute the essential points of his Theory of Mutation and may be formulated as follows:

1. New elementary species appear suddenly without any intermediate steps.
2. From the very moment of their origin, it is usually the case that new elementary species are completely constant.
3. Most of the new types correspond exactly, with regard to their characters, to elementary species and not to real varieties.
4. Elementary species usually appear in a large number of individuals at the same time, or at least during the same period.
5. New characters exhibit no special relation with regard to individual variability.
6. Mutations forming new elementary species occur in all directions. Changes take place in all the organs in almost every possible direction.
7. Mutation occurs periodically.

There are two methods of studying mutations in nature, and de Vries successfully used both of them. First he gathered mutants from the locality where he discovered the original stock, and second he took seeds from that locality in order to sow them under extremely favorable conditions of cultivation. His extensive use of the second method enabled him to obtain those mutants which in the open would have perished early because of extreme changes in their degree of vitality. De Vries knew that the discovery of a mutant never coincides with the genesis of the mutation. To learn more about this process of genesis, he considered it absolutely necessary, wherever possible, to produce mutations at will. All the mutations in de Vries's experiments arose suddenly and unexpectedly, and their manner of appearance is thus identical with the genesis of single variations, sports, and heterogenetic changes. The constancy of the mutations was proved by sowing the seeds of *Oe. gigas*, *albida*, *oblonga*, *rubrinervis*, and *nanella*. However, the constancy of *Oe. lata*, which is wholly pistillate, remained questionable, as was that of *Oe. laevifolia* and *Oe. brevistylis* which were encountered only in the original locality. *Oe. scintillans* was exceptional in that it produced *oblonga* and typical *Lamarckiana*.

The question of the size, or range, of the leaps caused by a mutation was discussed along with that on constancy. De Vries claimed that variability proceeded by way of a series of jerks, pushes, or jumps—he in fact spoke of a stepwise variability (*eine stossweise Variabilität*) rather than a jumplike variability (*eine sprunghafte Variabilität*), the adjective commonly employed by other writers—whereby these individual jerks, or pushes, produce changes that are quite small but nonetheless uniform. It follows that mutations do not necessarily produce greater differences than those found in extreme variants, i.e., products of the type of variability that is caused by environmental influences.

On more than one occasion, de Vries discussed the differences between elementary species and varieties. In his opinion elementary species differed from their nearest neighbors in almost all characters, whereas varieties differed from the maternal species, though usually only in a single character or, rarely, in a very small number of characters. The number of characters changed by a mutation thus determined whether he spoke of the genesis of a new elementary species or of a new variety. Simultaneous changes in several organs following a single mutation convinced him that mutations can occur in all directions. Deeply influenced by Darwin's theory of natural selection, he concluded that the mutability of *Oe. Lamarckiana* in fact amply satisfied all the requirements of that theory, since some of the new forms perished without offspring, and natural selection determined the vitality and distribution of the others.

From the fact that he could observe a high incidence of mutations solely in *Oe. Lamarckiana*, de Vries concluded that mutability must appear periodically in different species. Darwin had already assumed that species *in statu nascendi* were more flexible than others, thus admitting that variability was periodic. De Vries thought that his results warranted the assumption that species can remain quite unchanged for long periods but that they begin to produce new forms when certain conditions are satisfied. Whenever these conditions are met, elementary species appear simultaneously in a large number of individuals. He established that the seven principal species of *Oe. Lamarckiana* appeared with a frequency of approximately 1 to 3 percent. He thought that the only explanation of this fact lay in "the assumption of the presence of a tendency, in a latent condition, to those mutations in the apparently normal individuals of my cultures." Thus, he held the view that the characters of the new species which subsequently appeared in his cultures had already been present, in a latent form, at the beginning of his experiments on mutation in 1886, and had then become manifest on certain occasions. The faculty of mutation was thus a hereditary, latent character, and *Oe. Lamarckiana* was a species in which such latent characters were occasionally activated. He realized that this assumption was purely hypothetical and used the term premutation to refer to the genesis of these latent characters. His conclusion was that a mutation may follow every premutation even though it does not always do so, but that a premutation must precede every mutation, provided it is a progressive mutation, that is, one that gives rise to a character which is genuinely new. Moreover, de Vries distinguished between retrogressive and degressive mutations which led to the manifestation of tendencies already present. Although he was unable to pass a clear judgment on the causes of premutation, he assumed that it was primarily external causes, such as extremely favorable or unfavorable environmental factors, which were responsible for the genesis of premutation. "It is an important task for subsequent research," he said on one occasion, "to determine the conditions of this premutation and to bring it about, wherever possible, at will." It was knowledge of these conditions alone that could lead to the discovery of the Laws of Mutation, that is, to the control of mutability in the service of a greater understanding of the reciprocal relationships between organisms. Thus at the turn of the century he already clearly recognized the necessity for experimental research into mutation—a venture that could not proclaim its first triumphs until more than two decades later. He also raised the question whether the coefficient of mutation could be altered by artificial means so as to augment the number of mutants that were particularly rare. Probing more deeply into the causes of the genesis of mutation, he also raised the issue of the primary process of mutation, as a result of which we observe changes in the characters. He frequently stressed that all of the new characters we perceive in a mutant are no more than the collective expression of a single prior change. He justified this conclusion on the ground that the different characters of a mutant are without exception securely and inseparably combined. He also realized that this single internal change is governed by a process of interaction with all the other characters of the organism and that the visible result, the mutant, is determined partly by mutation and partly by the original characters of the organism.

His occasional lack of precision in the formulation of concepts notwithstanding, we have adhered closely to de Vries's own account of problems in the theory of mutation, even though he also made the mistake of considering all of the *Oenothera* mutants as cases of true mutation. The most important points of the Mutation Theory are here cited in order to show that, even though his experimental material was in some ways unsuitable, many of his conclusions are still valid today. Indeed, some have meanwhile proved correct beyond all doubt, or else still provide a focal point for research. Despite the multiplicity of the characters altered by a mutation, de Vries arrived at the conclusion that it was only a single unit character that was changed. But it is clear that elsewhere he expressed grave doubts about this view, which were extremely significant, since they contain a hint that de Vries was not completely satisfied with his interpretation of the *Oenothera* mutants. It seems, in fact, that he already anticipated the correct explanation of the mutations in *Oenothera*, for he says in the first volume of his *Mutationstheorie* (p. 232):

> This intimate union of the characters which appear together without exception, always simultaneously and unexpectedly, clearly stood in need of explanation. Two possibilities are available. First, it is conceivable that all these visible characters are merely the expression of a single transformation—that it is, in fact, a single elementary character which appears for the first time in every mutation. On the other hand, we might suppose that the elements of a species are transformed in groups during the course of a mutation. That the characters of plants are united in groups of varying magnitudes, and that it often happens that whole groups rather than single units react to external influences, or remain intimately bound together in crosses or in cultivation, can scarcely be subject to doubt and has frequently been affirmed on theoretical grounds.

By introducing an ordering principle into the multiplicity of the phenomena of variability, Korschinsky's Theory of Heterogenesis made an important advance, but it is the work of de Vries, the first great experimental studies in mutation, that heralds a new era in genetics. His work coincides with the rediscovery of Mendel's Laws of Heredity in the year of the birth of classical genetics. Ever since that time these two branches of genetics, the analysis of hereditary factors and the theory of the origin of inherited variations, have been closely linked together. While this historical survey has tried to show that the variability of organisms was viewed primarily with regard to its value for the origin of new, phylogenetically important forms, the emphasis was now to be placed upon questions about the nature of these changes, the manner of their origin and the frequency of their appearance, and thus upon questions about the nature of the process of mutation.

8

The Cytological Discoveries of the Nineteenth Century and Their Influence on the Idioplasm Theory of Heredity

The last three decades of the nineteenth century were of outstanding importance in the development of research into the cell, which finally led to the clarification of the cytological bases of heredity and replaced ancient hypotheses on reproduction and fertilization with theories that were confirmed by experiment.

Microscopic research had first been undertaken by Leeuwenhoek and Malpighi in the second half of the seventeenth century. However, progress in this branch of study had been impeded by chromatic aberration of the lens up to the time of the Swedish physicist Klingenstierna (1698-1765), who discovered the physical requirements for the construction of the achromatic lens. Achromatic objectives were then manufactured in accordance with Klingenstierna's specifications by the English instrument maker Dollond (1706-1761). At the beginning of the nineteenth century J. von Fraunhofer (1787-1826) in Germany, Chevalier in France, and Amici (1784-1860) in Italy undertook further work on the achromatic lens and its use in microscopes and telescopes; this enabled the first improved microscopes to be put on the market in the 1830s and opened an era of research that centered primarily on the study of the composition of the tissues of higher animals and plants and on protozoology. As early as 1828 the English botanist Robert Brown (1773-1858) discovered the movement of molecules within the cell, subsequently called Brownian movement, and in 1833 he discovered the cell nucleus.

Malpighi and Grew had already realized that wood is composed of cells, and the French botanist François de Mirbel (1776-1854) expanded this claim to include mosses, emphasizing that the cell was the basis of all structure in the vegetable kingdom. Amici (1824, 1830) discovered the pollen tubes and observed their function and growth.

Further progress in the field of microscopy was made by Hugo von Mohl (1805-1872) (Figure 34). Born in Stuttgart, he studied at the University of Tübingen and graduated with honors in medicine in 1828. He then took up full-time research in botany and studied the anatomy of palms, cycads, and other plants. Extremely skillful with the microscope, von Mohl was able to grind his own lenses and became an unflagging champion of improvement in the techniques of microscopy. In 1832 he was appointed Professor of Physiology at Berne University, where he taught human physiology and botany. He returned to Tübingen in 1835 as head of the department of botany; he also founded the first Faculty of Sciences in Germany there. He remained in this post until his death on April 1, 1872.

In his microscopic research von Mohl described the contents of the cell, the structure of the cell wall, and the flowing movements of the cytoplasm. He discovered that cells multiply by formation of new

34 Hugo von Mohl (1805-1872)

35 Matthias Jacob Schleiden (1804-1881)

dividing walls. He described the cellular structure of the spiral vessels of the bast, or phloem, and explained in 1839 the formation of spores in cryptogams, which occurs when a parent cell is divided into four. In 1841 F. Unger found that new cells are produced by division at the point of vegetation.

Important advances were made in histology thanks to the work of Matthias Jacob Schleiden (Figure 35). Born in Hamburg on April 5, 1804, he studied law at Heidelberg University from 1824 to 1827. As soon as he had taken his degree he returned to Hamburg, where he practiced as a lawyer. In 1833, two years after an unsuccessful suicide attempt, he abandoned his law practice and began studying medicine at

Göttingen; he later took up the study of botany in Berlin. At that time he undertook his first studies in cell theory—work that ultimately resulted in his general theory of the nature and contents of the cell. In 1839 he went to Jena, and shortly after graduating from there, he was installed as Associate Professor of Botany. In 1846 he refused the offer of a post at Giessen and was appointed honorary professor on the Faculty of Medicine at Jena. Four years later he became director of the Botanical Gardens in that city. In 1862 he resigned his chair at Jena and went to Dresden and later to Dorpat where he became Professor of Anthropology. In 1864 he returned to Dresden, and spent the remaining years of his life in Darmstadt, Wiesbaden, and Frankfurt, where he died on June 23, 1881.

It was Schleiden who helped the instrument maker Carl Zeiss obtain permission to open a workshop for optical instruments in 1846, and he greatly encouraged the manufacture and improvement of microscopes there. As a result of his investigations into the developmental history of the phanerogam embryo, he discovered the nucleolus in the embryo sacs of various phanerogamous plants. These same studies, however, led him to adopt a number of erroneous views, e.g., that the pollen tube represented the female element of a plant and that the embryo was formed from it. He also thought mistakenly that the cell originated from the nucleus and that cellular tissue was produced by crystallization within the living mucus. For Schleiden, the cell walls were the principal constituents of the cell, and he failed to appreciate fully the importance of the cell nucleus, though he did view the plant, considered as a whole, as a community of cells, as a kind of polypstem (*Polypenstock*). For his research in the field and for his textbooks, *Grundzüge der wissenschaftlichen Botanik . . .* (1842), *Die Botanik als induktive Wissenschaft behandelt,* and *Die Pflanze und ihr Leben* (1848), Schleiden was recognized as the founder of medical pharmacognosy.

His conception of the free formation of cells from fluids or intercellular material—he compared it to the process of crystallization (nucleolus ⟶ nucleus ⟶ cell)—was first adopted by Theodor Schwann (Figure 36).

Schwann was born on December 7, 1810, in Neuss in the Rhineland. He studied medicine at Berlin, graduated in 1834, and became an assistant under Johannes Müller at the Institute of Anatomy. In 1839 he published his celebrated work, *Mikroskopische Untersuchungen über die Übereinstimmung in der Struktur und dem Wachstum der Tiere und*

36 Theodor Schwann (1810-1882)

Pflanzen. The same year he was appointed Professor of Anatomy at the University of Louvain. In 1848 he accepted the offer of a post at Liège where he taught physiology and comparative anatomy. He resigned his chair in 1878 and died in Cologne on January 11, 1882.

Schwann applied Schleiden's theory—that tissues develop from cells—to the animal kingdom. He discovered that the notochord of the tadpole, the germ layers of the chick, and the embryological tissue of the pig all consist of cells that resemble those found in plants. He was thus the author of a general cell theory which he expounded in his classic study (1839).

At about the same time, J. E. Purkinje[1] and his co-workers (especially Valentin) in Breslau made the same discovery independently of the school of Müller. Animal cells in various epithelia—chorda dorsalis, cartilaginous tissue, bony and nervous tissue, and various glands—were described as "corpuscles," "granules," or "globules." In a lecture delivered in 1837 Purkinje emphasized the similarity between these nucleate "granules" and plant cells.

The son of Czech parents, Jan Evangelista Purkinje (Figure 37) was born in Libochovice near Litoměřice in Bohemia on December 17, 1787. After attending a seminary for Roman Catholic priests, he studied theology for three years and then took up the study of philosophy and medicine at Prague. His dissertation, *Zur Physiologie des Sehens* was inspired by Goethe's *Theory of Color.* Goethe was impressed with the dissertation and in 1823 recommended to the Prussian government to appoint Purkinje to a professorship in Breslau. There, however, Purkinje encountered numerous difficulties in setting up his institute and was unable to open it officially until 1840. He directed a great deal of pioneering work in different branches of biology; his studies in the physiology of the sensory organs rank among his greatest achievements. He was, in fact, the founder of experimental physiology and microscopic anatomy. In 1850 he accepted the chair of physiology at the University of Prague, where he founded the Institute of Physiology.

He subsequently became a writer, translated poems of Goethe and Schiller into Czech, and devoted himself to the cultural and political problems of his native land. He died on July 28, 1869.

Purkinje, who discovered the nucleus of the ovarian egg of the hen in 1825, employed the term "protoplasm" to describe the rudimentary, undifferentiated material of the cell (1838). In 1846 von Mohl used the same term to refer to the fluid parts of the cell. The zoologist Max Schultze pointed out in 1863 that plant protoplasm and the mucous matter in animal cells are essentially the same. Bischoff (1842) showed that for mammals the first stage in the development of the egg is its

[1]In 1959 the German Academy of Naturalists (Leopoldina) together with the Czechoslovak Academy of Sciences held a symposium on the life and work of Jan Evangelista Purkinje. Its transactions constitute a valuable contribution to Purkinje scholarship and emphasize the versatility of his thought.

37 Jan Evangelista Purkinje (1787-1869)

division into two equal parts, and Kölliker (1844) found that the same was true of cephalopods. With these discoveries it was realized that the nature of an organism is determined not by its construction from solidified cells but by the semifluid, mucous content of those cells. In 1841 he also showed that sperm is formed from cells in the testes and is therefore part of the parental organism and not parasitic.

Rudolf Albert von Kölliker (Figure 38) was born in Zürich on July 6, 1817. He first studied zoology and botany there and in 1839 continued his studies in Bonn and Berlin. In 1840 he traveled with his friends Nägeli and Remak to the islands of Föhr and Helgoland in the North Sea and subsequently to Naples and Messina in order to

38 Rudolf Albert von Kölliker (1817-1905)

undertake studies in developmental history. He graduated from Heidelberg University in 1842. In 1844 he became Professor of Physiology and Comparative Anatomy at Zürich and was offered a chair of Physiology and Comparative Anatomy at the University of Würzburg in 1847. He had wide practical experience in every branch of histology and embryology and helped pave the way to subsequent research in cytology. His view that the course of evolution is determined by sudden changes resulting from internal causes places him in the forefront of thinkers who recognized the importance of the cell nucleus in the developmental process. Kölliker died in Würzburg on November 2, 1905.

K. Nägeli made a number of important contributions to our understanding of the formation and development of cells. Studying the formation of pollen in phanerogams (Liliaceae) in 1842, he discovered that the nuclei of two daughter cells are derived by division of the nucleus of the parent cell. In these cases of nuclear division (*Tradescantia*), Nägeli identified the chromosomes, which he called "transitory cytoblasts."

The anatomist K. B. Reichert (1811-1883) and the embryologist R. Remak (1815-1865),[2] both pupils of Johannes Müller, decisively furthered our understanding of the first stages of embryonic development. Remak realized that the frog's egg is a cell that divides to form new cells. Therefore, as Reichert also discovered, all the products of the process of cleavage in the egg are in fact cells in which division proceeds from the nuclei. Thus, normal growth is caused by the continuous multiplication of cells.

Rudolph Ludwig Virchow (1821-1902), also a pupil of Müller, applied cellular theory to medicine. A statesman, pathological anatomist, and anthropologist, Virchow ranks among the leading scholars of the nineteenth century. In his *Cellular-Pathologie* (1858), he stated that each cell is an autonomous unit of living matter and that the membrane and nucleus are its most important constituents. Like Remak, he rejected spontaneous generation and free cell formation. Virchow is the author of the famous aphorism "*omnis cellula e cellula.*" (All cells come from [preexisting] cells.)

Our contemporary notion of the cell, its importance, function, and constituents, was largely determined, however, by the work of the zoologist Max Schultze (1825-1874). Schultze lectured at Halle for a while and then became a professor at Bonn. He undertook important studies in the microscopic anatomy of lower and higher animals, investigated the terminal branchings of the nervous system, and showed (1861) that animal cells consist of protoplasm and a nucleus but do not possess a membrane; that is, the membrane is not an essential constituent of the cell (Brücke 1862). For Schultze protoplasm and the

[2]Thanks to a cabinet directive issued by King Friedrich Wilhelm IV, Remak, a Jew, was permitted to lecture at the University of Berlin in 1847. This was the first time a Jew had been allowed to take up an academic appointment. Remak became associate professor in 1859.

nucleus were the bearers of all vital functions. He used the term "protoplasm" to describe the basic material of the cell and held that it exhibited a specific consistency that varied according to the kind of animal and type of cell. As early as 1868, T. H. Huxley characterized protoplasm as the "physical basis of life." The first volume of *Handbuch der physiologischen Botanik*, written by W. Hofmeister (1867), gives a comprehensive account of all that was known about the nature of the plant cell at that time.

Further progress in cytology, particularly where problems about the origin of the sex cells and fertilization are concerned, was largely dependent on two factors: the provision of more powerful microscopes, most of which were manufactured by the optical instruments firm founded by Zeiss in Jena in 1846, and the development of fixatives for preserving microscopic preparations.

Already in the 1830s, Jacobson used chromic acid, and H. Müller potassium bichromate, on microscopic preparations; Max Schultze introduced osmic acid in the 1860s. Hartung employed carmine for staining by about 1850, Waldeyer started working with hematoxylin in 1863, and at the same time Benecke was using aniline dyes for staining botanical preparations. Finally, when Wilhelm His originated the use of the microtome in 1870, it was possible to prepare extremely thin sections for the microscope. These new techniques gave rise to a profusion of studies in the biology of reproduction and fertilization. At the end of the century, the results of these investigations, together with the discoveries made in research on hybridization, presented a clear picture of the material basis of heredity.

Meanwhile, however, embryological studies led to a great deal of speculation about the causes of continuity between successive, isomorphous generations. For example, in *Letters to a Naturalist Friend* (*Unsere Körperform und das physiologische Problem ihrer Entstehung*, 1874), Wilhelm His (1831-1904) proposed seven principles for the hereditary transmission of characters:

1. The maternal embryo, or more strictly speaking the egg, is a substance which can be stimulated to grow.
2. Under certain circumstances which cannot at present be stated in general terms it is possible, as is shown in parthenogenesis, for internal causes to stimulate the egg to grow; it accordingly develops without having previously been fertilized.
3. When parthenogenesis does not occur, the egg requires contact with male semen to initiate growth.
4. Growth, a process regulated in terms of space and time, presupposes that the stimulation of growth is also a function of space and time.

5. If hereditary transmission is possible through the agency of semen, the effect exercised by the semen upon the egg must be a function of space and time.
6. If the egg embodies the conditions of hereditary transmission by the mother, the substance of the egg cannot be completely homogeneous. The capacity of the egg to be stimulated to grow must be different in different places either as a result of the irregular distribution of its mass or because of its varied constitution. The egg's capacity to be stimulated to grow must be a function of space.
7. When the law is given according to which the stimulating effect of the individual spermatozoids is temporally and spatially propagated, when the place and time of their entry into the egg are also given, and when the law is given according to which the egg's capacity for stimulation is spatially distributed, then the combination of these conditions determines the law governing the growth of the embryo and thus the whole of its subsequent development.

According to His, the fertilized egg possesses a stimulus to grow, and the whole range of possible hereditary transference, from both the paternal and the maternal sides, resides in this stimulus. As he says, "It is not the form that is transmitted, nor a specific formative substance, but a stimulus for structuring growth; not the characters, but the initiation of an analogous developmental process" (p. 152).[3]

In the thirteenth *Letter to a Naturalist Friend* His resumes his discussion of agency in the hereditary transmission of characters. He asks how it is possible for spermatozoa to transmit specific individual characters of the father, or of one of their ancestors, and finds no reason to suppose that the specific characters of the generative substances are directly influenced by each and every structure of the parental organism. In this connection he decisively rejects the inheritance of acquired characters. For His, the unbounded and continuous developmental process, as exemplified by successive generations, was due to two factors: the fact that the embryo is stimulated to develop as a result of the union of both generative substances and that the organization of the generative substance depends on the organization of the parents. He considered differences between children and their parents and the reappearance of characters from members of earlier generations as variations from a mean form that are caused by changing external conditions and sexual crossing.

[3]Anticipating His by twenty-five years, Virchow already expressed a similar view in 1849. He claimed that the ovum was the starting point of a series of processes initiated by the animation of the vital processes within the mother's body. As a result of the action of the spermatozoids the egg is stimulated for a second time and this further stimulation enables it to develop independently. During this process, moreover, a number of characteristics of the father's body are transmitted to the egg (Posner 1922).

The same period witnessed the beginning of a time of intensive cytological research which concentrated first on the continuity, structure, and function of cell nuclei, then on the details of the processes of fertilization, and finally on the minutiae of mitosis. Following one another in rapid succession, the works in which these researches were reported gave rise to various disputes concerning the authenticity of dates and authors of certain discoveries, and it is therefore sometimes difficult to appraise their relative merits.

In his historical essay entitled *Dokumente zur Geschichte der Zeugungslehre* (1918), Oscar Hertwig (Figure 39) provides a critical review of the important studies undertaken from 1870-1890 on mitosis and fertilization.

Hertwig was born on April 21, 1849, at Friedberg, Hesse. In 1868 he and his younger brother Richard began to study medicine at Jena under Ernst Haeckel. In 1871 the brothers went to Bonn, where they both graduated under Schultze and became assistants at the Institute of Anatomy. Oscar Hertwig's first work was in histology. Stimulated by a journey to the Mediterranean with Haeckel, however, he turned to the study of fertilization in the sea urchin egg. It was on the strength of this research that he was appointed lecturer at the University of Jena in 1875 and subsequently (1881) Professor of Anatomy. In 1888 he accepted the offer of an appointment at Berlin where a chair in General Anatomy, Histology, Theory of Development, and Comparative Anatomy had been established for him—a post he held until he retired in 1921. Hertwig died on October 25, 1922, in Berlin.

In a study that for a long time attracted little attention, A. Schneider (1873) recorded his observations of important stages in the karyokinesis of the eggs of Platyhelminthes. He saw that the dividing nucleus forms threads that gather together to produce a rosette in the equatorial plane and that one part of the threads finally moves toward one pole, the other part toward the opposite pole.

In the seemingly anucleate fertilized eggs of nematodes and mollusks, O. Bütschli (1848-1920) and L. Auerbach observed (1873, 1874, 1875) that two new nuclei were formed and fused together. Bütschli (1876) and E. Strasburger (1875) saw the vesicular nucleus transform into a nuclear spindle in a number of different experimental subjects. The studies undertaken by O. Hertwig (1875) on the maturation process of the sea urchin egg, *Toxopneustes lividus*, revealed that contrary to Haeckel's view the ovum does not pass

39 Oscar Hertwig (1849-1922)

through a moneron stage during the course of its development; in other words, there is no break in the series of nuclear generations during the maturation process. It was the substance of the nucleus, designated as nuclein and readily identified by staining techniques, that was recognized to be the essential factor. While Hertwig was observing the different stages of the maturation process of the egg in sea urchins, E. van Beneden (1845-1910) investigated the first stages of the development of the mammalian egg (1875, in the rabbit and in the bat), and Bütschli (1876) studied the formation and development of polar bodies in detail.

It was extremely important to understand the nature of the different events that occur during fertilization. Since the time of Spallanzani it had been realized that the spermatozoids play an essential part in fertilization, but precisely how fertilization is accomplished was not known until 1875. The clarification of this process was made all the more difficult by belief in the dogma that the spermatozoids were bearers of a catalytic substance which they imparted singly or in greater numbers, through contact or fusion with the yolk surface in order to stimulate development.

As late as 1873 Wundt, the physiologist, wrote in his *Lehrbuch der Physiologie des Menschen* (p. 750):

The essential condition for fertilization is most probably the penetration of the seminal particles into the contents of the egg, a phenomenon that has been demonstrated in the most varied classes of vertebrates. Penetration is effected through the canal, termed the micropyle, leading to the core of the egg and is possible chiefly on account of the mobility of the seminal elements. When these have penetrated into the egg, they very quickly lose their mobility and dissolve in the yolk. . . .

A theory—or even a well-founded hypothesis—to explain the nature of the operations whereby the seminal elements, having penetrated into the yolk, stimulate the developmental process has yet to be formulated.

We are indebted to O. Hertwig for the explanation of the most important events occurring during fertilization. He observed the fertilization of the sea urchin egg (*Toxopneustes lividus*) in the natural state and was able to record each stage of this process by means of stained microscopic preparations. In his habilitation dissertation (1875), he concluded that "fertilization depends on the fusion of sexually differentiated nuclei." This process had already been observed in mollusks and nematodes by Warneck, Bütschli (1875), and Auerbach (1874), and in mammals by van Beneden (1875); Hertwig now confirmed that it also occurred in amphibians and mollusks. H. Fol (1845-1892) was the first researcher to observe (1877), in *Toxopneustes lividus* and *Asterias glacialis*, the penetration of the egg by the spermatozoon, the formation of the fertilization cone, and the rise of the membrana vitellina.

At the same time, moreover, Strasburger (b. February 1, 1844, in Warsaw, d. May 15, 1912 in Bonn) became the first botanist to show that plants are fertilized in the same way as animals, namely, by the union of the egg nucleus with a sperm nucleus, known as the generative nucleus, from the pollen tube. While it is true that Strasburger does not consider the problem of fertilization in the first edition of his *Zellbildung und Zelltheilung* (1875), the second edition (1876) contains a special chapter on this subject. He assumed, however, that it is only the substance of the male nucleus, and not the nucleus itself, that penetrates the egg. One of the great botanists of his age, Strasburger, together with Noll, Schenck, and Schimper, edited in 1894 the *Lehrbuch der Botanik für Hochschulen* (*College Textbook of Botany*) the twenty-eighth edition of which was published in 1962.

With regard to studies in nuclear division and in the processes which it entails, Bütschli (1875) was the first to discover nuclear spindles in the animal kingdom, and Strasburger (1875) the first to discover them in the vegetable kingdom. With the aid of staining techniques W. Flemming (1843-1915) revealed the threads or rods situated in the central part of the nuclear spindle. He called them "chromatin" (1879), and W. Waldeyer (1888) characterized them as chromosomes. Strasburger and Bütschli referred to these chromosomes as the nuclear plate which divides into two halves which in turn move in opposite directions to form two daughter nuclei.

Further progress in understanding mitosis was achieved thanks largely to the work of W. Flemming, who studied the changes that occur in the chromatin threads of salamander larvae during the mitotic transformation of the mother nucleus into two daughter nuclei. He discovered that the chromatin threads divided longitudinally into two equal parts. As early as 1883, W. Roux had framed a hypothesis which explained the significance of this phenomenon (p. 15):

> The configurations of nuclear division are mechanisms that make it possible for the nucleus to divide not merely according to its mass but also according to the mass and constitution of its separate qualities. The essential process of nuclear division is the division of the mother grains; all other stages serve to guarantee the transfer of each daughter grain, derived by this division of the mother grain, into the center of both the daughter cells.

With regard to botanical subjects, L. Guignard (1883) and E. Heuser (1884) observed longitudinal division of the chromosomes in *Fritillaria*; this was also confirmed by Strasburger (1884). At about the same time, longitudinal division of the chromosomes was described in *Salamandra* (1885) by K. Rabl (1853-1917), in *Ascaris megalocephala* after painstaking investigation by van Beneden (1884), and also in *Ascaris* by M. Nussbaum (1883, 1884).

The discovery, during these investigations, that there are laws governing the number of chromosomes was extremely important. Flemming (1882), the first researcher to undertake work designed to ascertain these numerical laws, attempted to determine the number of chromosomes in *Salamandra.* Strasburger (1882) also counted chromosomes and established, for example, that twelve was the dominant number in Liliaceae. Investigating epithelial cells and the cells of connective tissue in Salamander larvae, Rabl (1885) found that there is a specific numerical law for each type of cell. This claim was corroborated by the work of Nussbaum and van Beneden, who found without exception that there are four chromosomes in the blastomeres and primitve spermacoblasts of *Ascaris.* Rabl (1885) also has the merit of being the first researcher to formulate clearly the hypothesis of the individuality and continuity of chromosomes. He assumed that the nuclear threads (chromosomes) contained in the resting nuclei at the final stage of nuclear division reappear as the same structures during the subsequent division of the nuclei. Rabl, Boveri, and van Beneden all claimed, however, that they were the first to make this discovery. Now while there is no doubt that Boveri used Rabl's findings as a starting point, he nonetheless modified Rabl's ideas in various ways with the result that he is often regarded as the originator of the theory of chromosome individuality. And we are indeed indebted to Boveri for producing the decisive evidence in favor of this theory and the centromeric theory of fertilization.

The son of a physician, Theodor Boveri (Figure 40) was born on October 12, 1862, in Bamberg. Keenly interested in music and the fine arts, he first studied the humanities but soon changed over to the natural sciences. He took up anatomy and biology in Munich, where he graduated under the anatomist C. von Kupffer in 1885. In 1887, after a number of visits to Naples, he was appointed lecturer in zoology and comparative anatomy at the University of Munich. From 1891 to 1893 he was an assistant to Richard Hertwig. He was elected Professor of Zoology and Comparative Anatomy in the University of Würzburg in 1893, a post he held until 1915. In 1913, on grounds of ill health,

40 Theodor Boveri (1862-1915)

Boveri turned down the offer of the post of director of the newly founded Kaiser Wilhelm Institute of Biology in Berlin-Dahlem. However, the department heads nominated by Boveri—Max Hartmann, Richard Goldschmidt, Hans Spemann, and Otto Warburg—were appointed, and together with Carl Correns established the great reputation of the Institute. Boveri died in Würzburg in 1915.

His research on a species of roundworm, *Ascaris megalocephala*, containing four chromosomes, revealed that in the nuclear divisions of the roundworm egg the same configuration of chromosomes always recurs in the nucleus before and after the resting stage; that is, an identical pattern of individual chromosomes invariably reappears in the daughter cells. Drawing on his own findings and those of Rabl, Boveri postulated the general thesis that chromosomes are independent individuals and preserve this independence in the resting nucleus. Although there was convincing evidence in favor of the theory of chromosome individuality, the theory was unfavorably reviewed, especially by O. Hertwig and R. Fick (1905, 1907). According to the latter, the individual chromosomes are decomposed and their elements variously recombined when the nucleus is at the resting stage. In this way chromosome individuality is not maintained. In the report on his work on *Ascaris megalocephala univalens*, which contains only two chromosomes, Boveri (1909) provided a decisive refutation of this view, and Fick subsequently came to agree with him. The morphological conception of chromosome individuality gradually had to give way to a conception framed more in chemicophysiological terms. Viewed as a problem of individual macromolecular structure and its identical propagation, this concept now constitutes a focal point in molecular-genetic research.

A proper understanding of the events that occur during mitosis was necessary to make further advances in the biology of fertilization. Most of these advances stemmed from research on *Ascaris megalocephala*. A. Schneider (1883) observed that the spermatozoon attaches itself to the egg and then penetrates it and that the sperm nucleus is preserved for a long time in the yolk. He mistakenly concluded, however, that the sperm nucleus is dissolved and that the two nuclei which subsequently appear are derived by division of the germinal vesicle. Nussbaum was later (1883, 1884) able to show that a single sperm nucleus will fertilize the egg.

Van Beneden (1883, 1884) made the most exact observations by showing that in *Ascaris* the egg nucleus and the sperm nucleus remain separated in the egg cell for an extended time. He also established that the fertilized egg contains in its nucleus two chromosomes from the father and two from the mother. Moreover, these divide longitudinally and move toward the two poles of the nuclear spindle, thus guaranteeing that each new cell contains four daughter chromosomes, of which two come from the egg nucleus and two from the sperm nucleus. He was also the first to ask how it is possible in *Ascaris* for the nuclei of the fertilized egg cell to contain four chromosomes, while the pronuclei have only two. Investigating oogenesis and spermatogenesis, he discovered certain processes and used the term "reduction" to describe them; thus we may consider him the originator of the theory of reduction division. He investigated the distinctive features of nuclear division, though he often misconstrued them, especially the nature of the polar bodies, and came to some false conclusions that were subsequently rectified by Boveri (1887, 1888).

O. Hertwig (1890) conducted yet another exact investigation into spermatogenesis in *Ascaris.* He confirmed and enlarged upon the sometimes incomplete conclusions of G. Platner's studies (1886, 1889) of the spermatocytes of *Paludina*, *Helix*, and *Limax* by accurately recording each stage of spermatogenesis as it occurred. He also discovered in *Ascaris megalocephala* the full details of the reduction process which occurs at the final stage of oogenesis and spermatogenesis with the formation of three polar cells and the division of the spermatocyte into four spermatids, respectively.

The explanation of chromosomal reduction posed the question of the importance of this phenomenon for subsequent generations. This problem is closely linked to the accounts given of the nature of polar bodies. Van Beneden explained the reduction process in the egg cell in terms of the expulsion of its male chromosomes as polar bodies. He claimed that the most important event in fertilization was the replacement of the male chromosomes which had been given off in the polar bodies by the chromosomes of another sperm cell which, while regenerating the egg, initiated its subsequent development. He also was the author of the theory of cell hermaphoditism, a doctrine which was refuted by the cogent arguments marshaled, above all, by O. Hertwig.

In the center of the aster in the dividing egg of the roundworm, van Beneden (1887) and Boveri (1888) discovered roundish particles which were also dividing. They considered these particles as foci that provide a stimulus for the astral rays, and the rays themselves as a means of apportioning the chromosomes to the daughter cells. Boveri inquired into the origin of the centrosomes and found that during fertilization the middle piece of the spermatozoon penetrates the egg together with the sperm nucleus. This middle piece contains the centrosome, which by division produces the centrosomes of the spindle of the first cleavage; from these all the centrosomes contained in the subsequent body cells of an individual are derived. Boveri's theory of fertilization thus combined two processes: the combination and transmission of parental chromosomes containing genetic material, and the operations of the organ that stimulates cell division, the sperm centrosome, which is necessary for the further development of the egg. Thus, whenever a cell divides, two cyclical processes occur: the regularly repeated divisions of the centrosome leading to the formation of nuclear spindles, and the equally regular formation of the chromosomes and the nucleus. Although both processes are largely autonomous, they are coordinated when the astral rays meet and form the spindle, which then actively disperses the dividing chromosomes toward the two poles. While those parts of Boveri's theory of fertilization that deal with the formation of the chromosomes and the nucleus are still accepted as true today, belief in the importance of the role of the centrosome in the process of division diminished when it was shown that eggs can be made to develop without having been fertilized. Thus Jacques Loeb (1899) was able to induce "chemical parthenogenesis" in the sea urchin egg with the aid of chemical agents. Moreover, it was subsequently realized that in the higher plants development can be stimulated without the aid of centrosomes.

Boveri (1889) finally managed to establish the parity, with regard to subsequent physiological development, of the paternal and maternal hereditary material. Using the shaking method developed by the Hertwig brothers in 1887, he reduced sea urchin eggs to fragments and then fertilized anucleate fragments which developed into normal (that is, haploid paternal) larvae. Conversely, eggs containing only the chromosome material of the egg nucleus also developed into normal (that is, haploid maternal) individuals. This demonstrated that the sets of chromosomes in the egg and sperm nuclei are matched not merely morphologically but also in their influence on subsequent physiological development and as bearers of hereditary traits.

A. Weismann (Figure 41) provided yet another explanation of the polar bodies. It stemmed from his theory of the germ plasm in conjunction with his hypothesis of the continuity of the germ plasm, together with a number of subsidiary hypotheses.

The son of a classical linguist, August Weismann was born on January 17, 1834, in Frankfurt am Main. From 1852 to 1856 he studied medicine at the University of Göttingen where he received his

41 August Weismann (1834-1914)

degree in 1856. He then became an associate at the surgical clinic of the University of Rostock and subsequently at the Chemical Institute. Later, while practicing medicine in Frankfurt, he became concerned with problems of theoretical biology and histology. He abandoned his medical practice and went to Paris. In 1861 he visited Leuckart in Giessen and accepted the post of personal physician to Archduke Stephen of Austria in Schaumburg, where he would also have time for zoological research. From 1861 to 1863 he worked on a book on the development of diptera. On the strength of this work he was appointed lecturer in zoology and comparative anatomy at the Faculty of Medicine at Freiburg in 1863. One year later, a chronic eye affliction compelled him to abandon work with the microscope, though he was able to resume his research for a few years after 1874. He was appointed associate professor in 1865, and director of the Zoological Institute in 1867. Like his close friend Haeckel, Weismann was one of the pioneers of the theory of descent, though in sharp contrast to Haeckel he was thoroughly opposed to the idea of acquired characters. Weismann died on November 5, 1914, in Freiburg.

Weismann drew a distinction between the actual idioplasm or germ plasm which transmits the total complex of hereditary factors required to preserve the species—an idea anticipated by Nussbaum (1880)—and the histogenous nucleoplasm, derived by decomposition of the germ plasm, which contains merely fragments of the latter. This distinction was based on the view of animal breeders—also adopted by Galton and still believed in some quarters today—that the hereditary contribution from ancestors to descendants diminished in mathematical proportion with the remoteness of the ancestors until it is finally no longer divisible. For this reason Weismann postulated a reduction of the ancestral plasms through the elimination of the polar bodies. In the course of this reduction, the quantity of ancestral plasm that is removed is equal to the quantity reintroduced during fertilization. He thus distinguished two kinds of cell division: ordinary mitosis in which a maternal chromosome divides into two halves and a kind of division in which the undivided maternal chromosomes form two distinct groups in the equatorial plate with the result that each daughter nucleus contains only half the total number of chromosomes. Weismann termed the first kind "equation division," the second "reduction division." He held that the histogenous nucleoplasm is expelled in the chromosomes of the first polar body, while the expulsion of the second polar body effects a reduction division whereby half of the various ancestral germ plasms are eliminated. Weismann thus postulated the reduction of ancestral plasms for each successive generation.

The fact that these ancestral plasms were reduced by half did not mean that maturation divisions were no longer significant. According to

Weismann, the latter in fact constituted an important cause of individual variations, e.g., the fact that the children of the same parents are never exactly alike. Since the halving of the chromosomes can take place in different ways, very different ancestral plasms are eliminated from the countless germ cells of a given individual in maturation division, depending on how the chromosomes are arranged in groups at the onset of division. Because of the qualitative differences between chromosomes in the two sexes, i.e., the differences in the ancestral plasms, new combinations of ancestral plasms always appear during fertilization. A characteristic feature of sexual reproduction, such new combinations cause children to exhibit, though always in different proportions, the characters of their parents and earlier ancestors.

O. Hertwig (1890) considered the problem of the significance of reduction division in a more straightforward fashion. Nägeli had already seen the problem which arises from the union of two idioplasms: It is simply that the volume of idioplasm would necessarily increase with each successive generation unless there were processes that permit the union of the parental idioplasms, yet prohibit the corresponding increase in the volume of idioplasm that such a union entails. Hertwig posed the question: "How, in view of the union of two idioplasms which occurs on every occasion of reproduction, is a summation of idioplasm avoided in successive generations?" He gave the answer, based on observed facts: "Because the paternal and maternal idioplasms are both divided into two equal parts just before fertilization."

If the idioplasm is considered as being contained in the cell nucleus, the question can be formulated in a more precise manner: "By means of what process are the volume of chromatin and the number of chromosomes maintained at the same level, as required by the empirically determined law governing the number of chromosomes for a given species of animals or plants, since a duplication of these parts must occur with the union of two nuclei which takes place during fertilization?" The answer is as follows: "The summation of the chromatin which would necessarily occur with the fusion of two nuclei during fertilization is counteracted in cell life by an opposite process known as reduction."

The understanding of the function and significance of reduction division was the first vital step toward clarifying the localization of the idioplasm as the bearer of hereditary characters. When putting forward his speculative idioplasm theory, Nägeli had not commented on this issue and simply took it for granted that idioplasm, in the form of cords within the cell membrane, covered the surface of the cells and was concentrated in the nucleus. Independently of one another Hertwig and Strasburger further developed in 1884 the idioplasm theory into a new theory of heredity on the basis of what they had learned from their study of fertilization.

Hertwig's thesis—that fertilization depends on the fusion of the egg and sperm nuclei—entailed that the nuclear substance, and not the protoplasm, is the vital fertilizing agent, that the nuclear substance acts as a structured, organized component, and that fertilization is therefore a morphological process susceptible of observation: "Since the transmission of the father's characters to the animal which arises from the egg is a process necessarily bound up with fertilization, a further obvious conclusion may be drawn from the theory we have formulated: the nuclear substances are also the bearers of the hereditary characters transmitted by parents to their descendants" (1884, p. 1).

One year later Weismann (1885) adopted the theory of nuclear idioplasm. According to his theory of the germ plasm, a part of the germ substance from which a new individual develops is unused and remains constant and thus forms the basis for all subsequent generations. He described this process as the "Continuity of the Germ Plasm" and held that the germ plasm is transmitted, thanks to the "germ track" (*Keimbahn*), from cell to cell until it cuts loose of the soma. F. Galton (1876) and G. Jäger (1876) had already outlined a view of this kind, but it was Weismann who first drew out all the implications of this conception. In his early works Weismann had not specified where the germ plasm was located, but in 1885 he pronounced that "the nuclear substance alone can be the bearer of hereditary tendencies." He, however, did not thereby abandon his speculative theory of the dissection of ancestral plasm, a theory that owed much to Galton. With the help of numerous subsidiary hypotheses Weismann was now able to formulate a theory of heredity. Hertwig (1918) subsequently gave a penetrating criticism of this theory. Weismann (1885, 1892) developed and enlarged the theory along speculative lines to a far greater extent than did Hertwig, who constructed his theory solely on the basis of facts established by experiment. The studies of van Beneden, Rabl, and Boveri provided good backing for Weismann's theory of heredity. Like Hertwig, he assumed that Nägeli's idioplasm could be identified with the chromatin material of the nucleus and that pangenes or determinants were located in the chromosomes of the cell nucleus, which Weismann described as "ids." Hereditary variations in certain organs or characters were attributed to changes in the relevant determinants. As Weismann says: "A determinant is nothing more than a living element of the germ substance, and its presence in the germ conditions the appearance and the specific development of a particular part of the body." Every determinant is a self-sustaining entity capable of growth and reproduction; determinants are present in specific patterns in the germ plasm. For a determinant to become operative, it must first be conveyed to the part of the body which it is to condition. Weismann claimed that this was effected by means of an unequal division which distributed the determinants in an uneven manner, depending upon the part of the organ in which they are to be effective. Thus the determinants can be either in an active or in a passive state. They become active when they are reduced to biophors, the ultimate particles of which they are composed, which are then able to pass through the nuclear membrane into the body of the cell. Biophors, which consist of groups of molecules, are able to reproduce by division; hereditary variations can result from changes in the configuration of these molecules or in the arrangement of the atoms that form a given molecule. The biophors do not leave the nucleus and enter the cell body until the determinant has attained a state of maturity; it is this movement of the biophors that has a continuous influence upon the quality of the cell. Weismann held that the number of biophors was extremely large "since the total of all bodily parts capable of independent, hereditary variation must represent an exact measure for the number of minute, vital particles that constitute the germ plasm."

It is particularly difficult to evaluate Weismann's theory of the germ plasm since the meanings of the concepts—especially the "id"—used to formulate the theory were modified as scientific knowledge increased. In the first edition of *Vorträge über Deszendenztheorie* (1902) an "id" refers to the total complex of determinants which contains, in specific and regular patterns, the hereditary factors for a whole individual; subsequently, however, the notion of an "id" is identified with the concept of a "chromosome." But the chromosomes, that is to say various groups of determinants that appear as compact units, represent nonequivalent complexes comprising a limited number of units; each chromosome merely contains a specific number of determinants.

The concept of the germ track that, according to Weismann, secures the continuity of the germ plasm throughout successive generations must be rejected, at least as far as the higher plants are concerned, for it is clear that many of the body cells in plants are potentially capable of forming a new individual having precisely the same complex of hereditary factors (Kölliker 1885). To belittle Weismann's importance in the growth of biological knowledge at the turn of the century by criticizing his notion of the germ track (which stemmed from his own investigations into the origin of sex cells in hydromedusae) would, however, be very much mistaken. Ever ready to revise earlier views in the light of new scientific advances, Weismann, who had been trained in several branches of biological research, possessed the qualities of a true scientist. In fact, many of his ideas provided the groundwork for the theories concerning the nature of the material basis of heredity which were developed after the turn of the century during the classical period of genetics.

At first sight Weismann's theory of determinants, with its assumption of the existence of minute, material particles, seems comparable to Darwin's provisional hypothesis of pangenesis. A crucial difference between the two theories becomes apparent, however, when we consider the relation of the determinants to the body cells. In Darwin's theory, the gemmules are derived from the body cells, and the nature of the gemmules is conditioned by the body cells. According to Weismann, however, it is the determinants that are the decisive factor, since they determine the characters of the body cells. Hence, changes in the body cells that are caused by use or disuse or as the direct result of external influences cannot be inherited, for the changed body cells are not capable of exercising a similar influence on the germ plasm. The supposition that such changes *are* inherited entails that the changed body cells are able to modify the appropriate, passive determinants of the germ plasm in such a way that they will be capable of producing "determinates" modified in the same direction. Weismann conducted experiments with mice in order to examine the heredity of acquired mutilations, which proved that such characters are not transmitted. From our present viewpoint, his experiments seem rather rudimentary and limited both regarding the question they were designed to answer and the manner in which they were conducted. But they were essential at a time when there was so much hazy talk about unexamined phenomena which were alleged to be cases of the inheritance of acquired characters. The same also holds true for Weismann's views, within the context of the germ plasm theory, on the causes of variability in organisms. To begin with, however, he thought that variability could be traced back to changes resulting from external influences in the developmental tendencies of the germs, though limits were set to such changes, since the developmental tendencies were regulated by the specific nature of the germs. A second cause of variability is seen in the phenomenon of amphimixis; that is, in the mingling of unlike germs during sexual reproduction. A third cause lies in germinal selection: hereditary variations which are determined by the varying nutriment supplied to the determinants.

In the thirty-first of his *Lectures on the Theory of Descent*, Weismann writes concerning germinal selection and its role in producing hereditary variations (Vol. 2, p. 300):

> We might expect on a priori grounds that not only the random fluctuations of nutrition within the germ plasm would cause its elements to vary in this or that direction, but that there would also be influences of a more general kind, in particular those of nutrition and climate, which, to begin with, would affect the body as a whole, but with it also the germ plasm, and which would thus occasion variation, either in all or only in certain determinants. . . .
>
> This is in fact the case; there is no doubt that external influences, such as those emanating from the environment in which a species lives, are able to cause direct variations in the germ plasm, that is to say permanent, because hereditary variations. We have already referred to this process and characterized it as 'induced germinal selection.'

The view that there is a hereditary substratum composed of tiny, material particles is by far the most important of Weismann's contributions which were substantiated and extended in the course of the genetical researches undertaken during the first decades of the twentieth century. Localized within the nuclear material, these particles cause the fertilized egg cell to become a complex, differentiated structure which, composed of innumerable tiny units, is capable of exercising a formative influence on every part of the developing organism. These determinants are arrayed in groups within the chromosomes, and each determinant provides the material substratum for a terminal product which appears during ontogenesis. Determinants can be modified through the direct action of external influences and can thus give rise to hereditary variations. When Weismann first developed these views they were considered hypothetical and speculative. Subsequently, however, many of these views could be formulated more precisely by means of exact research, a fact which shows the seminal importance of Weismann's fundamental work and unflagging efforts for the scientific studies of his time and also reveals how he pointed the way toward further developments.

We have now completed our brief account of cytological studies in the morphology of the cell, in mitosis, and in fertilization—studies which eventually led to the discovery of reduction division—together with our account of the theories of heredity, derived from this discovery, that were proposed by so many outstanding researchers between 1870 and 1890. This review must suffice to show how two different areas of research, hybridization and cytology, were gradually drawn together in a way that ultimately led to a reasoned statement, supported by experimental evidence, of the material basis of heredity. A crucial period during this process was the last decade of the nineteenth century, when researchers experimenting with hybridization regained the precision demanded by Gregor Mendel by determining the numerical proportions, in the descendants of hybrids of closely related families, of those characteristics which were incorporated in the hybridization.

9 The Rediscovery of Mendel's Laws of Heredity

One of the last great nineteenth-century experiments designed to establish the laws governing the inheritance of individual characters was undertaken by a zoologist, Wilhelm Haacke, who conducted experiments in the hybridization of mice. His work (1893-1895) approaches Mendel's results even more closely than Naudin's, although Haacke did not record any numerical data. He held that some of the characters he investigated were located in the cytoplasm, others in the nucleus.

Haacke (1855-1912) took his degree under Haeckel in 1878 and was an assistant at the Zoological Institute at Jena in 1878 and 1879. After a short period as an associate under Moebius at Kiel, he emigrated in 1881 to New Zealand, where he first worked in a museum at Christchurch and, one year later, became the head curator of a museum in Adelaide. From 1888 to 1892 he was director of the zoological garden at Frankfurt am Main, and he subsequently taught at the Darmstadt Institute of Technology.

It is evident that Haacke's views were sharply opposed to those of A. Weismann, whom he attacked along with O. Hertwig and other scientists of his time. In the preface to his treatise *Gestaltung und Vererbung* (1893), Haacke writes:

> On the occasion of these demonstrations . . . I am submitting a theory of formation and inheritance which has slowly been perfected during the course of twelve long years, years that I have spent upon the blue waves of the ocean and in the green pastures of New Zealand, in the gloomy forests of New Guinea and in the bright coral reefs of the Torres Strait, in broad open spaces under the Australian sun and in the lush beech forests of my native Germany, hunting and fishing, in museums and in zoological gardens, working with the microscope and examining specimens in breeding cages, at zoological exhibitions and in the auditorium—for such have been the fortunes of my life.

Haacke experimented with two breeds of mice: common white mice (climbing mice), and gray Japanese waltzing mice. He held that the waltzing factor was located in the polar body and its cytoplasm, the gray factor in the nucleus or chromosomes. During reduction division in the hybrids, the different kinds of cytoplasm which have been combined in the cross are separated as are also the different nuclear substances. When t denotes the cytoplasm in the waltzing mice, s its "gray" nuclear substances, k the cytoplasm in the climbing mice, and w its "white" nuclear substances, then during reduction division in the hybrids t is separated from k, and s from w, which four can be mutually united. It follows that there are four possible kinds of reduced germ cells in the hybrid mice: *ts*, *tw*, *ks*, and *kw*. All four kinds occur in both sexes. The fertilized egg cells must therefore contain one of the sixteen possible combinations of the hereditary materials of the parents:

1. *ts, ts*	5. *tw, ts*	9. *ks, ts*	13. *kw, ts*
2. *ts, tw*	6. *tw, tw*	10. *ks, tw*	14. *kw, tw*
3. *ts, ks*	7. *tw, ks*	11. *ks, ks*	15. *kw, ks*
4. *ts, kw*	8. *tw, kw*	12. *ks, kw*	16. *kw, kw*

Haacke continues (1893, *Biologisches Zentralblatt*, p. 532):

For we are concerned with "iterated variations from four elements to the second degree." Of these sixteen combinations, 1 gives gray waltzing mice that resemble one of their grandparents; 6 gives white waltzing mice that resemble one grandparent in respect to their morphological characters, the other in respect to their color; 2 also gives waltzing mice, though their coloring is a blend of that of their grandparents; the same holds true for 5. The remaining combinations signify, in part, pure climbing mice, like those in 16 which completely resemble one of their grandparents, or others whose coloring is a blend of that of their grandparents; in part, however, they signify hybrid mice: on the one hand those whose coloring is identical with that of one of their grandparents, and on the other hand those whose coloring is a blend of that of both grandparents. According to our hypothesis, if we now bred *inter se* the mice of combination 1, that is, gray waltzing mice, the offspring from such unions would all be gray waltzing mice; if we were to breed *inter se* the mice of combination 2, we should obtain both gray and white mice, whereas from the mice of combination 6 we should always obtain white waltzing mice; but in all these cases we always obtain waltzing mice. It is just as easy theoretically to specify what we should obtain were we to breed *inter se* the mice of the remaining combinations, and I shall therefore refrain from considering these cases in any detail. In theory, therefore, we can specify in advance what will happen even in cases where we assume that both "gray" and "white" chromosomes appear in the cytoplasm of waltzing mice as a result of the reduction division of their germ cells stemming from hybrid mice. I do not know whether the number of chromosomes present in mice has been recorded, but this number would enable us to establish the possible combinations. However, I do not believe that it often happens that different chromosomes appear in cytoplasm which has undergone reduction, for all of my experiments deny this possibility. From these experiments it follows that the germ cells of the mice must behave precisely as in our hypothetical example. When the first generation of gray waltzing mice are crossed with white climbing mice and the resulting hybrid mice bred *inter se*, the third generation yields: 1 further gray waltzing mice; 2 white waltzing mice; 3 gray climbing mice; 4 white climbing mice; 5 piebald waltzing mice; 6 piebald climbing mice; 7 gray hybrid mice; 8 white hybrid mice; 9 piebald hybrid mice. When each of these types is subsequently bred *inter se*, we see that all of these combinations can be produced. Pure gray waltzing mice produce always and only pure gray waltzing mice, pure white climbing mice produce always and only pure white climbing mice. Appropriate breeding experiments reveal that the mice with mixed characters can be analyzed in terms of white and gray waltzing mice, and white and gray climbing mice. In many cases we very quickly reobtain pure-bred animals, i.e., animals that transmit their characters without exception, never yielding a reversion to type.

Thus we see that Haacke, like Mendel, designed a hybridization project involving two pairs of differentiating characters. Haacke's mistake, however, was to assume that one pair of characters (waltzing, climbing) was located in the cytoplasm, while the other (white, gray) was located in the nucleus; he was also mistaken in considering both pairs of characters to be units. A cross involving three pairs of differentiating characters would have shown that this view was untenable. For Haacke, the nucleus functioned solely in respect to the inheritance of the chemical characteristics of the adult organism, irrespective of its own morphological characters, while the inheritance of the morphological characteristics of the developed organism turned exclusively on the centrosome and the plasm of the cell body whose elements, according to Haacke's hypothesis, resemble that of the centrosome. This view is connected with Haacke's Theory of Gemmaria. "Gemmaria" are the individuals from which the egg plasm is formed; they are composed of "gemmae" and have a firm texture.

A few years later G. von Guaita (1898) conducted some very similar experiments in cross-breeding at Weismann's Institute in Freiburg. His results were not as clear-cut as those of Haacke and they were interpreted throughout in the light of Weismann's theory of determinants.

Guaita crossed albino house mice with piebald Japanese waltzing mice. In the first hybrid generation he found a complete uniformity of color (gray) and of behavior (absence of the waltzing character). In the second hybrid generation, however, as in Haacke's experiments, segregation of the characters occurred. Guaita summarized his results as follows (p. 331):

When crosses are made using two varieties of the house mouse, the Japanese waltzing mouse and the albino house mouse, . . . the second generation (= F_1) always yields gray mice that resemble the wild mouse in size and temperament. Thus . . . a complete reversion to the ancestral form takes place.

When these gray mice of the second generation are crossed with one another, the third generation (= F_2) yields considerable variation with regard to coloring, temperament, and size: some of the offspring inherit the characteristics of their parents, that is, those of the house mouse (= F_1), others revert to the type of their grandparents, the albino mouse or the Japanese waltzing mouse, while yet others are of an intermediate form between these different types. The fourth generation mice (= F_3), obtained from crosses between the mice of the third generation (= F_2), yielded results that seemed identical with those of the third generation.

In a subsequent paper (1899) Guaita reported on specific, arbitrarily chosen forms up to F_6 without obtaining results that differed from those of his earlier experiments.

The defects of Guaita's experimental work lie in the fact that from the highly diverse F_2 generation he selected and bred only certain particularly conspicuous animals. In doing so, he failed to consider all of the possibilities inherent in his cross-breeding experiments, and his analysis, involving only a small number of animals, accordingly failed to reveal any of the laws governing the inheritance of the individual characters under study.

The same criticism applies to a work published by the plant breeder Wilhelm Rimpau (1891), who wrote about natural and artificial hybrids of cereals, particularly wheat and peas. He did not know of Mendel's essay but he definitely was familiar with Focke's treatise. Yet he could have easily become acquainted with Mendel's work, the more so as Focke liberally quotes from Mendel's experiments on pea crossings. According to Rimpau, the characters of the first-generation hybrids are usually intermediate between those of their parents. But several generations of the progeny of these hybrids showed a high degree of morphological variation. While a complete reversion to one or another of the parental types was what usually occurred, a combination of both parental types and all kinds of other intermediate forms were also encountered. Moreover, new characters which were not present in either of the parents were occasionally observed.

During the closing years of the nineteenth century, however, one American and two Swedish scientists experimented with cereals and leguminous plants and came very close to discovering the laws governing the inheritance of individual characters.

At the Agricultural Congress held in Stockholm in 1897, P. Bolin and H. Tedin from Svalöf reported on experiments, undertaken since 1890, in which they had crossbred different varieties of barley, peas, and vetches (Åkerman and MacKey, 1948). They too found no variation in F_1, but they discovered in the F_2 and subsequent generations all the possible combinations of the parental characters and were able to estimate in advance, with a fair degree of mathematical accuracy, the ratios in which the characters appeared.

H. Brücher (1950) gives details of a paper presented by Pehr Bolin at the same congress. The following quotation shows how close Bolin came to making Mendel's discovery: "With regard to the heredity of the 'typical characters' of the second and subsequent generations there seems to exist a certain regularity. For the forms appearing in these generations represent all the possible combinations of parental characters and can therefore be calculated in advance with mathematical accuracy."

At the Fifteenth Annual Convention of the Association of American Agricultural Colleges and Experimental Stations, W. J. Spillman (1901) gave a report on the crosses he had made since 1899 between winter and spring wheats. He also found no significant variation in the first generation, but he observed, in F_2, a remarkable segregation into distinct types, many of which were simple combinations of the characters of the parents. Similar crossings gave the same results, which led Spillman to conclude that each hybrid produces distinct types, and possibly does so in certain proportions. To quote him: "When these results were classified, they confirmed the above suggestion;. and if similar results are shown to follow the crossing of other groups of wheats, it seems entirely possible to predict, in the main, what types will result from crossing any two established varieties, and approximately the proportion of each type that will appear in the second generation" (p. 88).

Spillman also found that the type that appeared most frequently in F_2 resembled the F_1 plants most of all. He realized that quantitative studies provided the key to understanding the inheritance of individual characters. H. F. Roberts (1929), who analyzed Spillman's data, established that the characters under study (length of the ear, absence of awns, pubescence and color of the glumes, and so forth) segregated in the ratios of 1:2:1 and 3:1.

The definitive clarification of the laws governing the inheritance of individual characters was finally achieved thanks to the work of three men, a Dutchman, Hugo de Vries; a German, Carl Correns; and an Austrian, Erich Tschermak, who, independently of each other, undertook experiments in plant hybridization. It is probable that none of the three had heard of Mendel when they first began to experiment, and they all thought they had discovered something entirely new, namely the laws of segregation of hybrids. They did not become acquainted with Mendel's essay until their experimental work was well under way.

Hugo de Vries (Figure 42) was born in Haarlem on February 16, 1848. He graduated from the University of Leiden in 1870, worked as an assistant under Hofmeister at Heidelberg, and then taught for four years at a school in Amsterdam. In 1875 he went to Würzburg where he worked under Sachs, studying the most important agricultural plants. In 1877 de Vries received a degree from Halle University and, after lecturing there for a while, was appointed lecturer in plant physiology at Amsterdam. In 1878 he became an associate professor and three years later a professor. His early work was in respiration, cecidology,

42 Hugo de Vries (1848-1935)

43 Carl Erich Correns (1864-1933)

and osmosis, and he did not begin his study of variation until 1880. He remained in Amsterdam until 1918 and then lived in Lunteren, where he died on May 21, 1935.

Carl Erich Correns (Figure 43) was born in Munich on September 19, 1864. After receiving a thorough background in botany from Nägeli, he graduated from the University of Munich, where he had undertaken research into the structure of the cell membrane. His studies in anatomy and physiology under Haberlandt at Graz and under

Schwendener in Berlin and his work in the physiology of irritability under Pfeffer at Leipzig helped to broaden his experience. In 1892 he went to lecture at Tübingen, where he began his analysis of hereditary processes. Ten years later he was appointed associate professor at Leipzig, and in 1909, at the age of forty-five, he was appointed to a professorship at Münster. He left Münster in 1914 to become director of the Kaiser Wilhelm Institute of Biology in Berlin-Dahlem. He died on February 14, 1933, in Berlin.

Erich von Tschermak-Seysenegg (Figure 44) was born in Vienna on November 15, 1871. His early studies were at the College of Agriculture and Forestry in Vienna and at the University of Vienna. Then, from 1893 to 1895, he studied agriculture at the Halle-Wittenberg University, where he took his degree under the botanist Gregor Kraus. He decided to specialize in plant breeding and practiced in Stendal, Quedlinburg, and Ghent, where in 1898 he began a series of experiments in which he crossed flowers and peas. He pursued these experiments the following year on an estate owned by the Austrian imperial family at Esslingen near Gross-Enzersdorf and finally completed them in Vienna. He received his first academic appointment in 1900 at the College of Agriculture and Forestry, where in 1902 he became lecturer in the department of plant production. In 1903 he was made associate professor and three years later he was appointed to a chair in plant breeding—the first in Europe. Tschermak was made a full professor in 1909 and was thus able to influence deeply the subsequent development of plant breeding in Austria. He died in Vienna on October 11, 1962.

At this point we must also remember the work of another scholar, William Bateson, who was intimately associated with the rediscovery of Gregor Mendel's laws of heredity and was the first to translate his studies into English. He was without doubt the most fervent supporter of genetics in his country after 1900.

Bateson (Figure 45) was born at Whitby on August 8, 1861, and was educated at St. John's College, Cambridge. His first scientific work—in the field of comparative developmental history—consisted of morphological studies of the little-known worm *Balanoglossus*. From 1886 to 1887 he took part in an expedition to the western part of Central Asia, where he studied evolutionary problems in Crustacea and mussels. In the years that followed he devoted all of his energies to the study of variation, with particular reference to the problems of "discontinuous variation" in evolution. His far-reaching discoveries in this field are

44 Erich von Tschermak-Seysenegg (1871-1962)

recorded in *Materials for the Study of Variation Treated with Especial Regard to Discontinuity in the Origin of Species*, published in 1894.

45 William Bateson (1861-1926)

The two biometers Weldon and Pearson, who advocated the importance of continuous variation as a factor in evolution, opposed Bateson in a long and bitter controversy.

Around 1897 Bateson began to conduct experiments in hybridization with *Lychnis*, poultry, and butterflies. These experiments taught him that the hybrid progeny should be examined statistically, each character being considered separately. From 1899 to 1910 he experimented in the garden of Merton House at Granchester near Cambridge. He was appointed Professor of Biology at Cambridge in 1908 and founded the so-called Cambridge school of genetics. In 1910 he became director of the John Innes Horticultural Institution at Merton, which thanks to his efforts became a center of genetical research in England. Bateson died at Merton on February 8, 1926.

In 1899 the Royal Horticultural Society held an International Conference on Hybridization at which Bateson read a paper on hybridization and cross-fertilization as a method of scientific investigation. It is clear from this paper that he imposed the same requirement as Mendel had done more than thirty years previously—a requirement, moreover, that was to be formulated by de Vries, Correns, and Tschermak one year later. As Bateson says (p. 63):

> What we first require is to know what happens when a variety is crossed with its nearest allies. If the result is to have a scientific value, it is almost absolutely necessary that the offspring of such crossing should then be examined statistically. It must be recorded how many of the offspring resembled each parent, and how many showed characters intermediate between those of the parents. If the parents differ in several characters, the offspring must be examined statistically, and marshalled, as it is called, in respect to each of those characters separately. . . . All that is really necessary is that some approximate numerical statement of the result should be kept.

The statistical analysis of the offspring and the separate investigation of each individual character were the requirements laid down by Bateson; all attempts to understand the inheritance of alternative individual characters without meeting these requirements were doomed to failure. Bateson did not become acquainted with Mendel's essay until 1900. He did realize, however, that in some crossings, e.g., in *Matthiola incana*, the union between a hairy and a glabrous variety produces offspring that are either completely hairy or completely glabrous; whereas in other cases, for example, when a hairy variety of *Biscutella laevigata* is crossed with a glabrous variety, intermediate forms also occur and subsequently segregate.

At a meeting of the Royal Horticultural Society held during the year in which Mendel's laws of heredity were rediscovered, Bateson gave a report on all the experiments that were held to confirm those laws. In his report to the Evolution Committee of the Royal Society the following year, Bateson coined the term "allelomorphs" for the members of an antagonistic pair of characters, the term "heterozygote" for a zygote formed by the union of a pair of different allelomorphs, and the term "homozygote" for the case in which two allelomorphs of the same type are united.[1]

The science of heredity thus owes its name, "genetics," to Bateson. He first used this term in a letter to a friend, Professor Adam Sedgwick:

If the Quick Fund were used for the foundation of a professorship relating to Heredity and Variation the best title would, I think, be "The Quick Professorship of the study of Heredity." No single word in common use quite gives this meaning. Such a word is badly wanted and if it were desirable to coin one, "Genetics" might do. Either expression clearly includes variation and the cognate phenomena.

The word *genetics* was first printed in a review by Bateson (1906) of a book by Lotsy (published in *Nature*, Volume 74, and reprinted in *Scientific Papers of William Bateson*). In it, Bateson writes (p. 91):

As the moment is favourable, may it be suggested that the branch of science the rapid growth of which forms the occasion of Professor Lotsy's book should now receive a distinctive name? Studies in "Experimental Evolution" or in the "Theory of Descent," strike a wrong note; for, theory apart, the physiology of heredity and variation is a definite branch of science, and if we knew nothing of evolution that science would still exist. To avoid further periphrasis, then, let us say *Genetics*.

Bateson first publicly suggested that the new science be called genetics at the Third International Conference on Genetics in 1906 (report published in 1907) in his paper entitled "The Progress of Genetic Research." He writes:

Like other new crafts, we have been complelled to adopt a terminology which, if somewhat deterrent to the novice, is so necessary a tool to the craftsman that it must be endured. But though these attributes of scientific activity are in evidence, the science itself is still nameless, and we can only describe our pursuit by cumbrous and often misleading periphrasis. To meet this difficulty I suggest, for the consideration of this Congress the term *Genetics*, which sufficiently indicates that our labours are devoted to the elucidation of the phenomena of heredity and variation: in other words, to the physiology of descent, with implied bearing on the theoretical problems of the evolutionist and the systematist, the application to the practical proolems of breeders, whether of animals or plants. After more or less undirected wanderings we have thus a definite aim in view.

[1] A posthumous study of Bateson and Correns by O. Renner (1961) offers an illuminating comparison of the achievements of the two men in genetics; Renner also points out certain misconceptions concerning the interpretation and the limitations of their discoveries. He emphasizes, moreover, how understanding in scientific matters can be facilitated through the use of carefully chosen terminology; in this connection he says of Bateson: "Purity of the germ cells, homozygote and heterozygote, epistatic and hypostatic, allelomorph (subsequently abbreviated to allele)—all of these terms together with the name for the new science of heredity and variation, Genetics, were coined by Bateson. Wherever geneticists are assembled, Bateson is among them—in their technical language" (p. 8)

As early as 1909 Bateson published *Mendel's Principles of Heredity.* This book presents an account of the experiments pertaining to the inheritance of individual characters that were conducted by Bateson and his followers.

We shall always remember this great pioneer of genetics and esteem his work, which is no less important than that of the three rediscoverers; Bateson may safely be included among that group of men who together opened wide the gateway to further understanding in our science.

One year after Bateson had developed the principles for the study of the inheritance of individual characters, de Vries, Correns, and Tschermak published, in rapid succession, the results of their investigations.

De Vries was the first. In 1900 he published two papers: "*Sur la loi de disjonction des hybrides*" (Concerning the Law of Segregation of Hybrids) (*Compt. rend. de l'Acad. Sciences,* Paris) and "*Das Spaltungsgesetz der Bastarde*" (The Law of Segregation of Hybrids) (*Ber. Dtsch. Bot. Ges.* Vol. XVIII, received March 14, 1900). There is no mention whatever of Mendel in the French paper, even though it was sent to be printed twelve days later (March 26, 1900) than the German one in which we read (p. 85n): "This important study [Mendel, *Experiments in Plant Hybridization*] is so rarely cited that I myself did not become acquainted with it until I had concluded the majority of my experiments and deduced from them the principles given in the text."

With de Vrie's German paper as yet unpublished, Carl Correns submitted his first report entitled "*G. Mendels Regal über das Verhalten der Nachkommenschaft der Rassenbastarde*" (G. Mendel's Law Concerning the Behavior of the Progeny of Racial Hybrids) (*Ber. Dtsch. Bot. Ges.* Vol. XVIII, received April 24, 1900). As the title indicates, Correns had become acquainted with Mendel's essay, though not until his experiments had been all but terminated and "the law-governed behavior and the explanation thereof . . ." had been discovered.

Then the third and youngest of the three rediscoverers, Erich von Tschermak, published a study entitled "*Über künstliche Kreuzung bei* Pisum sativum" (Concerning Artificial Crossing in *Pisum sativum*) (*Ber. Dtsch. Bot. Ges.* Vol. XVIII, received June 2, 1900). Mendel is mentioned on several occasions in this paper, as is also the report by Correns which had just been published.[2]

[2] Tschermak (1956) states that he and Correns first became acquainted with Mendel's essay in the autumn of 1899 as a result of reading Focke's treatise, *Die Pflanzen-Mischlinge* (1881).

Roberts (1929) has made a detailed study of when the three rediscoverers first heard of Mendel or read his fundamental essay. He corresponded personally with the rediscoverers on this point. In a letter dated December 18, 1924, de Vries wrote to Roberts: "After finishing most of these experiments, I happened to read L. H. Bailey's 'Plant Breeding' of 1895. In the list of literature of this book, I found the first mention of Mendel's now celebrated paper, and accordingly looked it up and studied it. Thereupon I published in March 1900 the results of my own investigations in the *Comptes Rendus de l'Académie des Sciences* . . ." (Roberts 1929, p. 323).

However, in a footnote appearing in the fourth edition of his book *Plant Breeding* (p. 155), Bailey quotes from a letter from de Vries, who had written: "Many years ago you had the kindness to send me your article on Cross-Breeding and Hybridization of 1892, and I hope it will interest you to know that it was by means of your bibliography therein that I learnt some years afterwards of the existence of Mendel's papers."

These two quotations are obviously inconsistent. In his letter to Roberts, de Vries (1924) says that he first became acquainted with Mendel's essay in 1895 while reading Bailey's *Plant Breeding*, although the first edition of this book does not contain a bibliography. In his much earlier letter to Bailey, however (written in 1908 at the latest), de Vries confirms that he learned of the title of Mendel's essay in 1892 from reading the bibliography of Bailey's paper *Cross-Breeding and Hybridization*. We cannot therefore dismiss the possibility that de Vries had already read Mendel's essay at that early date.

Zirkle (1964) emphasizes that some years before de Vries became familiar with Mendel's essay, he had obtained segregation data with eleven different species in a series of species crosses. He started his experiments in hybridization as early as 1892 and obtained in 1894, after crossing *Silene alba* × *S. alba* var. *glabra*, a segregation ratio of 392 hairy plants to 144 smooth plants. In 1893 de Vries crossed two varieties of *Papaver somniferum* (Mephisto × Danebrog) and obtained, two years later, a segregation ratio of 158 plants with black spots to 43 white plants. He self-pollinated these F_2 plants in 1896. All of the recessives bred true, unlike the dominant forms of which some bred true while the remainder segregated again in the ratio of 1095 plants with black spots to 358 white plants. The question must be raised why de Vries delayed the publication of these segregation data. According to a letter he wrote to R. E. Cleland, the reason was that he did not believe in their universal significance.

T. J. Stomps (1954), one of de Vries's students, gives a different account of the matter. He claims that de Vries did not become acquainted with Mendel's essay until 1900, when Professor Beijerinck of Delft, knowing that de Vries was conducting experiments in hybridization, sent him the offprint dated 1865. It is clear that Beijerinck knew of the essay a number of years before the rediscovery. In fact, as he has himself observed, he would have been the first to rediscover Mendel (five years before de Vries, quote from Itallie-van Embden, 1940) had he not have been obliged to abandon his hybridization experiments at Wageningen in 1885 when he became a bateriologist at the Dutch Distillery Works in Delft.

De Vries proceeded in his investigations according to the principles enunciated in *Intracellulare Pangenesis*—a work he had published in 1889, five years after the appearance of Nägeli's *Mechanical-Physiological Theory of Descent.* He realized that species characters are composed of numerous more or less independent factors, most of which recur in different species, while many of them are present in almost all species. The diversity of nature is caused by the combination of a relatively small number of factors which determine the individual hereditary qualities each of which can vary independently of the others. Each cell must contain, in a latent form, all the factors that determine the character of the species. In Section 1 of *Intracellulare Pangenesis* de Vries says prophetically: "These factors are the units which the science of heredity has to study. Just as physics and chemistry may be traced back to molecules and atoms, so the biological sciences must penetrate these units in order to explain in terms of their combinations the phenomena of the living world."

De Vries postulated the existence of a specific particle for each hereditary quality, the manifestations of these particles being the visible phenomena of heredity; he called these particles "pangenes." He held that each pangene is composed of numerous molecules and is able to grow and reproduce; upon every cell division the pangenes become evenly distributed in the new cells. Pangenes can be active or inactive, but they can reproduce in either state. They are located in the chromatin threads of the nuclei, but most of the characteristics within them are latent (inactive), whereas within the plasm they become active. Thus de Vries supposed that hereditary qualities are transferred from the nucleus to the cytoplasm and advanced the hypothesis that the whole of the living protoplasm consists of pangenes that constitute its living elements. Pangenes are thus transported into the plasm, but they need never return to the nucleus for they can also reproduce in the plasm. This view differs from the transportation hypothesis of Darwin, for whom the ability of the gemmules to return to the nuclei was an important consideration. According to de Vries, the type of variation which gives rise to new species derives from the fact that on rare occasions the two new pangenes produced during cell division are different from one another and not, as is normally the case, identical to the original ones.

For de Vries, then, a particular form of material bearer corresponds to each individual character. According to most earlier theories of hybridization, species, subspecies, and varieties were indivisible units whose hybrid forms yielded still further hybrids, but de Vries now urged (1900, p. 84):

Instead we must adopt the principle of the crossing of species characters. The detailed features of species characters must be regarded as quite distinct elements and studied from that perspective. They must always be treated independently of each other until reasons for a different kind of treatment are found. In every hybridization experiment a single character, or a certain number of characters, must alone be considered; the remaining characters can be temporarily disregarded. Or rather it does not matter if the parents differ from one another in other respects. Probably for experimental purposes the hybrids whose parents differ only in a single character constitute the simplest cases (monohybrids as opposed to di- and polyhybrids).

If the parents of a hybrid differ only with respect to a single pair of characters, these characters or qualities are mutually antagonistic. The hybridization experiment is concerned only with these qualities.

De Vries deduced the following principles from his experiments:

1. Of the two antagonistic qualities the hybrid always shows one only and it is fully developed. Consequently the hybrid cannot be distinguished from one of its parents with regard to that trait. Intermediate forms do not occur.
2. The two antagonistic qualities separate during the formation of the pollen and egg cells. In the majority of cases they do so in accordance with the ordinary laws of probability theory.

Like Mendel, who had already formulated these two principles in his essay, de Vries characterizes the quality that is visible in the hybrid as dominant, and the latent quality as recessive; he found that the systematically progressive quality (?) is usually dominant, and the original character of the two when the derivation of both is known.

De Vries found:

Dominant	Recessive
Papaver somniferum, tall form	*P. s. nanum*
Antirrhinum majus, red	*A. m. album*
Polemonium coeruleum, blue	*P. c. album*

And in cases where the derivation is not known, for example:

Dominant	Recessive	Known since
Chelidonium majus	*C. laciniatum*	± 1590
Oenothera Lamarckiana	*O. brevistylis*	± 1880
Lychnis vespertina (hairy)	*L. v. glabra*	± 1880

During the formation of the pollen grains and egg cells the antagonistic qualities separate and behave independently of each other. This results in the following law: "The pollen grains and egg cells of monohybrids are not hybrids but have the pure character of one or other of the two parental types" (p. 86).

The hybrid therefore contains 50 percent dominant and 50 percent recessive pollen grains together with 50 percent dominant and 50 percent recessive egg cells. When d = dominant and r = recessive, fertilization gives:

$$(d + r)\ (d + r) = d^2 + 2dr + r^2,$$

or

$$25\%\ d + 50\%\ dr + 25\%\ r.$$

Thus 25 percent of the offspring are plants having the quality of their father, 25 percent are plants having the quality of their mother, and 50 percent are plants that are also hybrids.

The hybrids have the dominant character, and results are therefore obtained in which 75 percent of the specimens have the dominant character and 25 percent have the recessive character:

A. From artificial crosses:

Dominant	Recessive	Rec.	Year of crossing
Agrostemma Githago	*nicaeensis*	24%	1898
Chelidonium majus	*laciniatum*	26%	1898
Hyoscyamus niger	*pallidus*	26%	1898
Lychnis diurna	*L. vespert.* (white)	27%	1892
Lychnis vespertina (hairy)	*glabra*	28%	1892
Oenothera Lamarckiana	*brevistylis*	22%	1898
Papaver somnif. Mephisto	Danebrog	28%	1893
Papaver somnif. nanum (single)	double	24%	1894
Zea mays (starchy)	*saccharata*	25%	1898

B. From natural crosses, e.g.

Dominant	Recessive	Rec.	Year in which experiment was conducted
Aster Tripolium	*album*	27%	1897
Chrysanthemum Roxburghi (yellow)	*album*	23%	1896
Coreopsis tinctoria	*brunnea*	25%	1896
Solanum nigrum	*chlorocarpum*	24%	1894
Veronica longifolia	*alba*	22%	1894
Viola cornuta	*alba*	23%	1899

Average result based on all these experiments 24.93%.

Of the 75 percent of the plants having the dominant character, a number were fertilized with their own pollen, and the offspring were subsequently counted. In 1896 an experiment with *Papaver somniferum* Mephisto × Danebrog was conducted and yielded in the first hybrid generation (F_2):

Dominant (Mephisto)	24%
Hybrids	51%
Recessive (Danebrog)	25%

When the hybrids were self-fertilized, 77 percent of the plants in the following generation had the dominant character while 23 percent had the recessive character. This experiment was continued for two further generations and always gave the same result.

If a hybrid is fertilized with the pollen of one of its parents, the following result is obtained:

$$(d + r)\, d = d^2 + dr, \text{ and}$$
$$(d + r)\, r = dr + r^2.$$

In the first case all the plants exhibit the dominant character, though 50 percent of them are hybrids; in the second case 50 percent of the total are hybrids having the dominant character, while the remaining 50 percent of the plants have the recessive character.

The same laws are valid for the study of two or more antagonistic characters in dihybrids or polyhybrids. The prickly variety, *Datura Tatula*, was crossed with *D. stramonium inermis*, and the hybrids showed the dominant characters: they had blue flowers and bore prickly fruits. The following generation segregated in the ratio

blue : white = 72.0:28.0%
prickly : smooth = 72.6:27.4%.

From these results the composition of the offspring can be calculated. If *A* denotes one pair of antagonistic qualities, and *B* the other, the following result is obtained for the hybrids with regard to:

A	25% dominant	50% hybrids (*d* x *r*)	25% recessive
B	6.25%*d*, 12.5%*dr*, 6.25%*r*	12.5%*d*, 25%*dr*, 12.5%*r*	6.25%*d*, 12.5%*dr*, 6.25%*r*

The visible qualities were therefore distributed according to the following percentage ratio:

1. *A* dominant and *B* recessive	18.75%
2. *A* recessive and *B* dominant	18.75%
3. *A* dominant and *B* dominant	56.25%
4. *A* recessive and *B* recessive	6.25%

In a hybridization experiment in which *Trifolium pratense album* was crossed with *T. pratense quinquefolium*, the trifoliate leaves and the white flowers were recessive. The composition of the offspring of the hybrids (220 plants) was as follows:

		Expected
1. Red and trifoliate	13%	19%
2. White and quinquefoliate	21%	19%
3. Red and quinquefoliate	61%	56%
4. White and trifoliate	5%	6%

In other plants (*Antirrhinum majus*, *Silene armeria*, *Brunella vulgaris*) de Vries often found simple qualities, for example, the color of the flowers, to be composite. Such qualities could be analyzed into their component parts by means of hybridization, which invariably confirmed the above ratios.

Correns had studied the formation of xenia in maize since 1894 and was thus led to undertake experiments in hybridization with peas. Work on a lengthy publication about the foliose mosses prevented him from making a detailed study of the results obtained in his hybridization experiments until 1899. According to a letter to Roberts (1929) written on January 23, 1925, Correns came upon the explanation for the ratios he had obtained as he pondered on his results one morning in bed. He first learned of the existence of Mendel's essay in the autumn of 1899, as a result of reading Focke's *Pflanzen-Mischlinge* (1881).

Referring to de Vries's paper (in *Compt. rend. de l'Acad. des Sciences*), which contains no mention of Mendel, Correns (1900) observes at the beginning of his publication that de Vries was now in the same position in which he had been formerly: he had considered the results of his investigations original until he had become acquainted with Mendel's essay. The essay, Correns continues, contained the same result and the same explanation if the terms "egg cell" or "ovum nucleus" were substituted for "germ cell" and "germinal vesicle" and, say, "generative nucleus" for "pollen cell."

Apart from his study of *Zea mays,* Correns too at first worked chiefly with varieties of peas whose characters could be arranged in pairs such that each member of the pair pertained to a given trait, for example, the color of the cotyledons, the color of the flower, the color of the seed coat, and so forth. With many such pairs one of the characters is much more strongly pronounced than the other in the hybrid and may therefore, following Mendel, be termed the dominant character, while the other character may be termed recessive. Contrary to de Vries, Correns realized that it is not true that one member is always dominant in any given pair of characters in the hybrid, for there are cases in which intermediate forms appear.

The facts agreeing with the findings of Mendel and de Vries which Correns discovered from his experiments are as follows:

1. All of the individuals in the first generation of hybrids are alike and show the dominant character. In this particular case the cotyledons are yellow.
2. When these hybrids with yellow seeds are sown, plants are obtained whose pods contain seeds having green cotyledons and also seeds having yellow cotyledons in the ratio of 3 yellow to 1 green.
3. In the third generation the seeds having green cotyledons again produce only plants with green seeds. In subsequent generations, too, they behave like the pure variety which possesses this character.
4. The seeds having yellow cotyledons yield offspring that may be divided into two classes: A, third-generation plants that have only yellow seeds, and B, third-generation plants that have both yellow and green seeds in the ratio of 3 yellow to 1 green, as in the second generation. With respect to the total number of individuals, class A stands to class B in a ratio of 1:2.
5. In the fourth and fifth generations the class A plants with yellow seeds produce only plants whose seeds are yellow; i.e., they behave like the pure variety which possesses this dominant character.
6. In all of the following generations the class B plants with green seeds produce only plants with green seeds.

7. The class B plants with yellow seeds again yield two types of plants in the ratio of 1:2, whose seeds behave in the manner described in 5 and 6.

The numerical data recorded by Correns in two experiments with peas are as follows:

Experiment I (Correns 1900, p. 162)
Hybrid between the "green, late-bearing Erfurt pea" with green seed and the "purple-violet Kneifel pea" with yellow seed*

1st generation			51 yellow 19			
2nd generation		619 yellow 25				206 green (25%)
	(28%)		18			11
3rd generation	251 yellow		550 yellow 18		195 green (26.2%)	538 green
	7	8 (44%)	10		14	10
4th generation	224 yellow	216 yellow	225 yellow	70 green (23.8%)	370 green	307 green

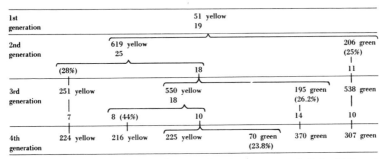

*Under identical treatment the plants produced an average of 43.3, 47.7, and 28.8 seeds in successive generations; this provides a good example of the effects of self-fertilization and also helps to explain the "giant growth" of many hybrids (*footnote added later by Correns*).

Experiment II (Correns 1900, p. 163)
Hybrid between the "green, late-bearing Erfurt pea" with green seed and the "bean pea" with yellow seed

1st generation				31 yellow 12	
2nd generation		775 yellow 21			247 green (24.2%)
	7 (33%)		14		20
3rd generation	292 yellow	462 yellow		149 green (23.6%)	670 green

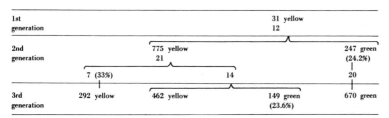

In order to explain these results Correns assumed, as had Mendel, that the recessive (green) character is suppressed in the hybrid by the dominant (yellow) character, with the result that all the seeds have yellow cotyledons. However, the herditary factor for the green color is preserved in a "latent" form, and there occurs prior to the formation of the reproductive nuclei a complete separation of the hereditary factors such that, in both sexes, one-half of the reproductive nuclei receives the hereditary factor for the recessive character while the other half receives the hereditary factor for the dominant character. The ratio of 1:1 suggests that the separation of the hereditary factors occurs during reduction division. Random combinations of the nuclei produce in the second generation three different classes of individuals: 25 percent have only the recessive character; 25 percent have only the dominant character; and 50 percent have both characters, even though the dominant character alone may be perceived.

Correns backcrossed maize hybrids with plants having the dominant character and then with plants having the recessive character. He found that in the first case, as is to be expected on theoretical grounds, only individuals having the dominant character appear; in the next generation one-half of these plants bred true with regard to the dominant character while the other half segregated into dominants and recessives in the ratio of 3:1.

He observed, as did Mendel, that the number of individuals containing both hereditary factors constantly diminishes with each successive generation. If in the second generation 50 percent of the plants were hybrids, then in the third generation there are only 25 percent, in the fourth only 12.5 percent, in the fifth 6.25 percent, and in the nth generation $100/(2^{n-1})$ percent.

Like Mendel, Correns crossed plants (*Zea mays*) which differed in two pairs of characters. From these crosses he logically derived nine different classes of individuals which, with regard to external distinguishing marks, could be arranged in four groups; the numbers of individuals in these groups appeared in the ratio of 9:3:3:1. In one particular case with *Zea mays* Correns obtained the following ratio (calculated on the basis of 1000):

565:191:176:68.

Mendel had concluded that "the pea hybrids form germ cells and pollen cells which, in their constitution, correspond in equal numbers to all the constant forms resulting from the combination of the characters united by fertilization," while Correns, using his own terminology, says: "The hybrid forms reproductive nuclei that unite in all possible combinations the hereditary factors for the individual parental characters, though never both hereditary factors for the same pair of characters. Each combination occurs with approximately the same frequency."

If the parents differ in one pair of characters (A, a), the hybrid forms two kinds of reproductive nuclei (A, a) which are identical to those of the parents. If the parents differ in two pairs of characters (A, a, B, b), the hybrid forms four different kinds of reproductive nuclei (AB, Ab, aB, ab), and 25 percent of the total number are of each kind.

If the parents differ in three pairs of characters (A, a, B, b, C, c), the hybrid forms eight different kinds of reproductive nuclei (ABC, ABc, AbC, Abc, aBC, aBc, abC, abc), and there are 12.5 percent of the total number of each kind. Correns goes on to say: "This is what I term the Mendelian Law; it includes also the *loi de disjonction* of de Vries. Everything else can be derived from this law" (p. 167).

Correns's results thus basically agree with the findings of Mendel and de Vries. However, he particularly emphasized two important points: First, that with a large number of pairs of characters it is not the case that one of the two characters is dominant, for intermediate forms are produced. Second, he explicitly stressed the limitations of the validity of the Mendelian Laws of Segregation (p. 166). He knew of this second point mainly as a result of his investigations into the hybrids in stocks (1900) which revealed that varieties having linked characters exist. He also noted from his experiments with *Matthiola* that reciprocal differences occur during hybridization which make the hybrid embryos increasingly resemble their mother.

Tschermak began his hybridization experiments on *Pisum sativum* in 1898 in the botanical garden at Ghent, his aim being to verify Darwin's experiments on the effects of cross- and self-fertilization in the vegetable kingdom. He also crossed different varieties of *Pisum sativum* in order to study the direct influence of foreign pollen on the shape and color of the seeds and, moreover, in order to survey, in successive generations of hybrids, the inheritance of the constant differentiating characters of the parents. Further, he "double-pollinated" some of the flowers to determine whether both kinds of pollen act simultaneously or whether one kind is predominant. Finally, he crossed the hybrids with their parental varieties or crossed pure varieties with hybrids, as Correns had done already. Tschermak became acquainted with Mendel's essay from Focke's reference in *Die Pflanzen-Mischlinge*, but not until his first experiments were completed.

His findings were as follows:

1. In the competition experiments with descendants of self-fertilized plants and descendants of different varieties which had been crossed it is only in certain cases, i.e., not invariably, that the latter show superior growth compared with the plants obtained by self-fertilization.
2. A direct influence of foreign pollen on the seeds may be determined in certain cases after crossing different varieties of peas.

3. Usually only one of the differentiating characters is fully developed in the hybrid. In agreement with Mendel's results, the round, smooth shape of the seed was dominant to the cubical, deeply wrinkled shape. The yellow cotyledon color was dominant to green (see also Correns). In individual cases the dominance is not complete, and "transitional forms" (prevalence) occur.
4. In the second hybrid generation, obtained by self-fertilization, dominant or prevailing and recessive characters appear in the ratio of approximately 3:1. (Tschermak does not give any exact numerical data for the segregation of specific characters in his first paper.)
5. If hybrids are pollinated with the pollen of the dominant parent, they exclusively produce seeds having the dominant character. Pollination of the hybrids with the pollen of the recessive parent gives yellow and green seeds approximately in the ratio of 57:43. Reciprocal crosses yielded essentially the same result.
6. "Double pollinations" of the hybrids using their own pollen and the pollen of one of the parental varieties can have results. One kind of pollen by no means excludes the other from the fertilization process.

Tschermak's first paper concerning his hybridization experiments with *Pisum* contains only data of a fairly general nature. In this respect it differs considerably from de Vries's paper, and even more from Correns's investigations; both of these researchers attempted to work out the statistical regularities of segregation for each successive generation, while Correns in particular (like Mendel thirty-five years earlier) presented extensive numerical data which had been obtained by experiment.

Tschermak's second paper, also written in 1900, is very comprehensive and does contain some numerical data regarding segregation, for example:
1. For outdoor hybrids that differed in one character pertaining to the seeds, 1087 yellow to 370 green = 2.9:1.
2. For outdoor hybrids which differed in two characters pertaining to the seeds, 435 yellow and smooth to 148 green and smooth to 116 yellow and wrinkled to 49 green and wrinkled = 9:3.0:2.4:1.

Gregor Mendel's laws of heredity were rediscovered by three different parties. In accordance with the principles enunciated in his *Intracellular Pangenesis*, de Vries had postulated the existence of specific particles, termed pangenes, for each hereditary quality and had urged that species characters be analyzed into quite distinct elements for the purpose of experiments in hybridization. His emphasis was on problems in the theory of descent and on the development of the "Mutation Theory," which he finally substantiated with an abundance of experimental data. Correns had been studying fertilization in maize and in peas and then discovered regularities when counting out the numbers of plants that had segregated. Tschermak, a plant breeder, had made experiments in hybridization in order to study the cases, described by Darwin, of increased growth following cross-pollination and further to study the formation of xenia in peas. It is clear that Tschermak must take the credit for being the first to recognize the importance of the laws of heredity in practical plant breeding. He has been the author of several detailed accounts of the history of the rediscovery of the laws of heredity (1951, 1956, 1960).

10

The Beginnings of the Chromosome Theory of Heredity

The rediscovery of Gregor Mendel's laws of heredity in 1900 marks the end of an era in the history of genetics: with the growth of human knowledge and experience and the rise of experimental methods in science, the fanciful notions concerning reproduction and heredity held at the beginning of this era and spanning more than twenty-five centuries gradually gave way to knowledge of the nature of heredity which could be provided with an ever firmer foundation. At the close of this era there arises the problem of the connection between the facts discovered in hybridization experiments and the cytological discoveries which so clearly proved that the chromosomes, located in the cell nucleus, played a part in the transmission of characters.

Shortly after the publication of the papers by de Vries, Correns, and Tschermak the cytologists E. B. Wilson (1902) and W. S. Sutton (1902), the zoologists T. Boveri (1904) and K. Heider (1906), the botanist W. A. Cannon (1902), and de Vries (1903) himself gave voice to the first ideas toward a "chromosome theory of heredity."

Mendel, however, had already assumed that the hereditary factors were distributed among the different germ cells, and he had realized that to the mosaic of characters there must correspond a compound of the germinal substance formed from representative particles. He writes: "The differentiating characters of two plants can finally, however, depend only upon differences in the composition and grouping of the elements that exist in vital interaction in the foundation cells of those plants" (1865, p. 42).

From the fact that the hereditary factors are transmitted by the father and by the mother in the same manner, Nägeli had already inferred that both parents supply an equal quantity of the hereditary substratum. Moreover, the considerable morphological disparity between the germ cells suggested that a distinct substance must be present to act as the bearer of the hereditary factors. This issue was clarified by the studies on fertilization undertaken by O. Hertwig and E. Strasburger. Nägeli's idioplasm is to be identified with the chromatin contained within the cell nuclei. The physiological equivalence of the male and female germ cells was demonstrated with the observation of the fusion of the two reproductive nuclei. The distinctive feature of the fertilized egg cell is that it unites paternal and maternal qualities: both parents supply equal numbers of the chromatin components (chromosomes) of the nucleus and, thanks to the mechanism of mitosis, these chromosomes pass on their combined hereditary materials from cell to cell in the course of ontogenesis. Since the individuality of the chromosomes is maintained throughout this process, the distinctive features of a given chromosome are transmitted on every occasion of cell division. The cell nuclei with their chromosomes represent, then, as Heider stated, a depository for hereditary factors, and the regular activation of these hereditary factors gives rise to the processes by which organisms are formed, though the interaction of the nucleus with the protoplasm is also important for the differentiation of the embryo.

As the bearer of hereditary traits, then, the cell nucleus is made up of a number of individuated components—the chromosomes. It was assumed that a given character, the color of the flower for example, is represented in the germinal substance by a hereditary factor located in a specific chromosome, while the hereditary factors for other characters were probably contained in a different chromosome. As each chromosome was probably composed of numerous, qualitatively different particles it was natural to formulate definite ideas concerning the structure of the hereditary substance in terms of tiny, individuated hereditary factors (pangenes, idioblasts, determinants) (Heider 1906). Boveri (1904) demonstrated in great detail that there are qualitative differences between chromosomes, and morphological studies (Montgomery 1901, Sutton 1902) also revealed that chromosomes can differ from one another in magnitude and in behavior. Thus Sutton found that the grasshopper *Brachystola magna* has six small chromosomes and sixteen large chromosomes—each of which is represented twice in the nucleus—which appear together in pairs. Montgomery was able to establish that in hemipters one of the partners in each pair of chromosomes is derived from the father, the other from the mother.

However, it is not always necessary for both sets of chromosomes to be present in order to bring about normal development. Boveri (1888, 1907) demonstrated this in his experiments on artificial parthenogenesis, in which the maternal set of chromosomes by itself produced normal development; in his further experiments on the fertilization of the eggs of the sea urchin (merogony) by means of a single male nucleus of anucleate fragments, *one* of the sets of parental chromosomes was found sufficient to bring about normal development. Boveri was also able to show that chromosomes are qualitatively different in yet another way—by observing the development of the double-fertilized eggs of sea urchins. Double-fertilization occurs in certain cells. When the chromosomes are distributed in the cleavage cells, irregular formation of the nuclear spindle produces abnormal combinations of chromosomes which then give rise to different or abnormal development in the embryos. This abnormal development indicates that the individual chromosomes are not equivalent and that they differ in the mechanical effects they have on development. From these studies of artificial parthenogenesis and double-fertilization it follows that the egg nucleus and the sperm nucleus contain a complete set each of the chromosomes necessary for normal development. To the paternal chromosomes A, B, C, D, therefore, there correspond matching chromosomes a, b, c, d, and each pair Aa, Bb, and so on is characterized by the possession of the same possibilities for development.

The abnormalities caused by the irregular distribution of the chromosomes led Boveri (1902) to consider the problem of the origin of malignant tumors in which multipolar mitoses are often found. He surmised that malignant tumors develop as a consequence of abnormal combinations of chromosomes, but he was unable to prove his hypothesis correct.

The importance of Boveri's work—especially the conclusions and discoveries drawn from the double-fertilized eggs of the sea urchin—led Wilson [in *Erinnerungen an Theodor Boveri* (Röntgen 1918)] to make the comment (pp. 75-76):

> One who, like the writer, had puzzled in vain over the riddle presented by the double-fertilized eggs of sea urchins could not read Boveri's complete and beautiful solution without a thrill; and it may be doubted whether a finer example of experimental, analytical and constructive work, compressed within such narrow limits—the paper on multipolar mitosis comprises but twenty pages and is without figures—can be found in the literature of modern biology. . . .
>
> This result, wholly new, was widely at variance with earlier conceptions of the chromosomes, and it is fundamental to our entire view of the cytological basis of heredity. As is often the case with discoveries of the first rank it gives at first sight little suggestion of its far-reaching importance or of the difficulties that had to be surmounted in its attainment.
>
> It brings forward the long-sought crucial evidence of the direct influence of the nucleus (chromosomes) in determination and development; all attempts to shake the force of that evidence have thus far proved unavailing. . . .

Boveri's classical experiments established the crucial connection between cytology and the study of hybridism and furnished, moreover, a rigorous proof of the importance of the chromosomes in development and formation. Boveri's contribution to cytology is as far-reaching as Mendel's contribution to the study of hybridism.

As research into hybridization has shown, the antagonistic pairs of characters maintain a certain independence in the hybrid; and we may conclude that the hereditary factor for these characters is located in the pairs of chromosomes within the cell nucleus so that the behavior of the pairs of chromosomes and of the pairs of hereditary factors during nuclear division, or in the hybrid progeny, is jointly regulated by the same precision system. During reduction division the operations of this system give rise to chromosome conjugation, in which a single paternal chromosome is united with a single maternal chromosome, and the formation of tetrads, which means that the mature reproductive cells contain (for example, in *Ascaris megalocephala* which has four chromosomes) four different possible combinations of chromosomes: AB, ab, Ab, aB. These possible combinations hold good for both sexes during the maturation period of the spermatozoa and of the egg; in the progeny of a couple of this type reproductive cells containing sixteen different combinations of the chromosome material taken from the grandparents can therefore be produced:

♀ ♂	♀ ♂
AB × AB	AB × Ab
ab × AB	ab × Ab
Ab × AB	Ab × Ab
aB × AB	aB × Ab
AB × ab	AB × aB
ab × ab	ab × aB
Ab × ab	Ab × aB
aB × ab	aB × aB

Animals and plants with a larger number of chromosomes will have a correspondingly greater—and ultimately incalculable—number of ways in which their chromosomes can combine. Reduction division causes the hereditary material contained in the chromosomes of the grandparents to be conveyed to the grandchildren in extremely diverse combinations and ensures that the complete set of chromosomes is preserved. Moreover, reduction division can divide the tetrad in two different ways—either during the first or during the second meiotic division—with the result that we speak accordingly of pre- or post-reduction division.

This process of division and combination, hitherto presented with reference to whole chromosomes, may be applied equally to the segregation of opposed pairs of characters. Mendel had in fact made just this assumption, which implies that the germ cells of the hybrids are not themselves hybrid but have the pure character of one or other of the hereditary factors. For in the hybrid Aa there must be formed, in both sexes and in equal proportions, germ cells of type A and of type a in order that the numerical ratios obtained by Mendel be produced. The segregation and distribution of the hereditary factors correspond to the formation of the tetrads and the distribution of the chromosomes in the germ cells during maturation division. Two types of germ cells are therefore produced in equal numbers in both sexes of the hybrid—one of these types contains only hereditary factor A, the other only hereditary factor a. On the assumption that all germ cells have an equal probability of being fertilized, the nuclei of these germ cells will unite at random, thus giving rise to four combinations of characters AA, Aa, aA, and aa, which represent (when A denotes the dominant character) the Mendelian segregation ratio of 3:1. The same result holds true if several pairs of opposed characters are united in the cross and if we assume that the individual pairs are distributed over different homologous chromosomes, that is, segregate independently of one another.

The first reflections that followed the rediscovery of Gregor Mendel's laws of heredity revealed, then, that these laws can be understood in terms of the behavior of the chromosomes during the formation of the germ cells. Hybrid research and cytology were thus united, and the ground was prepared for comprehensive analyses concerning the heredity of individual characteristics and their dependence upon one another and concerning the constitution and the composition of chromosomes in terms of qualitatively different elements. Moreover, the union of these two branches of research touched off the many and varied kinds of genetic investigation which heralded, at the beginning of the twentieth century, the era of classical genetics. The work undertaken during the first three decades of this century on the consolidation and improvement of the "chromosome theory of heredity" finally gave rise to the new problems which now mark a dynamic era in genetics.

Appendix

Translation of Gregor Mendel's letter to Carl von Nägeli*

July 3, 1870

Highly esteemed friend,
Please do not be angry with me for taking such a long time to express my gratitude for the living *Hieracia* you sent me, all of which arrived in good condition and are now growing magnificently. Construction work on the outlying dairy farms and other administrative matters occupied me to the exclusion of all else for many weeks, and when I returned to Brünn at Whitsuntide I met with urgent and time-consuming tasks there also. I did not become master of my own time until a very few days ago, but I am now in a position to resume my favorite occupation which I was obliged to discontinue because of an eye ailment, toward the end of June of last year.

I found myself in serious danger of having to abandon my experiments in hybridization completely—and this because of my own carelessness. Since diffused daylight was not wholly satisfactory for my work on the small *Hieracium* flowers, I availed myself of an illumination device (a mirror with convex lens) without suspecting the catastrophe I might have caused with it. After having worked a great deal in May and June with *H. Auricula* and *prealtum* I began to suffer from a strange weariness and exhaustion of the eyes which—despite my immediate efforts to spare my eyes as much as possible—developed into a serious condition and rendered me incapable of any exertion well into the winter. Since that time the malady has fortunately almost completely disappeared, with the result that I am again able to read for extended periods at a time and can also undertake fertilization experiments with *Hieracia* as well as can be done without artificial illumination.

With today's letter I am sending some living *Hieracium* hybrids and, where necessary, the ancestral forms are also enclosed.

Up to now the results of the experiments may be termed trifling, for they are too incomplete to allow any definite conclusions to be drawn from them. Some practical knowledge has nonetheless been gained, and I am taking the liberty of noting briefly the points that seem to me to be of some importance.

*The original letter is reproduced as Figure 26 (pp. 138-149).

To begin with I must mention that despite numerous attempts I have not yet succeeded in obtaining, by fertilization with foreign pollen, one single hybrid with some forms of piloselloids. This is the case, for example, with *H. aurantiacum.* With this species I have not as yet been able to neutralize the influence of its own pollen. *H. Pilosella* and *H. cymosum* also give rise to difficulties. With others, the varieties of *H. prealtum* for example, fertilization with foreign pollen succeeds more easily with the same treatment, and I have satisfied myself on more than one occasion that *H. Auricula,* when handled with care, is a completely reliable experimental plant. Last year I fertilized more than one hundred heads of this species with the pollen of *H. Pilosella, cymosum,* and *aurantiacum*; it is true that about one-half of them dried up because of injuries they had sustained and that only two to six seeds were obtained from each of the remaining heads, but the plants raised from those seeds are hybrids without exception. The tiny young plants of *H. Auricula* + *H. Pilosella* and *H. Auricula* + *H. cymosum* were, with the exception of a few specimens, unfortunately preyed upon by snails in the hothouse, but those of *H. Auricula* + *H. aurantiacum* were preserved and ninety-eight of them are now in the garden. They should flower next month.

Another type also seems to be very suitable for experiments, and I am enclosing it with the present consignment under the designation No. XII, for I am unable to classify it or to name it. I have found it in large numbers in areas where tree felling is in progress. My only attempt made last year, to fertilize it with pollen of *H. Pilosella* was wholly successful, for all of the twenty-nine plants obtained are hybrids.

I venture to remark here that I have hitherto used only one single type of *H. Pilosella* for fertilizations. However, I am not completely certain that a mistake was not made in last year's consignment for I subsequently noticed, at the time of flowering, that a different type of *Pilosella,* growing nearby, had invaded the territory of my experimental plant; I am therefore enclosing the plant, designated as *H. Pilosella* (Brünn), a second time. I shall not presume to express an opinion as to whether this form is related to *H. echioides* but will merely mention that it frequently occurs here, while the nearest known locality of *H. echioides* is some five miles away. The plant, designated as *H. prealtum* (?), which I dispatched to you, was found growing there last year; together with *H. echioides* and *H. prealtum,* and it is therefore beyond doubt that the supposition that it belongs to the *H. echioides-prealtum* series was correct. A comparison with the ancestral species in fact shows that it is more closely related to *H. echioides.*

I should be most grateful, my honored friend, if you would inform me at your convenience of your opinion on *Hieracium* No. XII, as this plant, with *H. Auricula,* is among the best experimental plants inasmuch as a fairly large number of hybrids may be obtained rather easily from it. This fact is important because the variations occurring among hybrid individuals cannot be evaluated unless a fairly large number of hybrids is derived from the same fertilization.

In point of fact variants have hitherto appeared in all cases in which several specimens of the hybrid were obtained from the same fertilization. I was most surprised, I must acknowledge, to observe that the influence of the pollen of one species upon the ovules of another species can produce divergent, even fundamentally different forms—all the more so since I had satisfied myself, by cultivating plants under experimental conditions, that the parental strains, when self-fertilized, yield only constant progeny. In *Pisum* as well as in other genera I had observed only uniform hybrids and I therefore expected the same in *Hieracium.* I must confess to you, my honored friend, how greatly I was deceived in this matter. As long as two years ago two specimens of the hybrid *H. Auricula* + *H. aurantiacum* flowered in my cultures. In one of them the paternity of *H. aurantiacum* was manifest at first sight; but not so in the other. Since at that time I was of the opinion that only one hybrid form could be produced by the two parental species, I considered this second specimen to be a fortuitous importation and disregarded it, for it had different leaves and a completely different, yellow flower color. On this account I enclosed in last year's consignment only the specimen which closely resembles *H. aurantiacum* in the color of the flower. However, when subsequently the hybrid resulting from the above fertilization and also the hybrid *H. Auricula* + *H. pratense* (var.) came to flower, each in three specimens and as many variants, the true facts of the case could no longer be misunderstood.

I see from your esteemed letter that the specimen of *H. Auricula* + *H. aurantiacum* and the parental species *H. Auricula,* which I dispatched to you, have perished; I am therefore replacing them and also include the long-misunderstood hybrid twin, designated *H. Auric.* + *H. aurant.* 868 *b.* Last year's three specimens are designated 869 *c, d, e.* The variant *c* is completely fertile.

One variant of the hybrid *H. Auricula* + *H. pratense* (var.), and the ancestral species *H. pratense,* died during the winter. The latter was not exactly a typical *H. pratense* for it bore some stellate hairs upon the leaves. The two specimens which you kindly sent me perished during the very first year in the garden; one atrophied without having flowered, the other during the flowering period. I have not yet been able to find the pure species in this locality.

The hybrid *H. No. XII* + *H. Pilosella* (Brünn) has just ceased blooming. There are very striking differences between the twenty-nine specimens available. Indeed, they all represent the transitional forms between the one ancestral species and the other, though nobody would take them for siblings if he found them growing in the natural state. I shall send you the whole collection as soon as the runners have taken root sufficiently, which should be the case in a few weeks. By that time I hope also to be able to report on this year's principal experiment with *H. Auricula* + *H. aurantiacum* which, in view of the fairly large number of specimens available, should provide me with some information.

This year *Hieracium* No. XII has been fertilized with *H. Pilosella vulgare* (München), and next year a comparison of the two hybrid series *H. No. XII* + *H. Pilosella* (Brünn) and *H. No. XII* + *H. Pilosella vulgare* (München) should not be without interest. Also, *H. Auricula* has been crossed with *H. Pilosella vulgare* (M.) and with *H. Pilosella* (Br.); it will very soon also be done with *H. Pilosella niveum* (M.). I have as yet seen only one flower of *H. Pilosella incanum*; it is to be hoped that others will appear. Twenty-five plants of the hybrid *H. prealtum* (Bauhini ?) + *H. aurantiacum*, of which I sent you two specimens last year, will come into flower. There are differences between them too, as far as may be seen at present. Of two specimens which were raised in pots and have long since ceased blooming, one is completely fertile, the other almost completely sterile.

Total sterility and complete fertility also occur in the hybrid series, *H. Auricula* + *H. aurantiacum*. The second generation of the hybrids *H. setigerum prealtum* (?) + *H. aurantiacum* and *H. prealtum* (Bauhini ?) + *H. aurantiacum* has flowered, as has the third generation of *H. prealtum* + *H. flagellare*. In these generations also the offspring did not vary. On this occasion I am unable to refrain from observing how astonishing it is that the behavior of *Hieracium* hybrids is directly opposed to that of *Pisum* hybrids. It is clear that we are concerned here only with individual phenomena which fall under a higher, more general law.

If one wishes to trace the development of the offspring of the hybrids having only partial fertility, it is necessary to shield the flowers most carefully from the influence of foreign pollen, for individual ovules—which would, under normal circumstances, remain unfertilized because of the very largely inferior quality of the pollen from the same plant—can be readily fertilized by pollen from other plants. I am also sending some double hybrids which are descended from *H. prealtum*

(Bauhini ?) + *H. aurantiacum*; I allowed this plant to flower among specimens of *H. Pilosella* (Brünn), but removed it from all other *Hieracia*. These hybrids should therefore be designated as (*H. prealtum* [Bauh. ?] + *H. aurantiacum*) + *H. Pilosella* (Brünn). They are, in many respects, very interesting forms.

If the stigmata in the flowers of partially fertile hybrids are covered with the pollen of different, though not too distantly related species, they always form more seeds than when they are kept isolated and are dependent on self-fertilization; this is due solely to the action of the foreign pollen, as can easily be shown to be the case by cultivating the seeds. With hybrids that are completely fertile, on the other hand, scrupulous isolation is unnecessary. Experiments with *H. prealtum* ? + *H. aurantiacum* showed that quantities of foreign pollen, even that of the two ancestral species, can be applied to the stigmata without impeding self-fertilization. All of the seeds yielded the original hybrid form.

I am also enclosing the hybrid *H. Cymosum* (München) + *H. Pilosella* (Brünn) in this consignment. It is the only hybrid I have hitherto obtained from *H. Cymosum*, although I have frequently tried to fertilize this species.

In the *Archieracia* it is very difficult to prevent self-fertilization. As yet only two hybrids have been obtained. The maternal parent of one of them is the species with light brown seeds which I sent you once as a dried specimen; the pollen was taken from a narrow-leaved *H. umbellatum*. The hybrid and the ancestral plants are enclosed. Among this year's seedlings, moreover, the fertilization of a form of *H. vulgatum*, also by the aforementioned *H. umbellatum*, can be seen to be a success. I search in vain for an *Archieracium* which would function as well as do *H. Auricula* and *Hieracium No. XII* among the piloselloids.

Of the *Archieracia*—for which I am indebted to your singular kindness—all except *H. glaucum* have been subjected to experiment. They are the following: *H. amplexicaule*, *pulmonarioides*, *humile*, *villosum*, *elongatum*, *canescens*, *hispidum Senatneri*, *picroides*, *albidum*, *prenanthoides*, *tridentatum*, and *gothicum*. In *H. amplexicaule* and *H. albidum* the artificially fertilized heads always dried up. I do not possess *H. alpinum*. From the seeds, labeled Breslau and München, which you were kind enough to send me, *H. nigrescens* and one other species were obtained, but it is not *H. alpinum*.

I must observe on this occasion that all of my *Archieracia* are growing well. It is true that *H. albidum* is rather delicate when raised in pots, particularly during the winter, but it keeps well in the ground.

The same holds good for the piloselloids, with the exception, however, of *H. pratense* and *H. Hoppeanum*; the latter died during the very first winter, both in pots and in the ground.

Last year I was not able to start any further hybridization experiments on account of my eye ailment. But one experiment seemed to me to be so important that I could not bring myself to postpone it to a later date. It concerns the view of Naudin and Darwin that a single pollen grain does not suffice to fertilize the ovule. Like Naudin, I used *Mirabilis Jalappa* as an experimental plant; but the result of my experiment is entirely different. From fertilization using only single pollen grains I obtained eighteen well-developed seeds, and from these seeds the same number of plants, of which ten are already in bloom. The majority of these plants are just as luxuriant as those that result from free self-fertilization. It is true that a few specimens are at present somewhat retarded in growth, but in view of the success displayed by the others the cause of this must lie in the fact that not all pollen grains are equally capable of effecting fertilization and that, moreover, in the experiment under consideration, the competition of other pollen grains was excluded. When several are competing, we may assume that only the strongest will be successful in accomplishing fertilization.

However, I intend to repeat these experiments; it should also be possible to prove directly by experiment whether two or more pollen grains can contribute to the fertilization of the ovule in *Mirabilis*. According to Naudin, at least three are required! With regard to the experiments of previous years, the ones with *Matthiola annua* and *glabra*, *Zea*, and *Mirabilis* were concluded last year. Their hybrids behave exactly like those of *Pisum*. The views concerning hybrids of these genera, which Darwin adopted from the reports of others and incorporated into his treatise *The Variation of Animals and Plants under Domestication*, need to be corrected in many respects.

Two experiments are still being continued. There are now about 200 uniform specimens of the hybrid formed from *Lychnis diurna* and *L. vespertina*. The first generation should flower in August.

The color experiments with *Matthiola* have now already lasted for six years and will probably be continued for some years yet. With the data already obtained, I hope eventually to get to the root of the matter. The lack of a reliable color chart was very troublesome in these experiments. I had ordered from Erfurt an assortment of *Matthiola annua* comprising thirty-six named colors, but it proved to be inadequate for my purposes. I have given very special attention to this experiment and I shall venture to report on it as soon as the 1500 plants in this year's culture have been examined. I should be able to do this when I send you the hybrid series *H.* No. XII + *H. Pilosella.*

Thanking you once again, my most honored friend, for the consignment you were kind enough to send me, I sign this letter with the expression of the highest esteem,

Your devoted,

Gr. Mendel

Brünn, 3 July 1870

Picture Credits

1 Assyrian relief from the time of Assurnasirpal II (883-859 B.C.), which shows the artificial pollination of the female inflorescences of the date palm, performed by priests wearing bird masks (winged demons).

2 Hippocrates (460-370 B.C.)
Hekler, *Bildniskunst der Griechen und Römer*, Stuttgart, 1912

3 Democritus (460-377 B.C.)
Schefold, *Die Bildnisse der antiken Dichter, Redner und Denker* Basel, 1943

4 Epicurus (c. 342-271 B.C.)
Supplied by Dr. D. Panos, Athens

5 Galen (?) (A.D. 129-c. 199)
American Journal of Archaeology, Vol. 59, 1955

6 Aristotle (384-322 B.C.)
Almquist, *Grosse Biologen*

7 Theophrastus (c. 372-287 B.C.)
Richter, *The Portraits of the Greeks*, Vol. 2 (London: Phaidon Press, 1965)

8 Avicenna (980-1037)
Deutsche Fotothek Dresden

9 Albertus Magnus (1193-1280)
Deutsche Fotothek Dresden

10 St. Thomas Aquinas (1225-1274)
Deutsche Fotothek Dresden

11 William Harvey (1578-1657)
Deutsche Fotothek Dresden

12 Anton van Leeuwenhoek (1632-1723)
Deutsche Fotothek Dresden

13 Marcello Malpighi (1628-1694)
Catalogus Illustratus Iconothecae Botanicae Horti Bergiani Stockholmiensi

14 Albrecht von Haller (1708-1777)
Deutsche Fotothek Dresden

15 Georges Louis Leclerc de Buffon (1707-1788)
W. A. Locy, *Die Biologie und ihre Schöpfer*

16 Pierre-Louis Moreau de Maupertuis (1698-1759)
Deutsche Fotothek Dresden

17 Karl Ernst von Baer (1792-1876)
H. Freund and A. Berg (eds.), *Geschichte der Mikroskopie*, Vol. I, *Biologie* (Frankfurt/Main: Umschau-Verlag, 1963)

18 Rudolf Jakob Camerarius (1665-1721)
Acta physicomedica Academiae Caesarae Leopoldino Carolinae

19 Carl von Linné (1707-1778)
Deutsche Fotothek Dresden

20 Josef Gottlieb Kölreuter (1733-1806)
H. F. Roberts, *Plant Hybridization before Mendel*

21 Count Giorgio Gallesio (1772-1839)
Supplied by Mr. S. Martini, Berne

22 Carl Friedrich Gärtner (1772-1850)
Botanical Museum, Berlin-Dahlem

23 D. A. Godron (1807-1880)
H. F. Roberts, *Plant Hybridization before Mendel*

24 Charles Naudin (1815-1899)
H. Iltis, *Charles Naudin, Der Züchter*, Vol. 1, pp. 248-250

25 Gregor Mendel (1822-1884)
Mendel Archives, Brünn

26 Gregor Mendel's letter to Carl von Nägeli, written July 3, 1870.
Supplied by Frau Dr. Correns, Munich

27 I. F. Schmalhausen (1849-1894)
Supplied by Professor A. E. Gaisinovich, Moscow

28 Gregor Mendel's letter to A. Kerner von Marilaun
Dörfler collection, Uppsala University Library

29 Carl von Nägeli (1817-1891)
Almquist, *Grosse Biologen*

30 Wilhelm Olbers Focke (1834-1922)
H. F. Roberts, *Plant Hybridization before Mendel*

31 Charles Darwin (1809-1882)
Autobiographie, Urania Verlag, 1959

32 Francis Galton (1822-1911)
Pearson, *The Life, Letters and Labours of Francis Galton* (London: Cambridge University Press, 1930)

33 Ernst Haeckel (1834-1919)
Archives, German Academy of Naturalists, Leopoldina, Halle

34 Hugo von Mohl (1805-1872)
H. Freund and A. Berg (eds.), *Geschichte der Mikroskopie*, Vol. I, *Biologie* (Frankfurt/Main: Umschau-Verlag, 1963)

35 Matthias Jacob Schleiden (1804-1881)
Almquist, *Grosse Biologen*

36 Theodor Schwann (1810-1882)
Almquist, *Grosse Biologen*

37 Jan Evangelista Purkinje (1787-1869)
Supplied by Professor Jírovec, Prague

38 Rudolf Albert von Kölliker (1817-1905)
Genetics, Vol. 22, 1937

39 Oscar Hertwig (1849-1922)
Supplied by Professor Paula Hertwig

40 Theodor Boveri (1862-1915)
Forscher und Wissenschaftler im heutigen Europa, Vol. 2, *Erforscher des Lebens* (Oldenburg/Hamburg: Stalling)

41 August Weismann (1834-1914)
Forscher und Wissenschaftler im heutigen Europa, Vol. 2, *Erforscher des Lebens* (Oldenburg/Hamburg: Stalling)

42 Hugo de Vries (1848-1935)
Dahlgren *Botanisk Genetik*

43 Carl Erich Correns (1864-1933)
Photo Verlag Scherl, Berlin

44 Erich von Tschermak-Seysenegg (1871-1962)
Photo Fayer, Vienna

45 William Bateson (1861-1926)
Supplied by Professor Nachtsheim, Berlin-Dahlem

Bibliography

Akakia, M.
1597 De morbis muliebribus. In: J. Spachius, Gynaeciorum sive de mulierum . . . affectibus & morbis libri. Argentinae.

Åkerman, A., and J. MacKey
1948 The breeding of self-fertilized plants by crossing. Some experiments during 60 years of breeding at the Swedish Seed Association. Svalöf 1886–1946. History and present problems. Carl Bloms Boktryckerei, Lund.

Albertus Magnus
De animalibus libri XXVI. Ed. from the Cölner Urschrift by H. Stadler. Beiträge zur Geschichte der Philosophie des Mittelalters, Aschendorffsche Verlagsbuchhandlung, Münster i. W. 1916.

Almquist, E.
1919 Linnés Vererbungsforschungen. Bot. Jahrb. **55**, 1–18.
— 1922 Linné und das natürliche Pflanzensystem. Beiblatt zu d. Bot. Jahrb. **58**, No. 1, 1–14.

Amici, G. B.
1824 Observations sur diverses espèces de plantes. Ann. Sci. Nat. Sér. 1, **2**, 65–67 (Trans. of his paper of 1823 in: Memorie di Matematica e di Fisica della Società Italiana della Scienza, Modena, **12**, 234–286).
— 1830 Note sur le mode d'action du pollen sur le stigmate; extrait d'une lettre de M. Amici à M. Mirbel. Ann. Sci. Nat. sér. 1, **21**, 329–332.

André, E.
1861 Une variété de l'érable à feuilles de frêne. Rev. hort., Paris, 269.

Aristotle
Opera omnia. Ed. Academia regia Boruss. (I. Bekker).

Ascherson, P., and P. Magnus
1891 Die Verbreitung der hellfrüchtigen Spielarten der europäischen Vaccinien sowie der *Vaccinium* bewohnenden *Sclerotinia*-Arten. Verh. zool.-bot. Ges. Vienna, 1891, **XLI.**

Auerbach, L.
1874 Organologische Studien, Vols. 1, 2. Breslau.

D'Azara, F.
1801 Essais sur l'Histoire Naturelle des Quadrupèdes de la Province du Paraguay. Pts. I and II. Paris.

Baer, K. E. v.
1827 De ovi mammalium et hominis genesi. Epistola ad Acad. Imp. scientiarium Petropolitanam. Cum tab. aenea (colorat.). Lipsiae, summptib: Leopoldi Vossii **4**, 40 p.
— 1865 Nachrichten über Leben und Schriften des Herrn Geheimrathes Dr. Karl Ernst v. Baer, mitgetheilt von ihm selbst. Buchdrukkerei der Kaiserlichen Akademie der Wissenschaften. St. Petersburg.

Bailey, L. H.
1892 Cross-breeding and hybridization. New York.
— 1896 Plant Breeding. Macmillan and Co., New York. 2nd. ed. 1902.

Ballantyne, J. W.
1894/95 Antenatal pathology and heredity in the Hippocratic writings. Transact. Edinburgh Obstetr. Soc. **20**, 51–66.

Balss, H.
1923 Präformation und Epigenese in der griechischen Philosophie. Arch. stor. sc. Rome, **IV**, 319–325.
— 1928 Albertus Magnus als Zoologe. Verlag der Münchener Drucke, Munich.
— 1934 Über die Vererbungstheorie des Galenos. Sudhoffs Arch. Gesch. d. Med. Naturw. **27**, 229–234.
— 1936 Die Zeugungslehre und Embryologie in der Antike. Quellen Gesch. Naturw. Med. **5,** 1–82.
— 1947 Albertus Magnus als Biologe. Wiss. Verlagsgesellschaft, Stuttgart.

Baltzer, F.
1963 Theodor Boveri und die Entwicklung der Chromosomentheorie der Vererbung. Zu Boveris 100. Geburtstag am 12. Oktober 1962. Die Naturw. **50**, 141–146.

Barthelmess, A.
1952 Vererbungswissenschaft aus Orbis Academicus, Problemgeschichten der Wissenschaft in Dokumenten und Darstellungen. Naturw. Abt. F. Gessner, ed. Verlag Karl Alber, Freiburg/Munich.

Bateson, W., and A. Bateson
1891 On variations in the floral symmetry of certain plants having irregular corollas. J. Linn. Soc., Bot., **28.**

Bateson, W.
1894 Materials for the study of variation treated with especial regard to discontinuity in the origin of species. Macmillan and Co., London.
— 1895 The origin of cultivated *Cineraria.* Nature **51**, 605–607.
— 1899 Hybridization and cross-breeding as a method of scientific investigation. J. Roy. Hort. Soc. **24**, 59–66.
— 1907 The progress of genetics since the rediscovery of Mendel's papers. Progr. Rei. Bot. **1**, 368–418.
— 1907 The progress of genetic research. Report of the third internat. Conf. 1906 on genetics. Spottiswoode & Co., London.
— 1909 Mendel's principles of heredity. Cambridge Univ. Press.
— 1928 Scientific papers of William Bateson, ed. by R. C. Punnett. 2 Vols. Cambridge University Press, London.

Bauhin, G.
1596 Phytopinax.
— 1620 Prodromos theatri botanici.

Bauhin, J.
1651 Historia plantarum universalis.

De Beer, Gavin
1964 Mendel, Darwin and Fisher (1865–1965). Notes and Records of the Roy. Soc. of London **19**, 192–226.

Behrens, J.
1896 Joseph Gottlieb Koelreuter. Verh. Naturwiss. Verein Karlsruhe **11** (1888–1895), 268–320.

Beneden, Ed. van
1875 La maturation de l'oeuf, la fécondation et les premières phases du développement embryonnaire des mammifères d'après des recherches faites chez le lapin. Bull. Acad. Roy. Sc. Belgique. Bruxelles 1875, Séance du 4 décembre No. 12.
— 1883 L'appareil sexuel femelle de l'Ascaride mégalocéphale. Arch. biol. **IV**, 95.
— 1883 Recherches sur la maturation de l'oeuf et la fécondation. Arch. biol. **IV**, 265.

Beneden, Ed. van, and Ch. Julin
1884 La spermatogénèse chez l'ascaride mégalocéphale. Bull. Acad. Roy. Belgique, 3 sér., **VII**, No. 4.

Beneden, Ed. van, and A. Neyt
1887 Nouvelles recherches sur la fécondation et la division mitosique chez l'ascaride mégalocéphale. Bull. de l'Acad. Roy. Belg. **XIV**, No. 8.

Beseler and Märcker
1887 Versuche über den Anbauwert verschiedener Hafersorten. Magdeburg. Ztg. Nr. 206, 207, 229.

Bischoff, T. L. W.
1842 Entwicklungsgeschichte des Kanincheneies. Braunschweig.

Blomeyer
1877 Vom Versuchsfeld des Landw. Instituts zu Leipzig. Fühlings Landw. Ztg. 402.

Du Bois, A. M.
1937 Die Entwicklung des Vererbungsgedankens. Ciba Zschr. Vol. 4 Nr. 45, Basel.

Bolin, P.
1897 Einige Beobachtungen über ungleiches Vererbungsvermögen gewisser Charaktere bei Gerstenkreuzungen. Sveriges Utsädesförenings Tidskr., Årg. VII, Vol. 4.

Bonnet, C.
1772 Betrachtungen über die organisierten Körper. Trans. by Goeze, 2 Vols. Lemgo.

Bose, N. K.
1963 Theory of heredity in the Manusamhita. Bull. Nat. Inst. Sci. India No. 21, 226–227.

Boveri, T.
1887 Zellen-Studien. Die Bildung der Richtungskörper bei *Ascaris megalocephala* und *Asc. lumbric.* Zschr. Med. Naturw., Jena, **21** (N. F. Vol. 14), 423–515.
— 1887 Über die Befruchtung der Eier von *Ascaris megalocephala.* S. ber. Ges. Morph. München **III**, 2, 71–80.
— 1887 Über den Anteil des Spermatozoon an der Teilung des Eies. S. ber. Ges. Morph. München.

— 1888 Die Vorgänge der Zellteilung und Befruchtung in ihrer Beziehung zur Vererbungsfrage, Verh. Münchener Ges. Anthrop.
— 1888 Zellen-Studien II. Die Befruchtung und Teilung des Eies von *Ascaris megalocephala*. Zschr. Med. Naturw., Jena, **22** (N. F. Vol. 15), 685–882.
— 1889 Zellen-Studien III. Über das Verhalten der chromatischen Kernsubstanz bei der Bildung der Richtungskörper und bei der Befruchtung. Zschr. Med. Naturw., Jena, **24** (N. F. Vol. 17) 314–401.
— 1889 Ein geschlechtlich erzeugter Organismus ohne mütterliche Eigenschaften. S. ber. Ges., Morph. u. Phys. München **5**.
— 1895 Über die Befruchtung und Entwicklungsfähigkeit kernloser Seeigeleier und über die Möglichkeit ihrer Bastardierung. Arch. Entw. mech. **2**, 394–443.
— 1902 Über mehrpolige Mitosen als Mittel zur Analyse des Zellkerns. Verh. Phys. Med. Ges., Würzburg, N. F. **35**, 67–90.
— 1902 Das Problem der Befruchtung. Fischer, Jena.
— 1904 Ergebnisse über die Konstitution der chromatischen Substanz des Zellkerns. Fischer, Jena.
— 1907 Zellen-Studien VI. Die Entwicklung dispermer Seeigeleier. Ein Beitrag zur Befruchtungslehre und zur Theorie des Kerns. Zschr. Naturw., Jena, **43**, 1–292.
— 1909 Die Blastomerenkerne von *Ascaris megalocephala* und die Theorie der Chromosomenindividualität. Arch. Zellf. **3**, 181–268.

Braun, A.
1851 Betrachtungen über die Erscheinung der Verjüngung in der Natur. Leipzig.

Brentjes, B.
1962 Gazellen und Antilopen als Vorläufer der Haustiere im alten Orient. Wiss. Zschr. Martin-Luther-Univ. Halle-Wittenberg **XI**, 537–548.
— 1962 Die Caprinae. Wiss. Zschr. Martin-Luther-Univ. Halle-Wittenberg **XI**, 549–594.
— 1962 Nutz- und Hausvögel im Alten Orient. Wiss. Zschr. Martin-Luther-Univ. Halle-Wittenberg **XI**, 635–702.
— 1962 Wildtier und Haustier im Alten Orient. Lebendiges Altertum Bd. 11, Akademie-Verlag, Berlin.
— 1964 Ein archäologischer Beitrag zur Entwicklungsgeschichte der Haustiere. Biol. Rdsch. **I**, 213–227.
— 1965 Zu einigen Theorien um das Auftreten des Pferdes im Alten Orient. Z. f. Tierzüchtung u. Züchtungsbiologie **81**, 348–354.
— 1965 Der geschichtliche Tierweltwechsel in Vorderasien und Nordafrika in altertumskundlicher Sicht. Säugetierkundl. Mitteilungen, BLV Bayer. Landwirtschaftsverlag, München, **13**, 101–109.
— 1965 Die Haustierwerdung im Orient. Die Neue Brehm-Bücherei, A. Ziemsen Verlag, Wittenberg Lutherstadt.

Brock, J.
1888 Einige ältere Autoren über die Vererbung erworbener Eigenschaften. Biol. Zbl. **8**, 491–499.

Brooks, W. K.
1883 The law of heredity. A study of the cause of variation and the origin of living organisms. 2nd ed. John Murphy and Co., Baltimore and New York.

Brosse, Guy de la
1636 Description du jardin royal des plantes médicinales.

Brown, R.
1831 Observations on the organs and mode of fecundation in Orchideae and Asclepiadeae. Transact. Linn. Soc. London.
— 1833 Observations on the organs and mode of fecundation in Orchideae and Asclepiadeae. Transact. Linn. Soc. **16**, 685.

Brücher, H.
1950 Pehr Bolin, ein skandinavischer Vorläufer bei der Wiederentdeckung der Vererbungs-Gesetze. Forsch. Fortschr. **26**, 313–314.

Brücke, M. E.
1862 Die Elementarorganismen, S.ber. Akad. Wiss., Math.-Nat. Kl., **XLIV**, II. Abt. 1861, Wien.

Buchanan Smith, A. D., O. J. Robinson, and D. M. Bryant
1936 The genetics of the pig. Bibliogr. Genet. **XII**, 1–160.

Bütschli, O.
1873 Beiträge zur Kenntnis der freilebenden Nematoden. Nov. Acta Leopoldina **36**.
— 1875 Vorläufige Mittheilung über Untersuchungen betreffend die ersten Entwicklungsvorgänge im befruchteten Ei von Nematoden und Schnecken. Zschr. wiss. Zool. **25** (ed. by Siebold, Kölliker, Ehlers), Leipzig.
— 1875 Vorläufige Mittheilung einiger Resultate von Studien über die Conjugation der Infusorien und die Zelltheilung. Zschr. wiss. Zool., Leipzig (W. Engelmann) **25**, 426–441.
— 1876 Studien über die ersten Entwicklungsvorgänge der Eizelle, die Zelltheilung und die Conjugation der Infusorien. Abh. Senckenberg. Naturf. Ges. Frankfurt a. M. **X**, 3. Heft, 213–462.
— 1877 Zur Kenntnis des Theilungsprocesses der Knorpelzellen. Zschr. wiss. Zool. **29** (ed. by Siebold, Kölliker, Ehlers), Leipzig.
— 1877 Entwicklungsgeschichtliche Beiträge. Zschr. wiss. Zool. **29**, 216–254.

Buffon, Georges Louis Leclerc de
1749 Histoire naturelle.

Bugge, G.
1929 Das Buch der großen Chemiker, Vol. I. Verlag Chemie, G. m. b. H., Berlin.

Camerarius, R. J.
1694 De sexu plantarum epistola. Trans. and ed. by M. Möbius. Ostwalds Klassiker der Exakten Wissenschaften Nr. 105, Verlag von Wilhelm Engelmann, Leipzig 1899.

Candolle, A. P. de
1824 Memoir on the different species, races, and varieties of the genus

brassica (cabbage), and of the genera allied to it, which are cultivated in Europe. Transact. Hort. Soc. **V**.

Candolle, C. de
1897 Remarques sur la Tératologie Végétale. Bibliothèque universelle et revue Suisse, Arch. physiques natur., IV. Période 3.

Cannon, W. A.
1902 A cytological basis for Mendelian laws. Bull. Tor. Bot. Club **29**, 657–661.

Capelle, W.
1938 Die Vorsokratiker. Die Fragmente und Quellenberichte. Kröners Taschenausgabe **119**, Alfred Kröner Verlag, Stuttgart.
— 1949 Theophrast über Pflanzenentartung. Museum Helveticum, **6**, 57–84, Benno Schwabe u. Co. Verlag, Basel.

Carrière, E. A.
1859 Sur une variété de cyprès. Rev. hort., Paris, 166.
— 1859 Sur le pêcher pleureur. Rev. hort., Paris, 419.
— 1863 Robinia pseudo-acacia Decaisneana. Rev. hort., Paris, 151.
— 1865 Production et fixation des variétés dans les végétaux. Paris.
— 1867 Traité général des Conifères ou description de toutes les espèces et variétés. Nouv. Edition, 1., 2, Paris (chez l'Auteur, Rue de Buffon, 53).
— 1870 Amygdalos monstrosa. Rev. hort., Paris, 549–561.
— 1878 Une page inédite à propos du lilas de perse. Rev. hort., Paris, 50° année, 451–455.
— 1886 Arbres et arbustes nouveaux ou peu connus. Rev. hort., Paris, 58° année, 398.
— 1887 Production spontanée d'un cérisier à fleurs roses. Rev. hort., Paris, 70.
— 1892 Groseillier à maquereaux sans épines. Rev. hort., Paris, 180–183.

Chabraeus
1666 Stirpium Sciagraphia.

Clos, D.
1889 Du nanisme dans le règne végétal. Mém. Acad. Sc. Série **9**, I.

Clusius, C.
1601 Rariorum plantarum historia.

Cohen, H. M.
1875 Das Gesetz der Befruchtung und Vererbung, begründet auf die physiologische Bedeutung der Ovula und Spermatozoen. Verlag C. H. Becksche Buchhandlung.

Cole, F. J.
1930 Early theories of sexual generation, The Clarendon Press, Oxford.

Coleman, W.
1965 Cell, Nucleus and Inheritance: An historical study. Proc. Amer. Phil. Soc. **109**, 124–158.

Colladon, J. A.
1821 Résumé travaux de la Soc. Cant. de Genève. Notice des séances Soc. Helv. Sciences Nat. Bibl. Universelle **17**, 828–829.

Colman, J. F.
1849 White variety of the swallow. The Zoologist **VII**, 2392.

Columella, Junius Moderatus
De re rustica libri XII et liber de arboribus. Ed. by J. M. Gessner, Mannheim 1781.

Conklin, E. G.
1937 Rudolph Albert v. Kölliker. Genetics **22**.

Cook, R.
1937 A chronology of genetics. U.S. Depart. Agric. Yearbook 1937.

Cordeaux, W. H.
1850 Varieties of the yellowhammer (Emberiza citrinella) and blackbird (Turdus merula). The Zoologist **VIII**, 2851.

Correns, C.
1899 Untersuchungen über die Xenien bei *Zea mays* (Vorläufige Mitteilung). Ber. Dtsch. Bot. Ges. **XVII**, 410–417.
— 1900 G. Mendels Regel über das Verhalten der Nachkommenschaft der Rassenbastarde. Ber. Dtsch. Bot. Ges. **XVIII**, 158–168 (received April 24, 1900).
— 1900 Über Levkojenbastarde. Zur Kenntnis der Grenzen der Mendelschen Regeln. Bot. Cbl. **LXXXIV**, 1–16.
— 1902 Über den Modus und den Zeitpunkt der Spaltung der Anlagen vom Erbsen-Typus. Bot. Ztg. **60**, 65–82.
— 1905 Gregor Mendels Briefe an Carl Nägeli 1866–1873. Ein Nachtrag zu den veröffentlichten Bastardierungsversuchen Mendels. Abh. d. Math.-Phys. Kl. Sächs. Ges. Wiss. **XXIX**, Nr. III, 189–265.
— 1922 Etwas über Gregor Mendels Leben und Wirken. Die Naturwissenschaften **10**, 623–631.

Le Couteur, J.
1843 Über die Varietäten, Eigenthümlichkeiten und Classification des Weizens. Translated by F. A. Rüder. Verl. d. J. C. Hinrichsschen Buchhandlung, Leipzig.

Cox, C. F.
1909 Charles Darwin and the mutation theory. Amer. Natural. **XLIII**, 65–91.

Crombie, A. C.
1965 Von Augustinus bis Galilei. Die Emanzipation der Naturwissenschaft. Kiepenheuer & Witsch, Köln-Berlin.

Darlington, C. D.
1959 Darwin's place in history. Basil Blackwell, Oxford.

Darwin, Charles
1859 The origin of species by means of natural selection. (With an introduction by Prof. Sir A. Keith). Everyman's Library, edited by Ernest Rhys, No. 811, J. M. Dent & Sons Ltd., London. Sixth ed. 1872.

— 1868 The variation of animals and plants under domestication. In two volumes. John Murray, London.
— 1871 The descent of man, and selection in relation to sex. John Murray, London.
— 1878 The effects of cross- and self-fertilisation in the vegetable kingdom. 2nd edition. John Murray, London.
— 1889 The life and letters of Charles Darwin. Including an autobiographical chapter. Edited by his son Francis Darwin. In two volumes. D. Appleton and Company, New York.
— 1893 The variation of animals and plants under domestication. Second ed. 2 vols. John Murray, London.

Darwin, E.
1794 Zoonomia; or the laws of organic life, Vol. I, J. Johnson, St. Paul's Churchyard, London.

Davis, P. D. E., and A. A. Dent
1966 Animals that changed the world. Phoenix House, London.

Detmer, W.
1887 Zum Problem der Vererbung. Pflügers Arch. G. Physiol. **41**, 203–215.

Dionis, P.
1698 Dissertation sur la génération de l'homme. Paris.

Dittrich, M.
1959 Getreideumwandlung und Artproblem. Fischer, Jena.

Dobzhansky, T.
1965 The Mendel Centennial. The Rockefeller University Press, New York.

Druery, C. T.
1903/1904 Plant variation under wild conditions. Hort. J. **28**.

Duchesne, M.
1766 Histoire naturelle des fraisiers. Didot le jeune et C. J. Panckoucke, Paris.

Dudley, P.
1724 Observations on some of the plants in New England, with remarkable instances of the nature and power of vegetation. Phil. Trans. Roy. Soc. London **33**, 194–200.

Duff, J.
1850 Varieties of the blackbird (Turdus merula). The Zoologist **VIII**, 2765.

Dunn, L. C.
1965 Ideas about living units, 1864–1909: A chapter in the history of genetics. Perspectives in Biology and Medicine **VIII**, 335–346, University of Chicago Press, Chicago.
— 1965 Mendel, his work and his place in history. Proc. Amer. Phil. Soc. **109**, 189–198.
— 1965 A short history of genetics. McGraw-Hill Book Company, New York.

Dunn, L. C., and W. Landauer
1934 The genetics of the rumpless fowl with evidence of a case of changing dominance. J. Genet. **29**, 217–243.

Dyer, W. T. Thiselton
1897 The cultural evolution of *Cyclamen latifolium*. Proc. Roy. Soc. London **61**, No. 371, 135.

Dzierzon, J.
1854 Der Bienenfreund aus Schlesien. Distributed by A. Bänder in Brieg.

Edwardson, J. R.
1962 Another reference to Mendel before 1900. J. Hered. **53**, 152.

Eichling, C. W.
1942 I talked with Mendel. J. Hered. **33**, 243–246.

Ernst, A.
1901 Ueber Pseudo-Hermaphroditismus bei *Nigella syncarpa*. Flora **88**, No. 1.

Fairchild, T.
1722 The City Gardener. London.
— 1724 An account of some new experiments, relating to the different, and sometimes contrary motion of the sap in plants and trees. Phil. Trans. Roy. Soc. London **33**, 127–129.

Faivre, E.
1868 La Variabilité des espèces et ses limites. Paris.

Fick, R.
1905 Betrachtungen über die Chromosomen, ihre Individualität, Reduktion und Vererbung. Arch. Anat. Physiol., Anat. Abt. Suppl.
— 1907 Vererbungsfragen, Reduktions und Chromosomenhypothesen, Bastardregeln. Erg. Anat. Entw.-Gesch. **16**.

Fisher, R. A.
1936 Has Mendel's work been rediscovered? Annals of Science **1**, 115–137.

Flemming, W.
1879 Beiträge zur Kenntnis der Zelle und ihrer Lebenserscheinungen. Part I, received Sept. 17, 1878. Arch. mikrosk. Anat. **XVI**.
— 1880 II. Teil eingereicht Dezember 1879. Arch. mikrosk. Anat. **XVIII**.
— 1882 III. Teil. Arch. Mikrosk. Anat. **XX**.
— 1882 Zellsubstanz, Kern und Zelltheilung. Verlag von F. C. W. Vogel, Leipzig.

Focke, W. O.
1881 Die Pflanzen-Mischlinge, ein Beitrag zur Biologie der Gewächse. Verlag Gebr. Borntraeger, Berlin.

Fol, H.
1877 Sur le commencement de l'hénogenie chez divers animaux. Arch. Sc. physiques natur. J. **58**.

Frank, F., and K. Zimmermann
1957 Färbungsmutationen bei der Feldmaus [*Microtus arvalis* (Pall.)]. Zschr. Säugetierk. **22**, 87–100.

Freund, H., and A. Berg (Eds.)
1963 Geschichte der Mikroskopie. Leben und Werk großer Forscher. Vol. I: Biologie. Umschau-Verlag, Frankfurt/Main.

Gärtner, C. F.
1827 "Correspondenz" Flora oder Botanische Zeitung, 10.Jahrgang, **I**, 79–80.
— 1828 "Correspondenz" Flora oder Botanische Zeitung, 11.Jahrgang, **II**, 553–559.
— 1833 Über Fruchtbildung und Bastard-Pflanzen. Flora oder Allgemeine Botanische Zeitung, 16.Jahrgang, **I**, 209–217.
— 1849 Versuche und Beobachtungen über die Bastarderzeugung im Pflanzenreich . . . K. F. Hering & Co., Stuttgart.

Gaisinovich, A. E.
1935 Gregor Mendel and his predecessors (in O. Sashre, Sh. Noden, G. Mendel′ Izbrannye raboty a rastitel′nykh gibridakh) Moscow.
— 1956/1957 Notizen von C. F. Wolff über die Bemerkungen der Opponenten zu seiner Dissertation. Wiss. Zschr. Friedrich-Schiller-Univ. Jena **6**, 121–124.
— 1961 K. F. Bol′f (C. F. Wolff) i uchenie o razvitii organizmov. Akademiia Nauk SSSR, Moscow.
— 1965 Pervoe izlozhenie raboty G. Mendelia v Rosii (I. F. Schmalhausen 1874). Biull. Mosk. ob-va ispyt. prirody. Otd. biol. **70**, No. 4, 22–24.
— 1966 An early account of G. Mendel's work in Russia (I. F. Schmalhausen, 1874), Proc. G. Mendel Memorial Symp. 1865–1965, ed. by M. Sosna. Academia, Prague.

Galen
Galeni opera. Ed. Kühn, Leipzig 1825.

Gallesio, Count Giorgio
1811 Traité du Citrus. Fantin, Paris.
— 1816 Teoria della Riproduzione vegetale. Capurro, Pisa.

Galton, F.
1865 Hereditary talent and character. Macmillan's Magazine **12**, 157.
— 1869 Hereditary genius. An inquiry into its laws and consequences. Macmillan and Co., London.
— 1876 The history of twins as a criterium of the relative powers of nature and nurture. London.
— 1876 A theory of heredity. J. Anthrop. Inst. London **V**, 329–348.
— 1877 Typical laws of heredity. Nature **XV**, 492, 512, 532 and Proc. Roy. Inst. Great Britain **VIII**, 282–301.
— 1889 Natural Inheritance. Macmillan and Co., London.
— 1897 A new law of heredity. Nature **LVI**, 235–237.
— 1897 The average contribution of each of several ancestors to the total heritage of the offspring. Proc. Roy. Soc. **61**, 401.
— 1910 Hereditary genius: An inquiry into its laws and consequences. Peter Smith, Gloucester, Mass.

Garth, J. C.
1849 Albino variety of the swallow (Hirundo rustica) and: Black variety of the bullfinch (Loxia pyrrhula). The Zoologist **VII**, 2568.
— 1850 White variety of the twite—Cream-coloured variety of the rook (Cornus frugilegus). The Zoologist **VIII**, 2953.

Gates, R. R.
1915 An anticipatory mutationist. Amer. Natural. **49**, 645–648.
— 1916 Huxley as a mutationist. Amer. Natural. **50**, 126–128.

Gaupp, E.
1917 August Weismann. Sein Leben und sein Werk. Fischer, Jena.

Gautier, A.
1886 Du méchanisme de la variation des êtres vivants. Felix Alcan, Paris.
— 1901 Les mécanismes de l'hybridation et la production des races. Rev. Viticult. **16.**

Gerlo, A.
1961 Lukrez, Gipfel der antiken Atomistik. Translated from the Flemish by Rose-Marie Seyberlich. Lebendiges Altertum, Populäre Schriftenreihe für Altertumswissenschaft 5, Akademie-Verlag, Berlin.

Gessner, C.
1551–1558 Historia animalium. Tiguri.
— 1560 Nomenclator aquatilium animantium. Tiguri.

Gigon, O.
1936 Zu Anaxagoras. Philologus edited by A. Rehm and J. Stroux, **XCI** (N. F. Vol. XLV), 1–41.

Glass, B.
1947 Maupertuis and the beginning of genetics. Quart. Rev. Biol. **22**, 196–210.

Gleditsch, J. G.
1749 Essai d'une fécondation artificielle, fait sur l'espèce de palmier qu'on nomme *Palma dactylifera folio flabelliformi.* Hist. Acad. Roy. Sc. Belles Lettres, Berlin, 103–108.

Gloede, F.
1859 Le fraisier quinquefolia. Rev. hort., Paris, 346–347.

Godron, D. A.
1863 Des hybrides végétaux, considérés au point de vue de leur fécondité et de la perpétuité ou non-perpétuité de leurs caractères. Ann. Sc. Natur. 4me Série, Botanique, **XIX**, 135–179.
— 1872 De l'espèce et des races dans les êtres organisés et spécialement de l'unité de l'espèce humaine, Vol. I, 2nd ed. Paris.
— 1873 Des Races végétales qui doivent leur origine à une monstruosité. Mém. Acad. Stanislas, Nancy, T. **VI.**

Goebel, K.
1886 Beiträge zur Kenntnis gefüllter Blüten. Jb. wiss. Bot. **17.**

Goethe, W. v.
Die Schriften zur Naturwissenschaft. Erste Abt. Bd. 9, S. 187–189, edited by Dorothea Kuhn. Hermann Böhlaus Nachfolger, Weimar 1954.

Götze, R.
1949 Besamung und Unfruchtbarkeit der Haussäugetiere. Verlag M. u. H. Schaper, Hannover.

Gordon, G.
1850 Variety of the common or house mouse. The Zoologist **8**, 2763–2764.

Goss, J.
1822 On the variation in the colour of peas occasioned by cross-impregnation. In a letter to the secretary. Transact. Hort. Soc. London **V**, 234–237.

Groenland, J.
1860 Sur deux monstruosités observées dans le genre Papaver. Rev. hort., Paris, 292–296.

Grüneberg, H.
1956 Genes in Mammalian Development. Inaugural Lecture Univ. College, London. H. K. Lewis & Co. Ltd., London.

Guaita, G. v.
1898 Versuche mit Kreuzungen von verschiedenen Rassen der Hausmaus. Ber. Naturw. Ges. Freiburg **10**, 317–332.
— 1899 Zweite Mitteilung über Versuche mit Kreuzungen mit verschiedenen Rassen der Hausmaus. Ber. Naturw. Ges. Freiburg **11**, 131–138.

Guignard, L.
1883 Sur la division du noyau cellulaire chez les végétaux. Compt. rend. Acad. Sc. Paris **97**, 646–648.

Haacke, W.
1893 Die Träger der Vererbung. Biol. Zbl. **13**, 525–542.
— 1893 Gestaltung und Vererbung—Eine Entwicklungsmechanik der Organismen. Leipzig.
— 1895 Über Wesen, Ursachen und Vererbung von Albinismus und Scheckung und über deren Bedeutung für vererbungstheoretische und entwicklungsmechanische Fragen. Biol. Zbl. **15**, 44–78.
— 1895 Die Bedeutung der Befruchtung und die Folgen der Inzestzucht. Biol. Zbl. **15**, 145–159.

Haeckel, E.
1866 Generelle Morphologie der Organismen. Allgemeine Grundzüge der organischen Formenwissenschaft, mechanisch begründet durch die von Charles Darwin reformierte Descendenztheorie. 2 Vols. G. Reimer, Berlin.
— 1876 Die Perigenesis der Plastidüle oder die Wellenzeugung der Lebenstheilchen. G. Reimer, Berlin.

Haedicke, W.
1936 Die Gedanken der Griechen über Familienherkunft und Vererbung. Diss. Halle, Akadem. Verlag, Halle.

Hagberg, K.
1946 Carl Linnaeus. Ein großes Leben aus dem Barock. Claassen u. Goverts, Hamburg.

Haller, A. v.
1757–1766 Elementa physiologiae corporis humani. 8 Vols., Lausanne.
— 1788 Alberts von Haller Grundriß der Physiologie für Vorlesungen. Nach der vierten lateinischen mit den Verbesserungen und Zusätzen des Herrn Prof. Wrisberg in Göttingen, vermehrten Ausgabe aufs neue übersetzt, und mit Anmerkungen versehen durch Herrn Hofrath Sömmerring in Mainz, mit einigen Anmerkungen begleitet und besorgt von P. F. Meckel, Professor in Halle. Haude und Spener, Berlin.

Hartmann, G. M.
1959 Der Materialismus in der Philosophie der griechisch-römischen Antike. Lebendiges Altertum Vol. 1, Akademie-Verlag, Berlin.

Harvey, W.
1628 De motu cordis et sanguinis in animalibus. Translated by Kenneth J. Franklin, Blackwell, Philadelphia 1958.
— 1651 Exercitationes de generatione animalium. London.

Heider, K.
1906 Vererbung und Chromosomen. Fischer, Jena.

Heimans, J.
1962 Hugo de Vries and the gene concept. The Amer. Nat. **96**, 93–104.

Heinisch, O.
1966 Biometrie und Genetik in historischer Sicht. Naturwiss. Rundschau **19**, 305–310.

Helbaek, H.
1959 Domestication of food plants in the old world. Science **130**, 365–372.

Helye, D.
1868 Sur la pélorie des mufliers. Rev. hort., Paris, 327.

Henschel, A.
1820 Von der Sexualität der Pflanzen. Breslau.

Henslow, G.
1903 Variations in animals and plants. Gard. Chron. Ser. III, **33**, 405 and **34**, 18–20.

Herbert, W.
1822 On the production of hybrid vegetables; with the result of many experiments made in the investigation of the subject. Transact. Hort. Soc. London **4**, 15–47.
— 1847 On hybridisation amongst vegetables. J. Hort. Soc. London **2**, 1–28 (Part I) and 81–107 (Part 2).

Herder, J. G.
1785–1792 Ideen zur Philosophie der Geschichte der Menschheit. Johann Friedrich Hartknoch, Riga and Leipzig.

Herre, W.
1958 Abstammung und Domestikation der Haustiere. Handbuch der Tierzüchtung **1**, Biologische Grundlagen der tierischen Leistungen, 1–58. Verlag Paul Parey, Berlin and Hamburg.

Hertwig, O.
1875 Beiträge zur Kenntnis der Bildung, Befruchtung und Theilung des tierischen Eies. I. Abh. Morph. Jb. **I**, 347–434 (Received August 1875). Also published separately as a *Habilitationsschrift* in October 1875.
— 1877 2. Abh. Morph. Jb. **III**, 1–83.
— 1877 Weitere Beiträge zur Kenntnis der Bildung, Befruchtung und Teilung des tierischen Eies (Received February 1877).
— 1878 3. Abh. Morph. Jb. **IV**, 156–213.
— 1884 Das Problem der Befruchtung und der Isotropie des Eies, eine Theorie der Vererbung. Zschr. Med. Naturw., Jena, **XVIII** (N. F. Bd. XI), 276–318.
— 1890 Vergleich der Ei- und Samenbildung bei Nematoden. Eine Grundlage für zelluläre Streitfragen. Arch. mikrosk. Anat. **36.**
— 1918 Dokumente zur Geschichte der Zeugungslehre. Eine historische Studie. Verlag F. Cohen, Bonn.

Hertwig, O., and R. Hertwig
1887 Über den Befruchtungs- und Teilungsvorgang des tierischen Eies unter dem Einfluß äußerer Agentien. Jena.

Hervé, G.
1912 Maupertuis génétiste. Rev. anthrop., Paris, **22**, 217–230.
— 1922 La génétique prémendélienne: Aristote et Réaumur. Rev. Antrop., Paris, **32**, 285–297.

Heuser, E.
1884 Beobachtungen über Zellkernteilung. Bot. Zbl.

Hippocrates
Medical Works, tr. by John Chadwick and W. N. Mann. 4 Vols. Blackwell, Philadelphia 1950.

His, W.
1870 Die Theorie der geschlechtlichen Zeugung I und II. Arch. Anthrop. **4**, 197–220 and 317–332, Verlag F. Vieweg u. Sohn, Braunschweig.
— 1872 Die Theorien der geschlechtlichen Zeugung III. Arch. Anthrop. **5**, 66–111.
— 1874 Unsere Körperform und das physiologische Problem ihrer Entstehung. Briefe an einen befreundeten Naturforscher. Verlag F. C. W. Vogel, Leipzig.

Hösch
1801 Versuch einer neuen Zeugungstheorie, Verlag Meyersche Buchhandlung, Lemgo.

Hofacker, J. D.
1827 De qualitatibus parentum in sobolem transeantibus. Tübingen.
— 1828 Ueber die Eigenschaften, welche sich bei Menschen und Thieren von den Eltern auf die Nachkommen vererben mit besonderer Rücksicht auf die Pferdezucht. C. F. Osiander, Tübingen.

Hoffmann, H.
1869 Untersuchungen zur Bestimmung des Werthes von Species und

Varietät: ein Beitrag zur Kritik der Darwinschen Hypothese. L. F. Rikersche Buchhandlung, Gießen.
— 1881 Rückblick auf meine Variations-Versuche von 1855–1880. Bot. Ztg. **39**, 345ff.

Hofmeister, W.
1867 Die Lehre von der Pflanzenzelle. Handbuch der physiolog. Botanik **1**, Verlag W. Engelmann, Leipzig.
— 1868 Allgemeine Morphologie der Gewächse. Handbuch der physiologischen Botanik **1**, 1. Abt., Verlag W. Engelmann, Leipzig.

Hommel, H.
1927 Moderne und hippokratische Vererbungstheorien. Arch. Gesch. Med., Leipzig, **XIX**, 105–122.

Hooke, R.
1665 Micrographia, London.

Hovorka-Kronfeld
1908 Vergleichende Volksmedizin **2**, 527–534. Verlag von Strecker und Schröder, Stuttgart.

Hughes, A.
1959 A History of Cytology. Abelard-Schuman, London and New York.

Humphreys, D.
1813 On a new variety in the breeds of sheep. Philos. Transact. Roy. Soc. London **I**, 88–95.

Hus, H.
1911 Jean Marchant; an eighteenth century mutationist. Amer. Natural. **XLV**, 492–506.

Huxley, T. H.
1868 The physical basis of life. Collected Essays.
— 1893 Darwiniana Essays. Macmillan and Co., London.

Iltis, A.
1954 Gregor Mendel's Autobiography. J. Hered. **45**, 231–234.

Iltis, H.
1923 Die Mendel-Jahrhundertfeier in Brünn (September 22–24, 1922). Studia Mendeliana, 5–30.
— 1924 Gregor Johann Mendel. Berlin.
— 1926 Gregor Mendels Selbstbiographie. Genetica, 329–334.
— 1929 Charles Naudin. Der Züchter **1**, 248–250.

Itallie-van Embden, W.
1940 Interview with Beijerinck.—Martinus Willem Beijerinck. His Life and His Work, Appendix J. The Hague.

Jäger, v.
1852 Gedächtnisrede auf das im Laufe des Jahres verstorbene geschätzte Vereinsmitglied, Med. Dr. v. Gärtner zu Calw. Jahreshefte d. Vereins f. vaterländ. Naturkunde in Württemberg **8**, 16–33.

Jäger, G.
1876 Zoologische Briefe. Verlag W. Braumüller, Vienna.

Jäggi, J.
1893 Die Blutbuche zu Buch am Irchel. Neujahrsblatt, herausgegeben von der Naturforschenden Gesellschaft auf das Jahr 1894, XCVI. Zürich.

Jahn, J.
1957/1958 Zur Geschichte der Wiederentdeckung der Mendelschen Gesetze. Wiss. Zschr. Friedrich-Schiller-Univ. Jena, Math.-Nat. Reihe, **7**, 215–227.
— 1965 W. O. Focke—M. W. Beijerinck und die Geschichte der "Wiederentdeckung" Mendels. Biolog. Rundschau **3**, 12–25.

Jakubíček, M., and J. Kubíček
1965 Bibliographia Mendeliana. Universitní Knihovna, Brno.

Jessen, K. F. W.
1864 Botanik der Gegenwart und Vorzeit in culturhistorischer Entwicklung. Ein Beitrag zur Geschichte der Abendländischen Völker. Republished by The Chronica Botanica, Waltham, Mass., USA, 1948.

Johannsen, W.
1917 Die Vererbungslehre bei Aristoteles und Hippokrates im Lichte heutiger Forschung. Naturwiss., Berlin, **5**, 389–397.
— 1923 Hundert Jahre Vererbungsforschung. Verh. Ges. Dtsch. Naturf. u. Ärzte **87**, 70–104, Verlag F. C. W. Vogel, Leipzig.

Jonston, I.
1660 Naenkeurige Beschryving van de Natur der vier-voetige Dieren. I. Schipper, Amsterdam I.

Kant, I.
1775 Von den verschiedenen Rassen des Menschen. Gesamtausgabe von Kants Schriften von Rosenkranz und Schubert **VI**, 313. Königsberg.
— 1785 Bestimmung des Begriffs einer Menschenrasse. Berlinische Mschr., Gedike and Biester (eds.), **VI**, 390.

Keeler, C. E., and S. Fuji
1937 The antiquity of mouse variations in the Orient. J. Hered. **28**, 93–96.

Keswani, N. H.
1963 The concepts of generation, reproduction, evolution and human development as found in the writings of Indian (Hindu) scholars during the early period (up to A.D. 1200) of Indian history. Bull. Nat. Inst. Sci. India No. 21, 206–225.

Keudel, K.
1936 Zur Geschichte und Kritik der Grundbegriffe der Vererbungslehre. Sudhoffs Arch. Gesch. Med. **28**, 381–416.

Kirchhoff, A.
1868 Caspar Friedrich Wolff. Sein Leben und seine Bedeutung für die Lehre von der organischen Entwicklung. Zschr. Med. Naturwiss., Jena, **IV**, 193–220.

Klatt, B.
1927 Entstehung der Haustiere. Handbuch der Vererbungswissen-

schaft, E. Baur and M. Hartmann (eds.) Verlag Gebrüder Borntraeger, Berlin.
— 1948 Haustier und Mensch. Richard Hermes Verlag, Hamburg.

Knight, T. A.
1799 An account of some experiments of the fecundation of vegetables. Philos. Transact. Roy. Soc. London, Part II, 195–204.
— 1823 Some remarks on the supposed influence of the pollen, in cross-breeding, upon the colour of the seed-coats of plants, and the qualities of their fruits. Transact. Hort. Soc. London **5**, 377–380.
— 1824 Notice of a new Variety of *Ulmus suberosa*, and of a successful method of grafting tender scions of trees. In a letter to the Secretary. Transact. Hort. Soc. London **V**.
— 1824 An account of a new variety of plum, called the Downton Imperatrice. In a letter to the secretary. Transact. Hort. Soc. London **V**.

Kölliker, A. v.
1841 Beiträge zur Kenntnis der Geschlechtsverhältnisse und Samenflüssigkeit wirbelloser Tiere. Berlin.
— 1844 Entwicklungsgeschichte der Cephalopoden. Zürich.
— 1852 Handbuch der Gewebelehre des Menschen. Verlag W. Engelmann, Leipzig.
— 1864 Über die Darwin'sche Schöpfungstheorie. Z. wiss. Zool. **XIV**, 174–186.
— 1872 Morphologie und Entwicklungsgeschichte des Pennatulidenstammes nebst allgemeinen Betrachtungen zur Descendenzlehre. Christian Winter, Frankfurt.
— 1885 Die Bedeutung der Zellenkerne für die Vorgänge der Vererbung. Z. wiss. Zool. **XLII**, 1–46.
— 1886 Das Karyoplasma und die Vererbung. Z. wiss. Zool. **XLVI**, 228–238.

Kölreuter, D. J. G.
1761–1766 Vorläufige Nachricht von einigen das Geschlecht der Pflanzen betreffenden Versuchen und Beobachtungen, nebst Fortsetzung 1, 2 und 3. Ostwalds Klassiker der Exakten Wissenschaften Nr. 41, Verlag W. Engelmann, Leipzig.

Korschinsky, S.
1901 Heterogenesis und Evolution. Flora **89**, 240–363.

Křiženecký, J.
1963 Mendels zweite erfolglose Lehramtsprüfung im Jahre 1856. Sudhoffs Arch. Gesch. Med. Naturw. **47**, 305–310, F. Steiner Verlag GmbH, Wiesbaden.
— 1965 Gregor Johann Mendel 1822–1884. Texte und Quellen zu seinem Wirken und Leben. Joh. Ambr. Barth Verlag, Leipzig.

Krumbiegel, I.
1957 Gregor Mendel und das Schicksal seiner Entdeckung. Sammlung Große Naturforscher, **22**, Wissenschaftl. Verlagsgesellschaft m. b. H., Stuttgart.

Kuckuck, H.
1959 Neuere Arbeiten zur Entstehung der hexaploiden Kulturweizen. Zschr. Pflanzenzücht. **41**, 205–226.

Kühn, C. G.
1830 Claudii Galeni Opera Omnia. Lipsiae.

Labus, J.
1929 Geschichtlicher Beitrag zu dem Problem der Vererbung von Krankheiten (Dissertation). Buchdruckerei von K. Henn, Freiburg i. Br.

Lamarck, J. B. de
1809 Philosophie Zoologique. Paris.

Landauer, W., and Tso Kan Chang
1949 The ancon or otter sheep. History and genetics. Journ. Hered. **XL**, 105–112.

Lawrence, B.
1967 Early domestic dogs. Zeitschr. f. Säugetierkunde **32**, 44–59.

Lawrence, W.
1819 Lectures on physiology, zoology and the natural history of man. J. Callow, med. Bookseller, London.

Laxton, T.
1866 Observations on the variations effected by crossing in the colour and character of the seed of peas. Rep. Intern. Hort. Exhib. Bot. Congr., 156.
— 1872 Notes on some changes and variations in the offspring of cross-fertilized peas. J. Roy. Hort. Soc. London **3**, 10–14.

Lecoq, H.
1862 De la Fécondation naturelle et artificielle des végétaux et de l'hybridation, considérée dans ses rapports avec l'horticulture, l'agriculture et la sylviculture. 2ème édition. Librairie Agricole de la Maison Rustique, Paris.

Leeuwenhoek, A. van
1679 Letter to Viscount Brouncker, November 1677. Phil. Trans. **XII**, 1040–1043.

Lehmann, E.
1916 Aus der Frühzeit der pflanzlichen Bastardierungskunde. Arch. Gesch. Naturw. Techn. 78–81.

Leibniz, G. W.
Hauptschriften zur Grundlegung der Philosophie. Übersetzt von Dr. A. Buchenau. Durchgesehen und mit Einleitungen und Erläuterungen herausgegeben von Dr. Ernst Cassirer. Band II. (Principal philosophical writings. Transl. by Dr. A. Buchenau. Revised edition with an introduction and commentary by Dr. Ernst Cassirer. Vol. II). Verlag der Dürr'schen Buchhandlung, Leipzig, 1906.

Lenz, H. O.
1856 Zoologie der alten Griechen und Römer. Becker'sche Buchhandlung, Gotha.

— 1859 Botanik der alten Griechen und Römer. Verlag Thienemann, Gotha.

Leonardo da Vinci
Notebooks, edited by Edward MacCurdy. Braziller, New York 1955.

Lesky, E.
1950 Die Zeugungs- und Vererbungslehren der Antike und ihr Nachwirken. Abh. der Geistes- u. Sozialwiss. Klasse der Akad. d. Wiss. u. der Literatur in Mainz, Jg. 1950, Nr. 19. Distributed by Franz Steiner Verlag G.m.b.H., Wiesbaden.

Lindley, J.
1824 A notice of certain seedling varieties of amaryllis, presented to the Society by the Hon. and Rev. William Herbert, in 1820, which flowered in the Society's garden, in February 1823. Transact. Hort. Soc. London **V**.

Linnaeus, C.
1735 Systema naturae, ed. 1. Lugduni Batavorum.
— 1743 Oratio de telluris habitabilis incremento. Lugduni Batavorum.
— 1744 Peloria (Daniel Rudberg). Amoenitates Academicae, Volumen Primum. Cornelium Haak, Lugduni Batavorum.
— 1753 Species plantarum, ed. 1. Holmiae.

Lippmann, E. O. v.
1911 Zur Geschichte der Vererbungstheorien. Mitt. Gesch. Med. Naturw., **X**, 384–385.

Loeb, J.
1899 On the nature of the process of fertilization and the artificial production of the normal larvae etc. Am. J. Physiol. **III**.

Logan, J.
1736 Some experiments concerning the impregnation of the seeds of plants. Philos. Transact. Roy. Soc. London **39**, 192–195.

Lotsy, J. P.
1908 Vorlesungen über Descendenztheorien **II**, 575. Fischer, Jena.

Lucas, P.
1847 Traité philosophique et physiologique de l'hérédité naturelle dans les états de santé et de maladie du système nerveux. Paris.

Lucretius
Titus Lucretius Carus: On the nature of things. Second edition. Translated by Cyril Bailey. Oxford University Press 1960.

Luria, S.
1963 Anfänge griechischen Denkens. Lebendiges Altertum Vol. 14, Akademie-Verlag, Berlin.

Lynch, I.
1900 The evolution of plants. J. Roy. Hort. Soc. **XXV**, Part I, 34–37.

Macfarlane, J. M.
1895 A comparison of the minute structure of plant hybrids with that of their parents, and its bearings on biological problems. Trans. Roy. Soc. Edinburgh **37**, 203–286.

Malpighi, M.
1686 Opera Omnia, 2 Vols. London (expanded edition, Leiden 1687).

Mangelsdorf, P. C., R. S. MacNeish, and W. C. Galinat
1964 Domestication of corn. Science **143**, 538–545.

Marchant, J.
1719 Observations sur la nature des plantes. Hist. Acad. Roy. Sc., Amsterdam.

Martin, R.
1928 Lehrbuch der Anthropologie in systematischer Darstellung. 2nd. ed., Vol. 2, Kraniologie, Osteologie. Fischer, Jena.

Martini
1871 Der mehrblütige Roggen. Verlag A. W. Kafemann, Danzig.

Martini, S.
1961 Gregor Mendel als Agronom und als Förderer der Landwirtschaft. Schweiz. Landw. Monatshefte No 1. Bern.
— 1961 Giorgio Gallesio, pomologist and precursor of Gregor Mendel. Fruit Varieties and Horticultural Digest, East Lansing, Mich.
— 1966 Albrecht von Haller (1708–1777) als Förderer der Forstwissenschaft und der Landwirtschaft in der Schweiz. Schweizer. Landw. Monatshefte **44**, 321–327.
— 1966 Conrad Gessner (1516–1565) als Förderer des Gartenbaus und der Landwirtschaft. Schweizer. Landw. Monatshefte **44**, 208–212.

Mason, S. F.
1962 History of the sciences. Revised edition. Collier, Riverside, N.J.

Masters, M. T.
1869 Vegetable teratology. London.
— 1886 Pflanzen-Teratologie. Verlag H. Haessel, Leipzig.

Mather, Cotton
1721 Religio Philosophica; or, The Christian Philosopher. London.

Maupertuis, P. M. de
1768 Oeuvres de Maupertuis. Nouvelle Edition. Vol. I: Essai de Cosmologie, 1–78, Vol. II: Vénus physique, 3–133, Système de la nature, 135–216, Lettres sur divers sujets, 217–372, Lettre sur le progrès des sciences, 373–431. J. M. Bruyset, Lyon.

May, W.
1917 Lucrez und Darwin. Naturwiss. **5**, 276–279.

Mayer, C. F.
1929 Die Personallehre in der Naturphilosophie von Albertus Magnus. Kyklos, Leipzig, **2**, 191–257.
— 1953 Genesis of Genetics. Acta Geneticae et Gemellologiae **II**, No. 3, 237–332.

Meckel, J. F.
1812 Handbuch der pathologischen Anatomie. Vol. 1. C. H. Reclam, Leipzig.

Meehan, T.
1875 Change by gradual modification not the universal law. 23rd Meeting 1874. Amer. Ass. Advanc. Sc. 7–12.
— 1884 Variation in Halesia. Proc. Nat. Acad. Sc. Philadelphia.
— 1885 Persistence in variations suddenly introduced. Proc. Nat. Acad. Sc. Philadelphia.

Megenberg, C. v.
Das Buch der Natur. In Neu-Hochdeutscher Sprache bearbeitet und mit Anmerkungen versehen von Dr. H. Schulz. Verlag u. Druck von Julius Abel, Greifswald 1897.

Mendel, G.
1865 (1866) Versuche über Pflanzenhybriden. Verh. Naturf.-Ver. Brünn **IV**, 3–47. Neudruck in Ostwalds Klassiker der Exakten Wissenschaften Nr. 121, Akademische Verlagsgesellschaft m. b. H., Leipzig 1940.
— 1870 Über einige aus künstlicher Befruchtung gewonnene *Hieracium*-Bastarde. Verh. Naturf.-Ver. Brünn **VIII**, 26–31. Neudruck in Ostwalds Klassiker der Exakten Wissenschaften Nr. 121, Akademische Verlagsgesellschaft m.b.H., Leipzig 1940.

Metzger, J.
1841 Die Getreidearten und Wiesengräser in botanischer und ökonomischer Hinsicht. Akad. Verlagsbuchh. F. Winter, Heidelberg.

Meyer, E. H. F.
1856 Geschichte der Botanik, **3**. Gebr. Bornträger Verlag. Königsberg.
— 1857 Geschichte der Botanik, **4**. Gebr. Bornträger Verlag, Königsberg.

Meyer, H.
1918 Das Vererbungsproblem bei Aristoteles. Philologus **75**, N. F., 29, 323.

Millardet, M. A.
1894 Note sur l'hybridation sans croisement ou fausse hybridation. Mem. de la Soc. des Sciences de Bordeaux **IV**, 347–372.

Miller, P.
1721 Letter to Mr. Richard Bradley. Dated Oct. 6, 1721. Published by Bradley in "Treatise of husbandry and gardening" **I**, 330–332, London 1726.
— 1731/68 The Gardeners Dictionary, London.

Mitterer, A.
1947 Die Zeugung der Organismen, insbesondere des Menschen. Nach dem Weltbild des Hl. Thomas von Aquin und dem der Gegenwart. Herder Verlag, Wien-Freiburg.
— 1956 Die Entwicklungslehre Augustins im Vergleich mit dem Weltbild des Hl. Thomas von Aquin und dem der Gegenwart. Herder Verlag, Wien-Freiburg.

Mode, H.
1960 Das Frühe Indien. Hermann Böhlaus Nachf., Weimar.

Moewes, F.
1913 Vorläufer Mendels. Naturw. Wschr., N. F., **XII**, 541–542.

Mohl, H. v.
1835 Ueber die Vermehrung der Pflanzenzelle durch Theilung. Dissert., Tübingen.
— 1846 Ueber die Saftbewegung im Inneren der Zellen. Bot. Ztg., Berlin.

Montgomery, T. H.
1901 A study of the chromosomes of the germ cells of metazoa. Transact. Amer. Philos. Soc. **XX**, New Series, 154.

Moore, T.
1859/60 The octavo nature-printed British ferns: being figures and descriptions of the species and varieties of ferns found in the United Kingdom. **I.** *Polypodium* to *Lastrea* 1859; **II.** *Athyrium* to *Ophioglossum* 1860. Bradbury and Evans, London.

Morel, B. A.
1857 Traité de dégénérescences physiques, intellectuelles et morales de l'espèce humaine etc. Paris.

Morgan, T. H.
1932 The rise of genetics. Science **76**, 261–267, 285–288.

Morison, R.
1669 Hortus regius blesensis.
— 1680 Plantarum Historiae universalis oxoniensis . . . 2. Oxonii.

Morsier, G. de, and M. Cramer
1957 Jean-Antoine Colladon et la découverte de la loi de l'hybridation en 1821. Gesnerus **14**, 113–123.

Motulsky, A. G.
1958 Josef Adams (1756–1818): The forgotten founder of medical genetics. Proc. X. Internat. Congr. Genet. Montr. **II**, 198. Univ. of Toronto Press.

Müntzing, A.
1958 Vererbungslehre. Methoden und Resultate. Fischer, Stuttgart.
— 1959 Darwin's views on variation under domestication in the light of present-day knowledge. Proc. Amer. Philos. Soc. **103**, No. 2, 190–220.

Munting, A.
1672 Waare Oeffeninge der Planten. Amsterdam.

Murr, J.
1896 Strahllose Blüten bei heimischen Kompositen. Dtsch. Bot. Mschr. **XIV**, Nr. 12.

Mursima, C.
1820 Caspar Friedrich Wolffs erneuertes Andenken (Berlin, 3. März 1819). Morphologische Hefte, Stuttgart, **I**, 2 cf Vol. **I**, 1817, pp. 252–256.

Nachtsheim, H.
1929 Die Entstehung der Kaninchenrassen im Lichte ihrer Genetik. Zschr. Tierzücht. **14**.
— 1949 Vom Wildtier zum Haustier. 2nd ed. Paul Parey, Berlin.

Nägeli, C.
1865 Ueber den Einfluss äusserer Verhältnisse auf die Varietätenbildung im Pflanzenreiche. Sitzungsber. Königl. Bayr. Akad. d. Wiss., München, **II**, 228–284.
— 1865 Die Bastardbildung im Pflanzenreiche. Sitzungsber. Königl. Bayr. Akad. d. Wiss., München, **II**, 395–443.

Nägeli, C. v.
1866 Über die abgeleiteten Pflanzenbastarde. Sitzungsber. Königl. Bayr. Akad. d. Wiss., München, **I**, 71–93.
— 1866 Die Theorie der Bastardbildung. Sitzungsber. Königl. Bayr. Akad. d. Wiss., München, **I**, 93–127.
— and A. Peter, 1885: Die Hieracien Mitteleuropas. Monographische Bearbeitung der Piloselloiden mit besonderer Berücksichtigung der mitteleuropäischen Sippen. R. Oldenbourg, München.
— 1884 Mechanisch-physiologische Theorie der Abstammungslehre. R. Oldenbourg Verlag, München-Leipzig.

Nägeli, K.
1842 Zur Entwickelungsgeschichte des Pollens bei den Phanerogamen. Orell, Füssli u. Comp., Zürich.

Nasse, C. F.
1820 Von einer erblichen Neigung zu tödtlichen Blutungen. Archiv für medizinische Erfahrung im Gebiete der praktischen Medizin und Staatsarzneikunde, ed. by Dr. Horn, Dr. Nasse, Dr. Henke, 385–434. G. Reimer, Berlin.

Nathusius, H. v.
1860 Die Rassen des Schweines. Berlin.

Natus, P.
1675 A phytological observation concerning oranges and lemons, both separately and in one piece produced on one and the same tree at Florence. Phil. Trans. **X**, 313–314.

Naudin, C.
1855 Réflexion sur l'hybridation dans les végétaux. Rev. Horticole, Paris, 4me Série, **4**, 351–354.
— 1856 Nouvelles recherches sur les caractères spécifiques et les variétés des plantes du genre *Cucurbita*. Ann. Sc. Natur., 4me Série, **6**, 5–73.
— 1861 Sur les plantes hybrides. Rev. Horticole, 4me Série, 396–399.
— 1861 La fécondation artificielle et la pratique horticole. Rev. hort., Paris, 465–467.
— 1863 Nouvelles recherches sur l'hybridité dans les végétaux. Ann. Sc. Natur. Botanique 4me Série, **XIX**, 180–203.
— 1864 De l'hybridité considérée comme cause de variabilité dans les végétaux. Comptes Rendus de l'Académie des Sciences **59**, 837–845 and Ann. Sc. Natur. Botanique, 5me Série, **3**, 153–163, 1865.
— 1865 Nouvelles recherches sur l'hybridité dans les végétaux. Nouv. Arch. mus. d'hist. nat. **I**, 25–174.

Němec, B.
1966 Mendel's discovery and Mendel's time. Proc. of the G. Mendel Memorial Symp. 1865–1965, ed. by M. Sosna, 3–10. Academia, Prague.

Nobis, G.
1963 Abstammung, Domestikation und Rassebildung unserer Haushunde. Naturw. Rdsch. **16**, 306.
Nordenskiöld, E.
1926 Die Geschichte der Biologie. Ein Überblick. German translation by G. Schneider. Fischer, Jena.
Notter, F.
1827 De qualitatibus parentum in sobolem transeuntibus, praesertim ratione rei equariae. Dissertatio, Praeside J. D. Hofacker, Tübingae, apud Ch. Fr. Osiandrum.
Nussbaum, M.
1880 Zur Differenzierung des Geschlechts im Tierreich. Arch. mikrosk. Anat., Bonn, **XVIII.**
— 1883 Über Befruchtung. Vorläufige Mitteilung. Sitzungsber. d. niederrhein. Ges. f. Nat. u. Heilk. Aug. 1883.
— 1884 Über die Veränderungen der Geschlechtsprodukte bis zur Eifurchung; ein Beitrag zur Lehre der Vererbung. Arch. f. mikrosk. Anat. **23.**
Olby, C. R.
1965 Francis Galton's derivation of Mendelian ratios in 1875. Heredity **20**, 636–638.
— 1965 A memorial to Mendel in Brno. New Scientist 880–881.
— 1966 Origins of Mendelism. Constable & Co. Ltd., London.
Orel, V.
1965 Gregor Mendel und die Landwirtschaft. Sborník (Acta universitatis agriculturae, Brno) **4**, 533–538.
— 1965 Gregor Mendel—as a person and as a discoverer. Anthropologia **II/3**, 5–9.
— 1966 Die Publizität der klassischen Arbeit Gregor Mendels vor der Wiederentdeckung im Jahre 1900. Folia Mendeliana No. 1, 23–29.
—, **J. Rozman, V. Veselý,** 1965 Mendel as a beekeeper. Moravian Museum, Brno.
Ostenfeld, C. H., and C. Raunkiär
1903 Kastrerings forsøg med *Hieracium* og andre Cichorieae. Bot. Tidsskr. **25**, 3 papers.
Pearson, K.
1900 Mathematical contribution theory of evolution: On the law of reversion. Proc. Roy. Soc. London **66.**
— 1901 Note on Mr. Bateson's paper: Heredity, differentiation, and other conceptions of biology: A consideration . . . Proc. Roy. Soc. London **69**, 450.
— 1901 On the principle of homotyposis and its relation to heredity, to the variability of the individual and to that of race. Philosoph. Transact. Roy. Soc.
— 1902 On the fundamental conception of biology. Biometrica **I**, 320–344.
— 1903 The law of ancestral heredity. Biometrica **II**, 221–236.
— 1904 A Mendelian view of law of ancestral heredity. Biometrica **III**, 109–112.

— 1909 Theory of ancestral contributions in heredity. Proc. Roy. Soc. London **81**.
— et al., 1901 On the principle of homotyposis and its relation to heredity. Proc. Roy. Soc. London **68**, 1–5.
— 1914–1930 The life, letters and labours of Francis Galton. 3 vols. Cambridge University Press, Cambridge.

Penzig, O.
Pflanzen-Teratologie. **I**, 1890, **II**, 1894. Angelo Ciminago, Genoa.

Pépin
1861 Une nouvelle variété de Troene. Rev. hort., Paris, 284.

Plate, L.
1932/33 Vererbungslehre. **I.** Mendelismus 1932. **II.** Sexualität und allgemeine Probleme 1933. Fischer, Jena.

Platner, G.
1886 Die Karyokinese bei den Lepidopteren als Grundlage für eine Theorie der Zellteilung. Internat. Mschr. Anat. Histol. **III**, H. 10.
— 1889 Beiträge zur Kenntnis der Zelle und ihrer Teilungserscheinungen. 1. Zellteilung u. Samenbildung in der Zwitterdrüse von *Limax agrestis*. 2. Samenbildung u. Zellteilung bei *Paludina vivipara* und *Helix pomatia*. Arch. mikrosk. Anat. **33**, 125–144.
— 1889 5. Samenbildung und Zellteilung im Hoden der Schmetterlinge. Arch. mikrosk. Anat. **33**, 192–213.

Plato
Works, tr. by B. Jowett, selected by Irwin Edman, 4 vols. in 1. Tudor Publ. Co., New York 1954.
— Platonis opera quae exstant-omnia. Paris 1578. Henricus Stephanus. 3 vols.

Platt, R.
1959 Darwin, Mendel, and Galton. Medical History **III**, 87–99.

Pliny (Gaius Plinius Secundus):
Natural History, 9 vols., Harvard University Press, Cambridge, Mass.
— Natural History. Ed. by Paul Turner. Carbondale, Ill., U. of Southern Illinois 1962.
— Naturalis historia. Ed. D. Detlefsen. Berlin 1866.

Poletyka, J. de
1754 De morbis haereditariis, dissertatio medica inauguralis. Lugduni Batavorum apud Joannem Heyligert.

Posner, C.
1922 Rudolf Virchow und das Vererbungsproblem. Arch. Frauenk. **VIII**, 14–23.

Prescott, W. H.
1844 History of the Conquest of Mexico. 2 vols., Everyman's Library, Dutton Pub. Co., New York.

Puissant, A.
1886 Genista Andreana. Rev. hort., Paris, 372–373.

Punnett, R. C.
1925 An early reference to Mendel's work. Nature, London, **116**, 606.
Purkyně-Symposion der Deutschen Akademie der Naturforscher Leopoldina in Gemeinschaft mit der Tschechoslowakischen Akademie der Wissenschaften am 31. 10. und 1. 11. 1959 Halle/Saale. (Purkinje-Symposium: German Academy of Naturalists, Leopoldina, and the Czechoslovak Academy of Sciences, 31 Oct. 59 and 1 Nov. 59). Ed. by R. Zaunick. Nova Acta Leopold. N.F. **24**, No. 151, Johann Ambrosius Barth, Leipzig 1961.

Purkině, J. E.
1825 Subjectae sunt symbolae ad ovi avium historiam ante incubationem. Ed. I, Bratislava. Ed. II, Leipzig 1830.

Purkinje, J. E.
1838 Über den Bau der Magen-Drüsen und über die Natur des Verdauungsprocesses. Ber. Vers. dtsch. Naturf. Aerzte in Prag im September 1837, 174–175.
— 1838 Körniger Überzug, welcher die plexus choroideos aller Hirnhölen beim Menschen und wohl bei allen Klassen der Rückgrathsthiere umgibt. Ber. Vers. dtsch. Naturf. Aerzte in Prag im September 1837, 178–179.
— 1838 Über die gangliöse Natur bestimmter Hirntheile. Ber. Vers. dtsch. Naturf. Aerzte in Prag im September 1837, 179–180.
— 1840 Über die Analogieen in den Struktur-Elementen des thierischen und pflanzlichen Organismus. Übers. Arb. Veränder. schles. Ges. vaterl. Cultur 1839, 81–82 (= Opera omnia II).

Quételet, A.
1835 Sur l'homme et le développement de ses facultés, ou essai de physique sociale. Paris.
— 1871 Anthropométrie. Paris.

Rabl, K.
1885 Über Zellteilung. Morph. Jb., Leipzig, **X**, 214.

Radl, E.
1909 Geschichte der biologischen Theorien. Part II. Verlag W. Engelmann, Leipzig.

Raikhov, B. E.
1947 An outline of the History of Evolutionary Theory in Russia up until the time of Darwin (Part II, Caspar Friedrich Wolff). Moscow and Leningrad (Russian).
— 1952 Russische Biologen als Evolutionisten bis Darwin, **I.** Moscow and Leningrad.
— 1964 Caspar Friedrich Wolff. Zool. Jb. Syst. **91**, 555–626.

Ramsbottom, J.
1937/38 Presidential Address. Linnaeus and the species concept. Proc. Linn. Soc. 150th session, 192–219.

Réaumur, R. A. Ferchault de
1749 Art de faire éclore et d'élever en toute saison des oiseaux domestiques de toutes espèces, etc. Paris.

Reed, C. A.
1960 A review of the archeological evidence on animal domestication in the prehistoric Near East. In: Braidwood-Howe, Prehistoric investigations in Iraqi Kurdistan. Chicago.

Reichert, K. B.
1840 Das Entwicklungsleben im Wirbelthier-Reich. Verlag von August Hirschwald, Berlin.
— 1843 Beiträge zur Kenntniss des Zustandes der heutigen Entwicklungsgeschichte.

Reinhardt, L.
1912 Kulturgeschichte der Nutztiere. Verlag E. Reinhardt, Munich.

Remak, R. R.
1852 Über extracellulare Entstehung thierischer Zellen und über die Vermehrung derselben durch Theilung. Müllers Arch. f. Anat., Physiol., wiss. Medizin 47–57.

Renner, O.
1924 Die Botanik vor Mendels Auferstehung. Naturwiss. **12**, 752–757.
— 1959 "Botanik" in Geist und Gestalt, vol. 2. Naturwissenschaften. Biograph. Beitr. z. Gesch. d. Bayer. Akad. Wiss. vornehmlich im 2. Jahrhundert ihres Bestehens. C. H. Becksche Verlagsbuchhandlung, Munich.
— 1961 William Bateson und Carl Correns. Sitzber. Heidelberg. Akad. Wiss. Math.-naturw. Kl. **60/61**, Springer-Verlag, Heidelberg.

Ribot, T.
1873 L'Hérédité psychologique, ses phénomènes, ses lois, ses causes, ses consequences. LaGrange, Paris.

Richter, O.
1924 Biographisches über Pater Gregor Mendel aus Brünns Archiven. Reprint from the memorial publication in honour of the 100th birthday of J. G. Mendel, issued by the Czechoslovak Eugenics Society in Prague.
— 1932 Gregor Mendels Reisen. Verh. Naturf. Ver. Brünn **63**, 1.
— 1941 75 Jahre seit Mendels Großtat und Mendels Stellungnahme zu Darwin's Werken auf Grund seiner Entdeckungen. Verh. Naturf. Ver. Brünn **72.**

Rimpau, W.
1877 Züchtung neuer Getreidearten. Landw. Jb.
— 1891 Kreuzungsprodukte landwirtschaftlicher Kulturpflanzen, 1–39. Verlag P. Parey, Berlin.
— 1899 Monstrositäten am Roggen. Dtsch. Landw. Pr. **XXVI**, 878, 901.

Roberts, H. F.
1919 The founders of the art of breeding. J. Hered. **10**, 99–106.
— 1929 Plant hybridization before Mendel. Princeton Univ. Press.

Robillard, J.
1890 Les semis du Robinia Decaisneana. Rev. hort., Paris, 518–519.

Romanes, G. J.
1881–1895 Hybridism. In: Encyclopaedia Britannica **12**, 422–426.

Röntgen, W. C.
1918 Erinnerungen an Theodor Boveri. Mit Beiträgen von W. Boveri, H. Beeg, H. Spemann, F. Baltzer, E. B. Wilson, A. Leiber, W. Wien, W. C. Röntgen. Verlag J. C. B. Mohr, Tübingen.

Rostand, J.
1956 L'atomisme en biologie, 202–208. Gallimard, Paris.
— 1958 Un précurseur de Johann Mendel. Le pharmacien suisse Colladon, "grand-père" de la génétique. Le Figaro littéraire, Aug. 30.

Roth, E.
1885 Die Thatsachen der Vererbung in geschichtlich-kritischer Darstellung. 2nd ed. Verlag A. Hirschwald, Berlin.

Roux, W.
1883 Über die Bedeutung der Kerntheilungsfiguren. Eine hypothetische Erörterung. Verlag W. Engelmann, Leipzig.

Royal Society of London
1879 Catalogue of Scientific Papers (1864–1873) **3**, 378.

Roze, E.
1895 Le *Chelidonium laciniatum* Miller. J. Botanique, Paris, **IX**, 296–307, 338–342.

Rückert
1894 Zur Eireifung bei Copepoden. Anat. Hefte **IV**, 2.
— 1895 Über das Selbständigbleiben der väterlichen und mütterlichen Kernsubstanz während der ersten Entwicklung des befruchteten *Cyclops*-Eies. Arch. mikrosk. Anat. **45**.

Rümker, K. v.
1889 Anleitung zur Getreidezüchtung auf wissenschaftlicher und praktischer Grundlage. Verlag Paul Parey, Berlin.

Rütimeyer
1861 Die Fauna der Pfahlbauten. Basel.

Sabine, J.
1824 Account and description of five new Chinese Chrysanthemums; with some observations on the treatment of all the kinds at present cultivated in England, and on other circumstances relating to the varieties generally. Transact. Hort. Soc. London **V**.
— 1824 Further account of Chinese Chrysanthemums; with descriptions of several new varieties. Transact. Hort. Soc. London **V**.

Sachs, J.
1875 Geschichte der Botanik vom 16. Jahrhundert bis 1860. Munich.

Sageret, M.
1826 Considérations sur la production des hybrides, des variantes et des variétés en général, et sur celles de la famille des cucurbitacées en particulier. Ann. Sc. Natur., Paris, Prem. Série, **8**, 294–314, Crochard, Libraire-Editeur.

Sajner, J.
1963 Gregor Mendels Krankheit und Tod. Sudhoffs Arch. Gesch. Med. Naturw. **47**, 377–382, Steiner-Verlag GmbH, Wiesbaden.

Salisbury, R. A.
1808 A short account of nectarines and peaches naturally produced on the same branch. Transact. Hort. Soc. London **1** (3rd ed. 1820) 103–106. W. Bulmer, London.

Scaliger, Julius Caesar
1566 Commentarii, et animadversiones, in sex libros de causis plantarum Theophrasti. Lugduni.

Schelver, F. J.
1812 Kritik der Lehre von den Geschlechtern der Pflanze. Heidelberg.
— 1814 Erste Fortsetzung seiner Kritik der Lehre von den Geschlechtern der Pflanze. Carlsruhe and Heidelberg.

Schiemann, E.
1932 Entstehung der Kulturpflanzen. Handbuch der Vererbungswissenschaft, edited by E. Baur and M. Hartmann. Verlag Gebr. Borntraeger, Berlin.
— 1943 Entstehung der Kulturpflanzen. Ergebnisse der Biologie **19**, 409–552, Springer-Verlag, Berlin.
— 1948 Weizen, Roggen, Gerste. Systematik, Geschichte und Verwendung. Fischer, Jena.

Schleiden, M. J.
1842 Grundzüge der wissenschaftlichen Botanik nebst einer methodologischen Einleitung zum Studium der Pflanze. Leipzig.
— 1848 Die Pflanze und ihr Leben. Verlag W. Engelmann, Leipzig.

Schmalhausen, I. F.
1874 On Plant Hybrids. Observations made from the Flora of St. Petersburg. Dissertation presented to the Faculty of Sciences of the Imperial University of St. Petersburg for the degree of Master of Botany. Printed by B. Demakov, St. Petersburg.

Schneider, A.
1873 Untersuchungen über Plathelminthen. Vierzehnter Ber. Oberhess. Ges. Natur-Heilk., Gießen.
— 1883 Das Ei und seine Befruchtung. Breslau.

Schrötter v. Kristelli
1870 Johann Ev. Purkinje. Alm. Akad. Wiss., Wien, **20**, 182–200.

Schultze, M.
1861 Ueber Muskelkörperchen und das, was man eine Zelle zu nennen habe. Reichert und du Bois Arch. Anat. Physiol., Verlag Veit et comp., Leipzig.
— 1863 Das Protoplasma der Rhizopoden und der Pflanzenzellen. Verlag W. Engelmann, Leipzig.

Schurig, M.
1720 Spermatologia Historico-Medica, h. e. Seminis Humani Consideratio Physico-Medico-Legalis, qva ejus natura et usus, insimulqve opus generationis et varia de coitu aliaqve huc pertinentia, v.g. de castratione, herniotomia, thimosi, circumcisione, recutitione, & infibulatione, item de hermaphroditis & sexum mutantibus, raris & selectis observationibus, annexo Indice locupletissimo. Johannes Beck, Frankfurt/Main.

Schwanitz, F.
1957 Die Entstehung der Kulturpflanzen. Series Verständl. Wiss. **63**, Springer-Verlag, Berlin.

Schwann, T.
1839 Mikroskop. Untersuchungen über die Übereinstimmung in der Struktur und dem Wachstum der Tiere und Pflanzen. Reprinted in Ostwalds Klassiker der exakten Wissenschaften No. 176, Verlag W. Engelmann, Leipzig.
— 1838 Über die Analogie in der Structur und dem Wachsthume der Thiere und Pflanzen. (From a letter to Prof. E. H. Weber.) Frorieps neueste Nachrichten No. 91 (No. 3, Vol. V; January).
— 1838 Fortsetzung der Untersuchungen über die Uebereinstimmung in der Structur der Thiere und Pflanzen. Frorieps neueste Nachrichten No. 103 (No. 15, Vol. V; February).
— 1838 Nachtrag zu den Untersuchungen über die Uebereinstimmung in der Structur der Thiere und Pflanzen. Frorieps neueste Nachrichten No. 112 (No. 2, Vol. VI, April.)

Scott, W. B.
1894 On variations and mutations. Amer. J. Sc. (Silliman), 3rd series **48.**

Seidlitz, C. v.
1865 Über die Vererbung der Lebensformen, Eigenschaften und Fähigkeiten organischer Wesen auf ihre Nachkommen in Bezug auf Physiologie und praktische Heilkunst, p. 14. Kaiserliche Hofbuchhandlung H. Schmitzdorff, St. Petersburg.

Sen, J.
1963 Some concepts of organic evolution of the ancient Hindus. Bull. Nat. Inst. Sci. India, No. 21, 184–188.

Seton, A.
1824 On the variation in the color of peas from cross-impregnation. Transact. Hort. Soc. London **V**, 236.

Shirreff, P.
1873 Improvement of the cereals and an essay on the wheat-fly. Print. for private circulation by William Blackwood and Sons, Edinburgh and London (German transl. by R. Hesse 1880, Verlag L. Hofstetter, Halle/S).

Shull, G. H.
1914 Duplicate genes for capsule-form in *Bursa bursa pastoris.* Zschr. indukt. Abstamm.-Vererb.lehre **12**, 97–149.

Siebert, H.
1861 Roger Bacon, sein Leben und seine Philosophie. Inaug.-Dissert. Marburg, Printed by C. L. Pfeil.

Sirks, M. J.
1915 Geschichtliches über Pelorienblüten. Naturw. Wschr. **30.**
— 1920 Praemendelistische erfelijkheidstheorien. Genetica **2**, 323–346.
— 1922 Francis Galton, 1822–16 Février 1922. Genetica **4**, 71–78.
— 1959 Leeuwenhoek on dominance in rabbits (1683). Genetica **XXX**, 292.

Solms-Laubach, H. Graf zu
1900 Cruciferenstudien. Bot. Ztg. **58.**
Spallanzani, L.
1785 Esperimenti che servono nella storia della generazione di animali e piante. Publicatio: Barthelmi Ciro, Genova.
— 1786 Versuche über die Erzeugung der Tiere und Pflanzen. Leipzig.
Spencer, H.
1864 The Principles of Biology. Williams and Norgate, London and Edinburgh.
— 1876 System der synthetischen Philosophie, Vol. III. Die Prinzipien der Biologie, Vols. I and II, transl. by B. Vetter. E. Schweizerbart'sche Verlagsbuchhandlung, Stuttgart.
Spillman, W. J.
1901 Quantitative studies on the transmission of parental characters of hybrid offspring. Proc. 15th ann. Convent. Assoc. Amer. Agric. Coll. Exper. Stat. held at Washington, D.C., November 12–14. U.S. Dept. Agric., Off. Exper. Stat. Bull. No. 115, 88–98.
— 1902 Quantitative studies on the transmission of parental characters to hybrid offspring. Proc. 15th ann. Convent. Assoc. Amer. Agric. Coll. Exper. Stat. Washington, Off. Exper. Stat. Bull. No. 115.
Sprengel, C. K.
1793 Das entdeckte Ceheimnis der Natur im Bau und in der Befruchtung der Blumen. Ed. Paul Knuth, Ostwalds Klassiker der exakten Wissenschaften, No. 48, 4 Vols. Leipzig 1894.
Stein, E.
1950 Dem Gedächtnis von Carl Erich Correns nach einem halben Jahrhundert der Vererbungswissenschaft. Die Naturwissenschaften **37,** 457–463.
Steinmann, G.
1908 Die geologischen Grundlagen der Abstammungslehre. Verlag W. Engelmann, Leipzig.
Stern, C.
1950 Boveri and the early days of Genetics. Nature **166,** 446.
Stiebitz, F.
1930 Über die Kausalerklärung der Vererbung bei Aristoteles. Sudhoffs Arch. Gesch. Med. **23,** 332–345.
Stomps, T. J.
1929 Aus dem Leben und Wirken von Hugo de Vries. Tübinger Naturw. Abh., Stuttgart, Nr. 12, 7–16.
— 1954 On the rediscovery of Mendel's work by Hugo de Vries. J. Hered., Washington, **45,** 293.
Strasburger, E.
1875 Über Zellbildung und Zelltheilung. H. Dabis, Jena.
— 1875 Über Vorgänge bei der Befruchtung. Tgbl. der 48. Versamml. dtsch. Naturf. Ärzte, Graz, 18–24 Sept. 1875.
— 1876 Über Zellbildung und Zellteilung. 2nd ed. Nebst Untersuchungen über Befruchtung.

— 1882 Über den Bau und das Wachstum der Zellhäute. Fischer, Jena.
— 1882 Über den Theilungsvorgang der Zellkerne und das Verhältnis der Kerntheilung zur Zelltheilung Arch. mikr. Anat. **XXI**, 476–590.
— 1884 Die Kontroversen der indirekten Kernteilung. Arch. mikrosk. Anat. **XXIII**.
— 1884 Neue Untersuchungen über den Befruchtungsvorgang bei den Phanerogamen als Grundlage für eine Theorie der Zeugung. Jena.

Stubbe, H.
1938 Genmutation. I. Allgemeiner Teil. Handbuch der Vererbungswissenschaft Vol. II F. Verlag Gebrüder Borntraeger, Berlin.

Studnicka, F. K.
1927 Joh. Ev. Purkinjes und seiner Schule Verdienste um die Entdeckung tierischer Zellen und um die Aufstellung der "Zellen"-Theorie. Acta Soc. Sc. Natur. Moravic., Brno, **IV**, Fasc. 4.
— 1927/28 Joh. Ev. Purkinjes und seiner Schule Verdienste um die Entdeckung der tierischen Zellen und um die Aufstellung der Zellentheorie. Anatom. Anz. **64**, 140–144, Fischer, Jena.

Stur, J.
1931 Zur Geschichte der Zeugungsprobleme. Arch. Gesch. Med. **24**, 312–328.

Sutton, W. S.
1902 On the morphology of the chromosome group in *Brachystola magna*. Biol. Bull. Marine biol. Labor **IV**, 1.
— 1902 The chromosomes in heredity. Biol. Bull. Marine biol. Labor. **IV**, 231–248.

Swingle, W. T.
1913 The date palm and its utilization in the southwestern states. Bureau of Plant Industry, U.S. Dept. Agric. Circ.

Tecoz, R. M.
1959 Un précurseur suisse de Mendel. Bull. Soc. Vaud. Sci. Nat. **67**, 127–132.

Theophrastus
Naturgeschichte der Gewächse. Transl. and commentary by K. Sprengel. Johann Friedrich Hammerich, Altona 1822.
— Vol. I, Historia plantarum. Vol. II, De causis plantarum. Ed. by Wimmer. Teubner Verlag, Leipzig 1854.
Sancti Thomae Aquinatis Opera Omnia. Parmae 1855. Typis Petri Fiaccadori.
S. Thomae Aquinatis Opera Omnia. Ed. by R. P. Mandonnet. 5 vols. P. Lethielleux, Paris 1927.

Thomson, M. A.
1888/89 The history and theory of heredity. Proc. Roy. Soc. Edinburgh **16**, 91–116.

Timoféeff-Ressovsky, N. W.
1966 Gregor Mendel. Proc. of the G. Mendel Memorial Symp. 1865–1965, ed. by M. Sosna, 47–55. Academia, Prague.

Tournefort, J. P.
1689 Schola botanica sive Catalogus plantarum quas ab aliquot annis in Horto regio parisiensi studiosis indigitavit.

Tschermak, E.
1900 Über künstliche Kreuzung bei *Pisum sativum*. Ber. Dtsch. Bot. Ges. **XVIII**, 232–239 (received June 2, 1900).
— 1900 Über künstliche Kreuzung bei *Pisum sativum*. Zschr. Landw. Versuchswes. in Österreich **III**, 465–555.
— 1937 Erinnerungen an die Wiederentdeckung der Mendel'schen Vererbungsgesetze vor 37 Jahren. Der Züchter **9**, 145–146.
— 1951 Historischer Rückblick auf die Wiederentdeckung der Gregor Mendelschen Arbeit. Verh. Zool. Bot. Ges. Wien **92**, 25–35.
— 1956 Gregor Mendels Versuchstätigkeit und die Zeit der Wiederentdeckung seiner Vererbungsgesetze. Novant'Anni delle Leggi Mendeliane. Ed. by L. Gedda, 113–117. Istituto "Gregorio Mendel," Roma.
— 1960 60 Jahre Mendelismus. Geschichte der Wiederentdeckung der Mendel'schen Vererbungsgesetze und ihre ersten Anwendungen auf Pflanzen, Tier und Mensch. Verh. Zool. Bot. Ges. Wien **100**, 14–25.

Turner, J.
1824 A description of some new pears. Transact. Hort. Soc. London **V**.

Unger, F.
1841 Genesis der Spiralgefäße. Linnea, Halle, **15**, 385–407.

Uschmann, G.
1955 Caspar Friedrich Wolff, ein Pionier der modernen Embryologie. Urania Verlag, Leipzig/Jena.
— 1959 Geschichte der Zoologie und der zoologischen Anstalten in Jena 1779–1919. VEB Gustav Fischer Verlag, Jena.
— and B. Hassenstein
1965 Der Briefwechsel zwischen Ernst Haeckel und August Weismann. Jenaer Reden und Schriften, Friedrich-Schiller-Universität, 7–68.

Valentin, G.
1836 Über Entwickelungsgeschichte der Pflanzengewebe. Übers. Arb. Veränder. schles. Ges. vaterl. Cultur im Jahre 1835, 87.
— 1836 Über den Verlauf und die letzen Enden der Nerven. Nova acta physico-med. Acad. Caes. Leopold.-Carol. Natur. Curios. **18**, 53–240.
— 1837 Feinere Anatomie der Sinnesorgane des Menschen und der Wirbelthiere. Repertorium f. Anatomie u. Physiologie **1**, Berlin 1836/1837, 141–147 and 300–316; **2**, 1837, 244–258. [Projected continuation until 1843 never published.]
— 1839 Grundzüge der Entwickelung der thierischen Gewebe. In: Rudolf Wagner, Lehrbuch der speziellen Physiologie Abt. 1, 132–139. Leipzig.

Vavra, M.
1965 Genetisch-Obstbauliche Analyse des Mendelschen Apfel- und Birnenzüchtungsplanes. Sborník (Acta universitatis agriculturae, Brno) **4**, 539–555.

Verlot, B.
1864 Mémoire sur la Production et la Fixation des Variétés dans les Plantes d'Ornement. J. Soc. Hort., Paris, **10**.

Vilmorin, H. L. de
1879 Note sur une expérience relative à l'étude de l'hérédité dans les végétaux. Mém. Soc. Nat. Agric. France, Paris.
— 1890 L'héredité chez les végétaux. Conférence de l'exposition universelle de 1889 Paris.
— 1894 Sur un Salpiglossis sinuata sans corolle. Bull. Soc. Bot. France **XXXI**, 216–217.

Vilmorin, L. L. de
1886 Notices sur l'amélioration des plantes par le semis et considérations sur l'hérédité dans les végétaux. New ed., Paris.

Vilmorin, M. L. de
1891 Bégonia Vernon ou B. semperflorens atropurpurea. Rev. hort., Paris, 84.

Viollet, J. B.
1860 Sur un nouveau muflier. Rev. hort., Paris, 446.

Virchow, R.
1858 Die Cellularpathologie in ihrer Begründung auf physiologische und pathologische Gewebelehre. Verlag A. Hirschwald, Berlin.

Virgil
Georgics, translated by Smith Palmer Bovie, University of Chicago Press, 1956.

Vries, H. de
1889 Intracellulare Pangenesis. Fischer, Jena.
— 1900 Sur la loi de disjonction des hybrides. C. R. Acad. Sc., Paris, **130**, 845–847.
— 1900 Das Spaltungsgesetz der Bastarde. Vorläufige Mittheilung. Ber. Dtsch. Bot. Ges. **XVIII**, 83–90 (received 11.3. 1900).
— 1900 Sur les unités des caractères spécifiques et leur application à l'étude des hybrides. Revue générale de botanique **12**, 257–271.
— 1901/03 Die Mutationstheorie. **1**, 1901, **2**, 1903. Verlag von Veit & Co., Leipzig.

Vrolik, G.
1827 Over een rankvormige ontwikkeling van witte lelieblomen. Nieuwe Verh. I. Kl. v.h. K. Nederl. Inst. van Wet. te Amsterdam Part **I**, 295–301.
— 1844 Ueber eine sonderbare Wucherung der Blumen bei *Digitalis purpurea.* Flora.

Vuillemin, M. P.
1893/95 Monstruosités provoquées par les variations du milieu extérieur chez le *Ranunculus repens.* Bull. Soc. Sc. Nancy **2**, 11/13.

Wakker, H. J.
1891 Eenige Mededeelingen over Pelorien. Ned. Kruidkund. Arch., 2. Serie, 5.

Waldeyer, W.
1887 Ueber die Karyokinese und ihre Bedeutung für die Vererbung. Dtsch. Med. Wschr. **13**, 925, 954, 975, 1001, 1018. Verlag G. Thieme, Leipzig.
— 1888 Über Karyokinese und ihre Beziehungen zu den Befruchtungsvorgängen. Arch. mikrosk. Anat. **32**.
— 1904 Caspar Friedrich Wolff. Festrede Sitzber. Preuss. Akad. Wiss. VI, Berlin.

Watermann, R.
1964 Vom Leben der Gewebe. Der Weg von der antiken Atomistik über die Zellenlehre bis zur modernen Molekularbiologie. Kölner Universitätsverlag, Cologne.

Weigert, C.
1887 VI. Neuere Vererbungstheorien. Schmidt's Jb. Med. **215**, 89–104, Verlag Otto Wigand, Leipzig.

Weiling, F.
1965 Die Mendelschen Erbversuche in biometrischer Sicht. Biometr. Zeitschr. **7**, 230–262.
— 1966 Hat J. G. Mendel bei seinen Versuchen "zu genau" gearbeitet?—Der χ^2-Test und seine Bedeutung für die Beurteilung genetischer Spaltungsverhältnisse. Der Züchter **36**, 359–365.
— 1966 Johann Gregor Mendel als Obstzüchter. Rhein. Monatsschr. f. Gemüse, Obst, Schnittblumen No. 12.

Weinstein, A.
1958 Did Nägeli fail to understand Mendel's work? Proc. X. Intern. Congress Genetics, Montreal, Vol. **II**, 339. Univ. Toronto Press, Toronto.
— 1965 The reception of Mendel's paper by his contemporaries. Natural History and Biology 997–1001.

Weismann, A.
1883 Über die Vererbung. Lecture, Jena.
— 1885 Die Continuität des Keimplasmas als Grundlage einer Theorie der Vererbung. Fischer, Jena.
— 1886 Zur Geschichte der Vererbungstheorien. Zool. Anz. 344–350.
— 1887 Über die Zahl der Richtungskörper und über ihre Bedeutung für die Vererbung. Fischer, Jena.
— 1892 Das Keimplasma. Eine Theorie der Vererbung. Fischer, Jena.
— 1892 Aufsätze über Vererbung und verwandte biologische Fragen. Fischer, Jena.
— 1895 Neue Gedanken zur Vererbungsfrage. Eine Antwort an H. Spencer. Fischer, Jena.
— 1902 Vorträge über Deszendenztheorie. 2 vols. Gustav Fischer, Jena.

Weldon, W. F.
1902 Mendel's laws of alternative inheritance in peas. Biometrica **I**, 228–254.
— 1902 On the ambiguity of Mendel's categories. Biometrica **II**, 44–45.

— 1903 Mr. Bateson's revision of Mendel's theory of heredity. Biometrica **II**, 286–288.
— 1903 Mendel's principles of heredity in mice. Nature **67**, 512, 610.
— 1904 Albinism in Sicily and Mendel's laws. Biometrica **III**, 107–109.

— and K. Pearson
1903 Inheritance in *Phaseolus*. Biometrica **II**, 499–503.

Werner, K.
1879 Die Psychologie, Erkenntnis- und Wissenschaftslehre des Roger Bacon. Sitzber. Phil.-Hist. Cl. Akad. Wiss. **93**, 467–576. Distributed by Verlag Gerold's Sohn, Vienna.

Wettstein, F. v.
1933 Joseph Gottlieb Koelreuter (zum zweihundertsten Geburtstag am 27. April 1933). Die Naturw. **21**, 309–310.

White, C. A.
1902 Saltatory origin of species. Bull. Torr. bot. Club.

Wichura, M.
1865 Die Bastardbefruchtung im Pflanzenreich, erläutert an den Bastarden der Weiden, 1–95. Verlag von E. Morgenstern, Breslau.

Widder, F.
1953 Die "laciniaten" Abänderungen des *Chelidonium majus* Linné. Phyton, Annales rei botanicae **5**, 153–162.

Wiegmann, A. F.
1828 Über die Bastarderzeugung im Pflanzenreiche. Verlag Friedrich Vieweg, Braunschweig.

Wiesner, J.
1892 Die Elementarstruktur und das Wachstum der lebenden Substanz. Alfred Hölder, Vienna.

Wilks, W.
1906 Gregor Johann Mendel. Report of the 3rd International Conference of Genetics, London.

Wilson, E. B.
1900 The cell in development and heredity. 2nd ed. The Macmillan Company, New York.
— 1902 Mendel's principles of heredity and the maturation of the germ-cells. Science **XVI**, 416.
— 1925 The cell in development and heredity. 3rd ed. The Macmillan Company, New York.

Wolff, C. F.
1759 Theoria Generationis. Halle.
— 1768 De formatione intestinorum. Novi Comment. Acad. Petropolitanae **XII** and **XIII**.
— 1896 Theoria Generationis (1759). Translated and edited by Paul Samasse. Ostwalds Klassiker der exakten Wissenschaften No. 84/85, Leipzig.

Wollny, E.
1885 Saat und Pflege der landwirtschaftlichen Kulturpflanzen. Verlag Paul Parey, Berlin.

Wriedt, C.
1925 Das Anconschaf. Zschr. ind. Abst. u. Vererb.lehre **XXXIX**, 281–286.

Wundt, W.
1873 Lehrbuch der Physiologie des Menschen. 3rd ed. Verlag F. Enke, Erlangen.

Ysabeau, A.
1859 Le Robinier d'Utterhart. Rev. hort., Paris, 548–550.

Ziegler, E.
1886 Können erworbene pathologische Eigenschaften vererbt werden und wie entstehen erbliche Krankheiten und Mißbildungen? Beitr. path. Anat. Physiol. **I**, 361–406.

Zirkle, C.
1932 Some forgotten records of hybridization and sex in plants (1716–1739). J. Hered. **23**, 433–448.
— 1935 The beginnings of plant hybridization. Univ. Pennsylvania Press, Philadelphia.
— 1935 The inheritance of acquired characters and the provisional hypothesis of pangenesis. Amer. Natural. **6**, 417–445.
— 1936 Further notes on pangenesis and the inheritance of acquired characters. Amer. Natural. **70**, 529–546.
— 1941 The Jumar or cross between the horse and the cow. Isis **33**, Part IV, 486–506.
— 1946 The early history of the idea of the inheritance of acquired characters and of pangenesis. Transact. Amer. Philos. Soc., Philadelphia, New Series, **XXXIV**.
— 1951 The knowledge of heredity before 1900. In: L. C. Dunn, Genetics in the 20th century. Macmillan Company, New York.
— 1951 Gregor Mendel and his precursors. Isis **42**, 97–104.
— 1964 Some oddities in the delayed discovery of Mendelism. Journ. Hered. **55**, 65–72.
— 1966 Some anomalies in the history of Mendelism. Proc. of the G. Mendel Memorial Symp. 1865–1965, ed. by M. Sosna, 31–37. Academia, Prague.

Supplementary Reading for the English Edition

Barnett, S. A. (ed.), 1958
A Century of Darwin. William Heinemann Ltd., London.

Blandino, G., 1969
Theories on the Nature of Life. Philosophical Library, New York.

Brink, R. A. (ed.), 1967
Heritage from Mendel. Proceedings of the Mendel Centennial Symposium sponsored by the Genetics Society of America 1965. University of Wisconsin Press, Madison, Wis.

Crew, F. A. E., 1966
The Foundation of Genetics. Pergamon Press, Oxford.

Darlington, C. D., 1964
Genetics and Man. George Allen and Unwin Ltd., London.

Darwin, F., 1968
Fr. Galton 1822–1911. Eugenic Reviews **60.**

Dunn, L. C., 1965
A Short History of Genetics. The Development of some of the main lines of thought: 1864–1939. McGraw-Hill Book Company, New York.

Facets of Genetics, 1970
Readings from Scientific American. W. H. Freeman, San Francisco.

Ford, E. B., 1965
Mendelism and evolution. (Science Paperbacks, first published in 1931.) Methuen & Co. Ltd., London.

Freeman, R. B., 1965
The works of Charles Darwin, an annotated bibliographical handlist. Dawson of Pall Mall, London.

Glass, B., O. Temkin, and W. Straus, Jr., 1959
Forerunners of Darwin, 1745–1859. The Johns Hopkins Press, Baltimore, Md.

Iltis, H., 1966
The life of Mendel. George Allen and Unwin Ltd., London.

Nordone, R. M. (ed.), 1968
Mendel centenary—Genetics, development and evolution. Catholic Univ. of America Press, Washington, D.C.

Proceedings of the Royal Society, 1966
From Mendel's factors to the genetic code. Royal Society, London.

Ravin, A. W., 1965
The evolution of genetics. Academic Press, New York and London.

Smith, J. M., 1966
The theory of evolution. 2nd ed. Penguin Books.

Stern, C., and E. R. Sherwood (eds.), 1966
The origin of genetics: A Mendel source book. W. H. Freeman and Company, San Francisco and London.

Sturtevant, A. H., 1965
A history of genetics. Harper and Row, New York.

Wichler, G., 1961
Charles Darwin: The founder of the theory of evolution and natural selection. Pergamon Press, New York-Oxford.

Whittinghill, M., 1965
Human genetics and its foundations. Reinhold Publishing Corporation, New York.

Index